Wild Edible Vegetables of Lesser Himalayas

Arshad Mehmood Abbasi • Munir Hussain Shah
Mir Ajab Khan

Wild Edible Vegetables of Lesser Himalayas

Ethnobotanical and Nutraceutical Aspects, Volume 1

Springer

Arshad Mehmood Abbasi (PhD)
Department of Environmental Sciences
COMSATS Institute of Information Technology
Abbottabad, Pakistan

Munir Hussain Shah (PhD)
Department of Chemistry
Quaid-i-Azam University
Islamabad, Pakistan

Mir Ajab Khan (PhD)
Department of Plant Sciences
Quaid-i-Azam University
Islamabad, Pakistan

ISBN 978-3-319-09542-4 ISBN 978-3-319-09543-1 (eBook)
DOI 10.1007/978-3-319-09543-1
Springer Cham Heidelberg New York Dordrecht London

Library of Congress Control Number: 2014949351

Printed on acid-free paper

Springer is part of Springer Science+Business Media (www.springer.com)

Preface

Major populace in the world does not have sufficient food, and millions of people are deficient in one or more micronutrients. Wild edible plants have played an important role in human life since time immemorial and a greater part of the rural communities depend on wild food resources including wild edible fruits and vegetables to meet their food requirements in the periods of food disaster. The diversity in wild plant species offers variety in family diet and contributes to household food security. Wild edible food species are known to be an excellent source of nutrients and serve as a food supplement and an appetizer. Throughout the world wild plants make an important contribution to the life of local communities. They play a significant role in a wide range of agricultural systems and have an important socioeconomic position through their use in medicines, dyes, poisons, shelter, fibers, and religious/cultural ceremonies. Ethnobotany deals with natural relationships of the plant kingdom with man and animals. It relates past and present interrelationships between human cultures and the plants, animals, and other organisms in their environment. Nutraceuticals are the substances that may be considered as food or part of food and provide health benefits to the consumers. Consumption of the fruits and vegetables has been strongly associated with reduced risk of cardiovascular disease, cancer, diabetes, Alzheimer disease, cataracts, and age-related functional decline.

This book provides a brief introduction of ethnobotanical and nutraceutical aspects of wild edible vegetables used in the region of Lesser Himalayas. A total of 50 wild edible vegetables species are included in this volume. A detailed description of each species is provided including photographs, botanical name, English/local name, family, flowering/fruiting period, status/habitat, parts used, distribution, ethnobotanical uses, cultural aspects, medicinal uses, and nutraceutical aspects. Medicinal uses include mode of preparation, way of application, and diseases cured; cultural aspects show species cultural index; nutraceutical aspects comprise proximate analysis of fats, proteins, fibers, carbohydrates, ash, moisture content, dry matter, and energy value; elemental analysis includes various essential and toxic metals; phytochemical screening involve total phenolic, flavonoids, flavonols, and ascorbic acid; and antioxidant potential in terms of DPPH scavenging activity, hydroxyl radical scavenging activity, H_2O_2 scavenging activity, Fe^{2+} chelating activity, ferric reducing antioxidant power, and phosphomolybdenum assay are also assessed for each species. The analytical data pertaining to the statistical distribution, correlations, and multivariate analyses of the nutrients, phytochemicals, selected metals, and antioxidant activities are also provided.

It was our intention to best convey maximum knowledge regarding ethnomedicinal and nutraceutical aspects in minimum words. Further work is desirable to assess detailed phytochemical profiling and mechanisms of antioxidant properties of these vegetables based on modern techniques. There is always space for improvements and perfection. The readers are best adjudicators to evaluate this effort. We would be very grateful for any comments and suggestions as to improvement in terms of clarity, vigor, and coverage of subject matter. It would be our greatest achievement if this book could attract the students/scientists of food science, agriculture, nutraceutical science, bioscience, biodiversity, applied ethnobotany, ethnoecology, ecology, wild life naturalists, tourists, and others who have some lovely feeling for nature.

Islamabad, Pakistan

Arshad Mehmood Abbasi
Munir Hussain Shah
Mir Ajab Khan

Contents

About the Authors

Arshad Mehmood Abbasi is an Assistant Professor of Environmental Sciences at COMSATS Institute of Information Technology, in Abbottabad, Pakistan since 2013. He received his Ph.D. in Plant Sciences with specialization in Ethnobotanical and Nutraceutical aspects of wild edible fruits and vegetables of Lesser Himalayas, Pakistan from Quaid-i-Azam University in Islamabad, Pakistan in 2013. His primary research and professional activities have been in Ethnobotany, Ethnomedicines, Medicinal Plants, Biodiversity, Forest Food Resources, Phytochemistry, Nutraceuticals, and Functional Foods. He is the editor of the *Journal of Ethnobiology and Ethnomedicines* and also authored the book entitled, *Medicinal Plant Biodiversity of Lesser Himalayas-Pakistan* published by Springer, USA in 2012. He has been the recipient of the *Outstanding Research Award* in the "Best Research Paper" category for the years 2010–2011, from the Higher Education Commission of Pakistan.

Mir Ajab Khan is a Professor of Plant Sciences at Quaid-i-Azam University in Islamabad, Pakistan. He received his Ph.D. in Biosystematics from Leicester University, England in 1984. His research and professional activities have been in Biosystematics, Chemotaxonomy, Taxonomy, Palynology, Anatomy, Medicinal Plants, Biofuels and Oil Yielding Plants, Ethnobotany, and Plant Biodiversity. He has published more than 300 research articles in journals of international repute in these areas and also authored a number of national and international books. He is also the recipient of a number of outstanding achievement and research awards by the Higher Education Commission of Pakistan.

Munir H. Shah is an Associate Professor of Chemistry at Quaid-i-Azam University in Islamabad, Pakistan. He received his Ph.D. in Analytical/Environmental Chemistry in 2007. His primary research and professional activities have been in Hydrogeochemistry, Atmospheric Chemistry, Clinical Chemistry, Toxicological Chemistry, and Chemometrics and has published several research articles on these topics. He was the recipient of the "Young Scientist in Chemistry" award, from the Third World Academy of Sciences in 2010 and the "Best Young Research Scholar" award, from the Higher Education Commission of Pakistan in 2011.

Chapter 1
Pakistan and Pakistani Himalayas

1.1 Pakistan

1.1.1 Geo-Climate

Pakistan has a long history of human settlements. Evidence of early civilizations dating from 3000 B.C. has been found in Harappa and Mohenjodaro. Pakistan is located in the north-west part of the Indo-Pakistan subcontinent about in the middle of the South Asian region. It lies between 24° and 37° north latitude and between 61° and 78° east longitude and stretches from the coast of Arabian Sea near the Delta of the river Indus so far north of the mountain ranges of Himalayas and Karakoram. From the deserted deserts of Thar to the forested valleys of Dir and Kohistan Pakistan, it covers a land area of 803,940 km^2 (0.67 % area of the world). About 58 % of the total land of Pakistan consists of mountain and plateau, whereas about 42 % are plains and deserts. Pakistan is surrounded by the neighboring countries; Iran to the west, Afghanistan to the north, China to the northeast, and India to the east and southeast, with about 1,000 km long coastline along the Arabian Sea (Food and Agriculture Organization of the United Nations 1987).

Pakistan has a great variety of landscapes with a diversified relief. It has all the majestic high mountain ranges, having enormous mass of snow and glaciers and 100 peaks of over 5,400 m in elevation. K-2 (8,563 m) is the second highest peak in the world. The great mountain masses the Karakoram, Himalaya and Hindu Kush form the watershed of the Indus River, which starts in the mountains of Ladakh, flows through Baltistan, bending southward through mountains into the immense plains of the Punjab, and falls into the Arabian Sea near Karachi. To the west, the Rocky Mountains which start from Chitral and extend through the Safed Koh in the Khyber Pakhtunkhwa (KPK) province converge with the semiarid Suleiman range in Balochistan. Torrential monsoon rains cause flash floods in some of these mountains. Their eastern drainage flows into the Indus. In the heartland of the country lies the flat alluvial flood plain which forms the drainage basin of this

A.M. Abbasi et al., *Wild Edible Vegetables of Lesser Himalayas*,
DOI 10.1007/978-3-319-09543-1_1

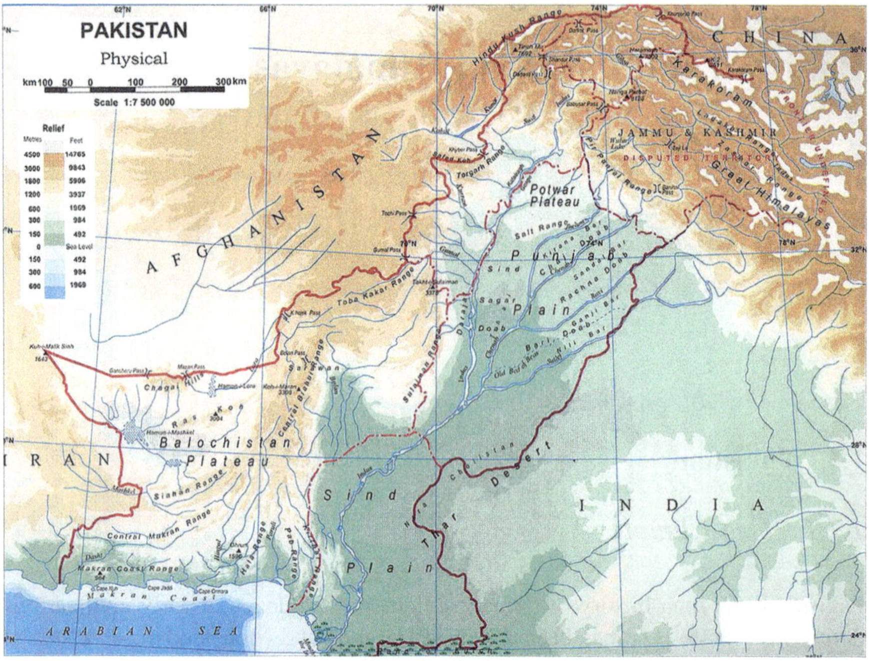

Fig. 1.1 Physical map of Pakistan (Source: http:// cn.bing.com/physical map of Pakistan)

mighty river. Summers are hot and dry, even in the lower mountain regions (Fig. 1.1).

The hot spell is interspersed by the monsoons which sweep into Pakistan from the Indian Ocean during early July to late September. The monsoon season is of immense importance to the biodiversity of the country. The moisture-laden winds end in heavy precipitation which is intense along the outer foothills of the great northern mountain barrier. The monsoon currents are blocked by the Nanga Parbat, after which a cold mountain desert emerges some 1,200 m above sea level. The direction of the monsoon largely bypasses the Baluchistan highland region, in particular the south-western corner of the province. This area is comprised of a vast desert plain called the Chaghai, which is an extension of the great Seistan depression, extending all the way into the Arabian Sea.

Rainfall in the country is highly erratic ranging from 30 mm annually in the Chaghai and Sibi deserts to as high as 1,350 mm in the Murree hills. In the Indus plain, the rain falls mainly during the monsoon season while the KPK is influenced by a Mediterranean climate with up to 60 % of the rainfall occurring in the winter months. In the foothill regions of the Hazara district and the Punjab, rainfall is more evenly distributed with a pronounced monsoon influence and only a quarter of the precipitation taking place during the winter months. Punjab and central parts of Northern Areas receive less than 250 mm of rainfall in a year. Northern Sindh, Southern Punjab, North-Western Balochistan, and the central parts of Northern

Areas receive less than 125 mm of rainfall. True humid conditions appear after the rainfall increases to 750 mm in plains and 625 mm in highlands. Pakistan can be divided into four broad temperature regions:

Hot summer and mild winter: 32 °C or more in summer and 10–21 °C in winter.
Warm summer and mild winter: 21–32 °C in summer and 10–21 °C in winter.
Warm summer and cool winter: 21–32 °C in summer and 0–10 °C in winter.
Mild summer and cold winter: 10–21 °C in summer and 0 and 10 °C in winter (UNEP).

1.1.2 Physiography

Pakistan has within its borders some of the world's highest and most stunning mountains such as Himalayas, Karakoram and Hindu Kush, Sulaiman, Toba Kakar, Kirthar, and Salt range. The Northern and Western Highlands produced by the mountain building movement extended from the Makran Coast in the south to the Pamir Plateau in the extreme north. The Northern and Western Highlands cover most of Balochistan, NWFP, Northern Areas, and parts of the Punjab. Five major physiographic entities of Pakistan are:

1. Mountainous North (The Himalayas, The Karakorams, and The Hindu Kush)
2. Koh-e-Safaid and Waziristan Hills
3. Sulaiman and Kirthar Mountains
4. Balochistan Plateau
5. Potowar Plateau (Salt Ranges, Kala Chitta Ranges, Margalla Hills)

1.1.3 Biodiversity

Pakistan has 225 Protected Areas (PAs), 14 national parks, 99 wildlife sanctuaries, and 96 game reserves. It is a world of rapidly shrinking wetlands, some of them of international significance of wondrous Juniper forests of Ziarat, Balouchistan province, and minute life forms. Pakistan covers a number of the world's eco-regions, ranging from the mangrove forests stretching from the Arabian Sea to the towering mountains of the western Himalayas, Hindukush, and Karakoram. One of the highest plateaus of the world "Deo Sai Plain" in Sakrdu presents a unique plant diversity of alpine zone. The Fairy meadows of Gilgit also present a wide range of plant biodiversity. The Khunjrab Pass between China and Pakistan is a famous place of alpine flora, where several plants of family Asteraceae are yet to be identified a part from the members of Caryophyllaceae, Primulaceae, and grasses. The flora and fauna of the country are composed of a blend of Palearctic and Indomalayan elements, with some groups also containing forms from the Ethiopian region (Biodiversity Action Plan for Pakistan 2000).

1.1.4 Eco-Regions

Under the Global 200, ecosystems have been ranked to carry out conservation through comparative analysis. It covers all habitats on the land masses and in the ocean. The Earth has been divided into 238 ecoregions, by the United Nation (UN), the National Geographic Society (NGS) with WWF. Out of them five are in Pakistan. The Global eco-regions of Pakistan are:

1. Rann of Kunth grasslands
2. Tibetan Plateau
3. Western Himalayan Temperate Forests
4. Indus Delta Ecosystem
5. Arabian Sea

1.1.5 Agro-Ecological Zones

The climate of Pakistan varies with altitude. It has some of the world's highest cold areas that occur above 5,175 m in the Himalayas and the hottest low areas in the Indus Plains with many intermediate ecological zones. Pakistan has ten major ecological zones. In the Arid and Semi-Arid areas, most parts are bare of vegetation. On the basis of physiography, climate, soil type, and agricultural land use Pakistan is divided into ten main agro-ecological zones, which are:

1. Indus Delta
2. Southern Irrigated Plain
3. Sandy Desert
4. Northern Irrigated Plain
5. Barani (rainfall)
6. Wet Mountains
7. Northern Dry Mountains
8. Western Dry Mountains
9. Dry Western Plateau
10. Sulaiman Piedmont

1.1.6 Forest Types

Most recently, data of land use including forest area have been reported by Forestry Sector Master Plan (FSMP), with the help of Landsite Satellite Thematic Mapper Images at a scale of 1:250,000 covering the whole of Pakistan. Although the landmass of Pakistan offers varied soil conditions but the area under forests in spite of very pleasant climatic conditions is hardly about 4.587 million hectors which is 5.2 % of the total land area, including the farmland trees and linear planting along roadsides, canal sides, and railway sides. Of the total forest area, about 3.311 million hectares (72.4 %) are managed and protected forests, while the rest 1.266 million hectares (27.6 %) are commercial forests (Hussain and Ilahi 1991; Sheikh 1987).

The type of forests that exist in Pakistan with relative share are moist and dry temperate: coniferous 40 %, scrub 28 %, tropical thorn 3 %, man-made irrigated 5 %, riverine 7 %, mangrove 8 %, and farm forests 9 %. At riverbanks and deltas, Riverain and Mangrove forests have emerged. On humid hills and mountains, pines and coniferous forests occur that change with altitude. The Dry sub-Tropical forests dominate up to an altitude of 1,000 m; the coniferous forests from 1,000 to 4,000 m. Above the tree line, dwarf alpine forests followed by the alpine pastures occur up to snow line.

The following forest types are found in Pakistan:

1.1.6.1 Tropical Forests

These forests are confined to the plain areas of Pakistan. They range from sea level to 300–400 m. These forests are divided into three subgroups (Beg 1975).

Tropical Littoral and Swamp Forests

These forests cover an area of 207,000 ha and occur in the Arabian Sea around the coast of Karachi and Pasni in Balochistan. The characteristics genera of these forests are *Avicennia*, *Rhizophora*, *Typha*, *Phragmites*, *Nymphaea*, *Nelumbium*, *Saccharum*, *Arundo*, *Tamarix*, *Acacia*, *Imperata*, *Paspalum*, *Sagittaria*, *Vitiveria*, *Salix*, *Potamogeton* and *Vallisnaria,* etc.

Tropical Dry Deciduous Forests

These are forests that occur at narrow strips in small areas of the Pakistan and mainly restricted to the Himalayan foothills and adjoining Siwalik and recent alluvial deposits. Such type of forests can be seen in the Rawalpindi foothills. The characteristics genera are *Salmila*, *Terminalia*, *Mallotus*, *Dalbergia*, *Carissa*, *Bauhenia*, *Lannea*, *Bombax*, *Sterculia*, *Flacourtia*, *Acacia Adhatoda*, *Gymnosporia,* and *Indigofera*.

Tropical Thorn Forests

These are low, open and pronouncedly xerophytic forests characterized by thorny, decedious, and hardwood species which occupy the whole of the Indus plain except the driest parts. The major representative genera are *Prosopis*, *Capparis*, *Zizyphus*, *Tamarix,* and *Salvadora*.

1.1.6.2 Montane Subtropical Forests

These forests lie above the tropical zone and extend from 450 to 1,500 m in different parts of the country. They exist at a transitional zone between the tropical and temperate vegetation, therefore contain mixture of species from both types. According to Champion et al. (1965) these forests are divided into two main zones

Sub-Tropical Broad-Leaved Evergreen Forests

These are xerophtic forests of thorny and small-leafed evergreen species. This type occurs on the foothills and lower slopes of the Himalayas, the Salt Range, Kalachitta, and the Sulaiman Range. The typical genera of these forests are; *Olea*, *Acacia*, *Reptonia*, *Dodonaea*, *Mallotus*, *Carrisa*, *Justicia*, *Bauhenia*, *Pisticia*, *Maytenus* and *Sageretia*. Total area of these forests is estimated to be 1,191,000 ha.

Sub-Tropical Chir Pine Forests

These are open inflammable pine forests sometimes with, but often without, a dry evergreen shrub layer and little or no under wood. This type consists of Chir pine (*Pinus roxburghii*) forests found between 900 and 1,700 m elevation in the Western Himalayas within the range of the south-west summer monsoon. The characteristics genera of these areas are *Pinus*, *Xylosoma*, *Machilus*, *Celtis*, *Quercus*, *Pyrus*, *Ficus*, *Punica*, *Ziziphus*, *Pisticia*, *Olea*, *Zanthoxylum*, *Berberis*, *Cotoneaster*, *Indigofera*, *Carrisa*, *Rubus*, *Dodonaea*, *Justicia*, *Mallotus*, *Daphne*, *Vitex*, *Clematis* and *Tylophora*, etc.

1.1.6.3 Montane Temperate Forests

These forests occupy altitudinal range from 1,360 to 3,000 m. These forests are best classified by the rainfall during the season of vegetative growth. The wettest sites are predominantly occupied by the broad-leaved evergreen species. These forests are further divided into two groups (Hussain and Ilahi 1991).

Himalayan Moist Temperate Forests

The evergreen forests of conifers, locally with some admixture of oak and deciduous broad-leaved trees, fall in this category. These forests extend all along the outer Himalayan ranges. Their undergrowth is rarely dense and consists of both evergreen and deciduous species. These forests occur between 1,500 and 3,000 m elevation in the Western Himalayas except where the rainfall falls below about 1,000 mm in the inner ranges, especially in the extreme north-west. These forests are divided into a lower and an upper zone. In the lower zone, *Cedrus deodara, Pinus wallichiana, Picea smithiana,* and *Abies pindrow* are the main conifer species in order of increasing altitude, with *Quercus incana* at lower altitudes and *Q. dilatata* above 2,130 m. In the upper zone *Abies pindrow, Populus, Salix,* and *Q. semecarpifolia* are the dominant tree species. There may be pockets of deciduous broad-leaved trees, mainly edaphically conditioned, in both the zones.

Himalayan Dry Temperate Forests

These are open evergreen forests with open scrub undergrowth. Both coniferous and broad-leaved species are present. This type occurs on the inner ranges and is mainly represented in the north-west. The representative genera of these forests are *Pinus, Quercus, Juniperus, Picea, Cedrus, Salix, Pisticia, Olea, Betula, Abies, Rosa, Ribes, Sorbus, Juniperus, Ephedra, Caranga, Salvia, Lonicera, Cotoneaster, Daphne, Berberis, Sorbaria, Artemesia, Astragulus, Thymus, Ferula, Allium, Tulipa, Nepeta* and *Ranunculus,* etc.

1.1.6.4 Sub-Alpine Forests

These forests occupy altitudinal range between 3,333 and 3,788 m and found throughout the Himalayas. In these forests trees are mainly deciduous, small sized and less in number. Evergreen conifers and mainly evergreen broad-leaved trees occur in relatively low open canopy, usually with a deciduous shrubby undergrowth of *Viburnum* and *Salix,* etc. *Abies, Betula, Pinus, Vibernum, Juniperus, Sorbus, Salix, Populus, Lonicera, Cononeaster, Rhododendron, Berberis, Bergennia, Euphorbia, Gentiana, Potentilla, Polygonum, Rumex, Ranunculus, Caltha, Anemone,* etc. are the characteristics genera of these forests.

1.1.6.5 Alpine Scrubs and Pastures

These forests consist of pasture and scrubby species and lie between 3,790 and 3,939 m altitude. The height of vegetation is generally less than a meter to 2 m. The herbaceous plants form mat-like structure. Growing season is very short due to predominant snow fall and strong winds. The characteristic genera are *Betula, Salix, Abies, Lonicera, Berberis, Cotoneaster, Juniperus, Rhododendron, Ephedra, Rumex, Rosa, Polyganum, Anemone, Senecio, Iris, Chrysanthemum, Gentiana, Arabis* and *Ranunculus,* etc.

1.1.6.6 Cold Deserts

They lie above the alpine zone at 4,550 m above sea level. Mostly they have no vegetation due to permanent snow cover. However, some herbaceous flora appears on the edges of these habitats after melting of the snow. The representative genera of such habitats are *Primula*, *Sibbalida*, *Sedium*, *Cerastium*, *Oxyria*, *Potentilla*, *Draba*, *Thlaspi*, *Arenaria*, *Luzula*, and *Waldheimia*.

1.2 Himalayas in Pakistan

Himalaya is derived from two Sanskrit words hima, "Snow," and alaya, "House" factually means "House of Snow." This word was coined by the ancient pilgrims of India who travelled in these mountains. The attractiveness of the Himalayas has lured visitors to this region since ancient times. Being the world's highest mountain chain, it constitutes the greatest fascination to climbers and trekkers all over the world. The Himalayan mountain system is the planet's highest and home to the world's highest peaks: the Eight-thousanders, including Mount Everest (Nepal/ Tibet) and K2 (Pakistan's Northern Areas). The Himalayas represent the tremendous power, beauty, and magnificence of Nature. Himalaya is the land that houses the world's largest non-polar ice masses.

The Himalayan system stretches across six countries: Bhutan, China, India, Nepal, Pakistan, and Afghanistan. Its widespread glacial networks feed Asia's major rivers including the Ganges, Indus, and Brahmaputra. More than a billion people rely on these glacier-fed water sources for drinking water and agriculture. The Himalayan range runs west to east, from the Indus river valley to the Brahmaputra river valley, thereby forming an arc 2,400 km long, which varies in width from 400 km in the western Kashmir-Xinjiang region to 150 km in the eastern Tibet-Arunachal Pradesh region.

1.2.1 Himalaya Range

The greater Himalaya range runs west to east, from the Indus river valley in northern Pakistan to the Brahmaputra river valley in northern India and Tibet, forming an arc 2,400 km long which varies in width from 400 km in Northern Pakistan to 150 km in the eastern Tibet. The Western Himalayas are situated in Kashmir valley and Northern Pakistan, to the south and east of Indus River, and is dominated by Nanga Parbat massif with highest peak rising 8,125 m above the sea level, Pakistan's second and world's ninth highest peak. Nanga Parbat peak is also famous for its unique Rupal Face which rises 4,600 m straight above its base and the mountain is considered as one of the most difficult and dangerous to climb,

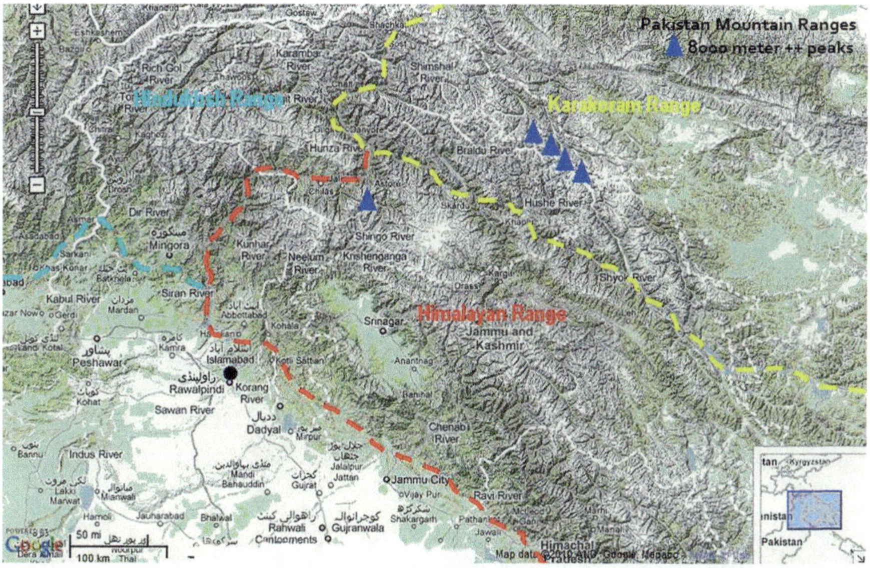

Fig. 1.2 Himalayas in Pakistan (Source: http:// cn.bing.com/physical map of Pakistan)

nicknamed as Killer Mountain. The Himalaya range occupies in Pakistan the regions of Kashmir, Kaghan, Kohistan, Deosai, and Chilas. The Western Himalya is also dominated by Deosai plains situated at the average height of about 4,114 m and covering an area more than 3,000 km^2. The place can be accessed from Skardu and is very well-known for its rich flora and fauna including summer flowers and endangered wild life species like brown bears. The Himalayan system in Pakistan is classified into three parallel and longitudinal zones (Fig. 1.2).

1.2.1.1 The Sub-Outer or Siwalik Himalayas

This zone extending over the southern part of Hazara-Murree hills, Rawalpindi hills, and Pabbi hills. Width of this zone varies from 8 to 48 km, with an average elevation of 909–1,212 m (Fig. 1.3).

1.2.1.2 The Lesser or Middle Himalayas

This zone occurs in the northern Hazara and Murree hills including main range, Dunga gali (4,546 m) in the north. Lesser or middle Himalayas are closely related to the grater Himalayas but of lower height. The average width of this zone is 80.5 km (Fig. 1.4).

Fig. 1.3 Panoramic view of a village in Lesser Himalayas, Pakistan

1.2.1.3 The Central or Great Himalayas

This zone comprises an average height of 6,061 m. Some of the height peaks of the central or great Himalayas are like Mount Everest (8,529 m) and Dhaulagiri (8,119 m) in Nepal and Nanga Parbat (8,066 m) in Pakistan.

1.2.2 Climate

Due to variation in the topography, altitudes aspects, and vegetation cover, the climate of Himalaya ranges shows tremendous variation. Even within the range, inner and outer parts contrastingly differ in the amount of rainfall, snowfall, and temperature conditions. The altitude varies from less than 500 m to over 5,000 m in the alpine regions which make with a shift in the climate also. The amount and time of rainfall, snowfall, and temperature are the major climatic factors.

Fig. 1.4 A view of Siri Paiya (Shogran)

1.2.2.1 Rainfall and Snowfall

It has been usually observed that the amount of rainfall reaches to its maximum up to 3,000 m within the moist part of the Lesser Himalayas. The climate of Lesser Himalayas falls into two major types

(a) Subtropical Continental Lowland including the plain and foot-hills zone.
(b) Subtropical Continental Highland including outer and middle Himalayas, Siwalik hills, Murree hills, and entire Hazara hills.

The climate of Haripur, Tarbela, Khanpur, Tanawal, Kahuta, Rawalpindi, and Islamabad appears to be likely dried subtropical. The average rainfall varies from 70 to 90 mm. There is a short monsoon period of 3 months during July, August, and September, which has monthly average of 100–200 mm. Snow is usually very rare phenomenon. In the lower Siwalik range covered by *Acacia-Olea* semi-evergreen forests, the climate is slightly mild. The upper Siwalik zone is the home of Chir pine; a true subtropical climate prevails. The climate data for Abbottabad, Mansehra, Goragali, Balakot, Bisham, Shinkiari, and Dadar shows that the average annual rainfall lies in between 100 and 130 mm. October and December appear to be the primary dry periods, which is followed by the secondary dry spell in May-June. A large part of the winter precipitation from the western disturbance is received in the form of snow.

The northern parts of Batagram, Balakot, Allai-Madakhel, Galiyat range, Dungagali range, entire Kaghan and Siran valleys, and middle parts of the Lesser Himalayas are characterized by cool temperate climate. The mean annual rainfall varies in between 100 and 130 mm. The precipitation from this source is mainly in the form of snow. Beyond Batakundi the monsoon rains are rare. The northern parts receive little rain but heavy snowfall in the winter. At the peaks monsoon fades and winter precipitation in the form of snow is more prominent as seen in Malika Parbat, Musa-ka-Musalla, and other peaks within the area.

1.2.2.2 Temperature

As a general rule the temperature decreases at a normal lapse rate of 0.5 ºC/100 m during an altitude rise. In the western Himalayas there is a decline of 3 ºC per every 303 m and 4 ºC in the eastern Himalayas. The mean annual temperature of Tarbela, Khanpur, Tanawal, Kahuta, Oghi, Balakot, Kakul, Rawalpindi, and Islamabad is around 21 ºC with summer temperature rising over 41 ºC. June and July are hot months while December and January are coldest. There is a definite cool winter with frost and occasional snowfall. The climatic data for Abbottabad, Mansehra, Ghoragali, Balakot, Bisham, Shinjiari, and Dadar indicate that the mean annual temperature flocculates in between 15 to 18 ºC. December and January are the coldest month and mean maximum temperature may be around 10 ºC. Frost and snowfall is common phenomenon. The mean annual temperature of the northern parts of Batagram, Balakot, Naran, Allai-Madakhel, Galiyat range, Dungagali range, entire Kaghan and Siran valleys, and middle parts of the Lesser Himalayas ranges from 4 to 10 ºC. In the wet temperate forests the winter accumulation of snow may be up to 7 m or more (Hussain and Ilahi 1991).

1.2.3 Hydrology

The important rivers of Himalayas in Pakistan are Siran, Dor, Haro, Kunhar, Jehlum, Soan, and Kurang. The mighty Indus strikes it on the west for 48 km and Jhelum River on the east for about 40 km. The Siran River arises in the north of the Bhogarmang valley then turns northwesterly in Heripur areas to join Indus at Tarbella. The total course of the Siran River is about 129 km. This river irrigates parts of Mansehra, Abbottabad, and Haripur. The river Dor originates at the northern end of Dungagali and flows through the Haripur plains and ultimately joins Siran near the north eastern end of the Gandgar range. The total length of Dor is 65 km and it irrigates parts of Abbottabad and Haripur. Haro rises at the southern end of Dungagali range where it has two branches. The eastern branch is called

Fig. 1.5 Kunhar river near Balakot

Dhund and western as Karral at the head of Khanpur tract (Fig. 1.5). The river Haro enters Rawalpindi district near Bhallan and then enters Attock district. The Kunhar River rises from the lake Lulusar at the end of Kaghan valley. After a very complicated course of 177 km, it joins the Jhelum at Pattan. In the Rawalpindi district river Jhelum skirts from the northern point near village Dewal covering a distance of about 112 km. The Soan River originates from the Murree hills and traverses deep valleys and then enters the plain near Pharwal fortress and finally joins Indus near Makhad. The Kurang River rises near Ghoragali, flows along Murree-Rawalpindi road, and joins the Soan near Sihala. It is also damed at Rawal. There are numerous tributaries of these rivers. They are popularly known as Kathas, Nallhas, Kassies, Leh, Kanshi, Lings, Sarin, Ghamlan, and Tamrah.

Lakes are confined to upper mountainous region in Kaghan valley. The three world famous lakes are Lulusar, Dudupat Sar, and Saiful Malook Sar. Sar means a lake. Lulusar is a crescent-shaped lake about 2.5 km long and 274 m wide, located 16 km west of Babusar Pass at an elevation of 3,384 m (11,167 ft). Dudupat Sar is a circular lake about half a kilometer in distance at 3,636 m (12,000 ft) height. Saiful Malooke Lake is located 10 km east of Naran. It is about half km long and 457 m broad at an altitude of 3,248 m (10,718 ft) (Hussain and Ilahi 1991).

1.2.4 Geology

The Himalayas ranges consist mostly of uplifted sedimentary and metamorphic rock. According to the modern theory of plate tectonics their formation is a result of a continental collision along the boundary between the Indo-Australian Plate and the Eurasian Plate. Geologically, Himalayas has been divided into two major zones, i.e., the inner or Hazara zone and outer or Kalachita zone (Tahirkheli 1982), with many formations and series as follow,

1.2.4.1 The Inner or Hazara Zone

This zone includes Salkhala series, Hazara slates, Tanawal formation, Tannaki conglomerates, Abbottabad formation, Galdanian formation, Hazara formation, and Mansehra granites.

1.2.4.2 The Outer or Kalachita Zone

This zone comprises Mianwali formation, Chak-Jabbi limestone, Samana Sukh formation, Kawagrah marl, Lockhart limestone, Patala formation, and Marghala hills limestone.

The main rock types of Lesser Himalayas are phyllites, schists, paragneisses, sandstone, quartizitic limestone, quartizitic sandstone, crystalline limestone, marble, minor conglomerate, acidic and basic igneous rocks, slates, calcareous and siliceous partings, magmatities, amphibolites, chlorite, kyanite, sillimanite, staurolite, granet and biolite grades, arenaceous, quartz, argillite limestone, dolomite, haematitic claystone, shaly siltstone, feldspars, black tourmaline, marl, and oolitic limestone.

1.2.5 Soil

Soil of the Himalayas in Pakistan falls into following major categories (Champion et al. 1965)

1.2.5.1 Alluvial Soils

Alluvial soils are water-transported soils which have been deposited along the water courses and in the valley floor. These soils are deep and show stratification of sand, clay, and loam.

1.2.5.2 Lateritic Soils

Lateritic soils are characteristic of tropical regions and develop on more basic rocks. They are usually acidic with poor minerals status. These soils not only affect the distribution of individual species, but also the vegetation as a whole.

1.2.5.3 Piedmont or Foothills Soils

They are present in the foothills and may be slightly acidic or neutral with high base status.

1.2.5.4 Montane Soils

These soils show features of North Temperate Zone having adequate mineral nutrients and humus. They are acidic in nature and usually have high organic matter.

1.2.5.5 Skeletal Soils

Skeletal soils occur in the high altitude area in the drier regions of Himalayas and western mountains. They, therefore, support xerophytic and degraded vegetation.

1.2.6 Vegetation

Vegetation is a general term for the plant life, which refers to the ground cover provided by plants, without specific reference to particular taxa, life form, structure, or any other specific botanical or geographic characteristics. Vegetation is the ecological aspect of the plant life growing together forming different groups on a unit of habitat. It is broader than the term flora which refers exclusively to species composition. Primeval redwood forests, coastal mangrove stands, sphagnum bogs, desert soil crusts, roadside weed patches, wheat fields, cultivated gardens and lawns—all are encompassed by the term vegetation.

The vegetation of the world can be divided into three main zones

- Polar
- Temperate
- Tropical

The vegetation of Himalayas falls within the subtropical, temperate, subalpine, and alpine zones. However, the *Acacia modesta forests* and *Tropical dry deciduous forests* lying in the foothills of the Lesser Himalayan ranges are also included (Fig. 1.6a, b). On the basis of climate, physiognomy, and altitude, Murree-Hazara hills can be divided into following zones (Naqvi 1976)

Fig. 1.6 (**a**) Vegetation of lesser Himalayas. (**b**) Vegetation of upper Himalayas

- *Sub-mountane zone*: This zone extends from foothills to height of 1,000 m throughout the outer Himalayas. It has two subzones, i.e.
  (a) *Acacia-Zizyphus* subzone dominated by *Acacia modesta* and *Zizyphus mauritiana* extending upto 700 m.
  (b) *Acacia-Olea* subzone, lying above the foothills. It is composed up of *Acacia modesta* and *Olea ferruginea*. The later gains much dominance in the upper limit of this range.
- *Montane zone*: This zone extends between 1,000 and 2,000 m, but may extend on warmer aspects up to 2,200 m. *Pinus roxburghii* is the dominating species.
- *Sub-alpine zone*: It extends between 2,000 and 3,000 m in outer Himalayas. They are predominantly coniferous in nature along with following associations.
  (a) *Pinus-Quercus association*: consisting of *Pinus wallichiana*, *Quercus incana*, *Q. dilatata* found between 1,800 and 2,200 m.
  (b) *Pinus-Abies association*: It is found in the middle and sub-alpine zone upto 2,600 m. The characteristics species beside blue pine is *Abies pindrow*.
  (c) *Mixed coniferous-Hardwood association*: It is the mixture of conifers and hard wood species. It is confined to the gentle slopes and more or less plain areas.
  (d) *Pinus-Abies-Picia association*: occurring in the upper portion of the sub-alpine regions in the outer Himalayas.
  (e) *The Timber vegetation*: The sub-alpine zone extends upto timber line which lies between 3,200 and 4,000 m in the outer Himalayas. *Pinus wallichiana*, *taxus wallichiana*, *Betula utilis*, and *Picea smithiana* are timber line trees in the Murree-Galiyat range.
- A*lpine zone*: This zone develops around 4,000 m in the outer Himalayas. Two habitat types are recognized.
  (a) *Alpine scrub*: composed entirely of deciduous shrubs reaching upto 2 m height, with some evergreen *Juniper*, *Rhododendron*, and *Ephedra*.
  (b) *Alpine Meadows and pastures*: are very common in the alpine zones. They may be natural or result due to extensive overgrazing. All plants are herbaceous.

1.2.7 Fauna

Knowledge of the fauna of the Eastern Himalayas Region is poor. Most of the information available is on the larger vertebrates that are easily observed and inventoried. Overall, more than 175 species of mammals and in excess of 500 species of birds are known from the region. The mammalian fauna in the lowlands is typically Indo-Malayan, consisting of Snow leopard, Sloth bear, Himalayan Ibex

Brown bear, Tibetan red fox, Asiatic black bear, Tibetan wolf, Marcopolo Sheep, Ermine, Alpine weasel, Stone Martin, Golden Marmot, Large-eared Pika, Hispid hare, Cape hare, Common field mouse, Royle's mountain vole, Flying squirrel, Lesser shrew, and Migratory hamster. Common birds are Lammegier vulture, Golden eagle, Himalayan griffon vulture, Eurasian black vulture, Marsh harrier, Eurasian sparrow hawk, Eurasian kestrel, Lesser kestrel, Saker falcon, Peregrine falcon, Himalayan Snow cock, Snow partridge, Chukar, Grey heron, Common sandpiper, Hill pigeon, Snow pigeon, Northern eagle owl, Eurasian cuckoo, Common swallow, Magpie, White-winged duck, Alpine cough, and Raven (WWF, Pakistan).

References

Beg, A. R. (1975). *Wildlife habitats of Pakistan. Bulletin monograph 5* (pp. 25–30). Peshawar: Pakistan Forest Institute.

Biodiversity Action Plan for Pakistan. (2000). Government of Pakistan, World Wide Fund for Nature (WWF), Pakistan and International Union for Conservation of Nature and Natural Resources (IUCN), Pakistan.

Champion, H. G., Seth, S. K., & Khattak, G. M. (1965). *Forest types of Pakistan*. Peshawar: Pakistan Forest Institute.

Food and Agriculture Organization of the United Nations. (1987). *Pakistan's experience in rangeland rehabilitation and improvement* (pp. 70). Rome, Italy: FAO.

Hussain, F., & Ilahi, I. (1991). *Ecology and vegetation of lesser Himalayas Pakistan* (1st ed., pp. 1–185). Peshawar, Pakistan: Jadoon Printing Press.

Naqvi, H. H. (1976). *Vegetation zonation of Murree-Hazara Hills. Final Project Rep*. Islamabad, Pakistan: University Grants Commission.

Sheikh, M. I. (1987). *Forest and forestry in Pakistan* (pp. 25). Peshawar: Pakistan Forest Institute.

Tahirkheli, R. A. K. (1982). *Geology of the Himalayas, Karakorum and Hindukush in Pakistan. Geological Bulletin of the University of Peshawar* (Sp. Issue, Vol. 15, pp. 1–51).

UNEP. Environment Assessment Programme for Asia and the Pacific, United Nations Environment Program (http://www.unep.org./environmentl governance/UNEP in the regions Asia and pacific)

Chapter 2
Ethnobotany and Ethnomedicine

2.1 Ethnobotany

2.1.1 History

Since the beginning of human civilization, people have used plants as food and medicine. Indian Sanskrit and Rigveda show thousands of plant species used as food, medicine, agriculture tools, and for the religious purposes. Perhaps as early as Neanderthal man, plants were believed to have healing power. The earliest uses are found in Babylonian circa 1770 B.C. in the code of Hammurabi and in ancient Egypt circa 1550 B.C. In fact, ancient Egyptians believed medicinal plants to have utility even in the after life of their pharaohs. Plants have been recorded from the Giza pyramids and can be found on display in a dark corner of Cairo Museum. The herbal medicines occupy distinct position right from the primeval period to present day. The ethnobotanical pharmacology is as old as man himself. In Indo-Pak first record of plant medicine was compiled in Rig Veda (4500–1600 B.C.) and Ayurveda (2500–600 B.C.). This system traces its origin to Greek medicine, which was adopted by Arabs and then spread to India and Europe. Presently about 80 % population of the world depends on the traditional system of healthcare (Ahmad 1999).

2.1.2 Definition

The term "Ethnobotany" was first used by Harshberger in the year 1895 at a lecture in Philadelphia. Ethnobotany is the study of the interrelations of man and plants/vegetation, but now it is understood as:

Total natural relationships of the plant kingdom with man and animals. It deals with past and present interrelationships between human cultures and the plants,

A.M. Abbasi et al., *Wild Edible Vegetables of Lesser Himalayas*,
DOI 10.1007/978-3-319-09543-1_2

animals, and other organisms in their environment. Like its parent field, ethnobotany makes apparent connection between human cultural practices and the subdisciplines of biology.

2.1.3 Aims of the Ethnobotany

Forests along with a great variety of useful plants are disappearing without any notice on the whole earth, such as in Hawaii one half of the indigenous floras are threatened with immediate extinction. Unfortunately, Hawaii is not unique among oceanic islands in its devastating loss of native species. Yet only a single conservation organization, Seacology (in Berkeley, California), is solely focused on island preservation. Currently, 12.5 % of all plant species on earth are threatened with immediate extension. Another problem is the availability of the knowledge about the different plant species existing in the far off places from the developed countries. Old people are using the herbs and other plants found in their areas in their daily life especially as medicines. As facilities are reaching in each and every part of the world, people are relying on the modern ways of the life and leaving their indigenous culture along with plants and their characteristics. With the demise of these old age people the indigenous cultures and knowledge about the plants is also disappearing. So ethnobotanists have not only to record the fruitful knowledge about the medicinal uses of certain plant species, but they also have to record and preserve the indigenous cultures of areas of their observation.

In our attempt to discover new medicines from the diverse life around us, we are going forward with significant advantages, but also significant challenges. At this moment, when we have never been better equipped technologically to evaluate natural products, the plant species are themselves disappearing. College courses and international conferences in ethnobotany are multiplying throughout the world, yet at this climax of interest, an increasing number of aged healers are dying, with their knowledge left unrecorded. The key objectives ethnobotanical investigations are focused on are as follows (Hamilton 2002):

(a) To find primitive ideas and conceptions of plant life
(b) To study the effects of a given plant environment on the lives, customs, religion, thoughts, and everyday practical affairs of the people
(c) Documentation of the traditional uses of the plants as food, for medicines, for material culture, and for ceremonial purposes
(d) To find the extent of indigenous people knowledge of the parts, functions, and activities of the plants
(e) To assess that into what categories are plants names and words that deal with plants grouped in the language of the people studied and what can be learned concerning the working of the folk mined by the study of these names?

2.1.4 *Significance*

Today ethnobotany is widely accepted as a science of human interactions with plants and its ecosystem. The fundamental structure of ethnobotanical research is to examine the dynamic relationship between human population, cultural values, and plants. Ethnobotany is more than simply a study of plants useful to people, for it's devoted to understanding the limitations and behavioral consequences of human population's action on their plant environment. Ethnobotanical data is highly significant for

- Economic botanists to explore new plant resources
- Getting fresh ideas for environment planners
- Selection of plant species for plantation and drugs development by pharmacologists, phytochemists, and clinicians
- A new source of history of plant names and for locating germplasm for agriculturists

Ethnobotanical study has become interdisciplinary study (besides botany and ethnology) involving archeology, sociology, folklore, mythology, linguistics, literature, forestry, ecology, agriculture, medicinal science, economics, phytochemistry, pharmacology, veterinary medicine, etc. Ethnobotany is also involved with use of microbes, lichens, mosses, and other lower plants. Ethnobotanists have innovated identification of food materials from the analysis of human feces: proved to be very useful tool to study prehistoric nutrition. The study of indigenous food production and local medicinal knowledge may have practical implications for developing sustainable agriculture and discovering new medicines. Ethnobotany also encourages an awareness of the link between biodiversity and cultural diversity, as well as a sophisticated understanding of the mutual influence (both beneficial and destructive) of plants and humans.

2.1.5 *Future Prospects*

Ethnobotany is a growing field, but the field ethnobotanists are still at a lower level and did not get higher attention, support, and respect than molecular biologists, because interest in this field is newly emerging. This field is rising now at the sky of science. Now scientific journals and societies have begun to publicize the studies of the ethnobotanists to peers, other scientists, and policy makers world-wide.

Ethnobotany issues are the focus of much public attention. Due to increased public interest and policy making in conservation, companies are looking to plants for new approaches to food, medicines, and energy sources. Departments of many Universities are opening positions for ethnobotanists. In 1992 Convention on Biological Diversity specifically addressed issues of traditional knowledge. It binds the signatory nations to three goals.

1. To respect, preserve, and maintain traditional knowledge
2. To promote wide application of traditional knowledge
3. To encourage equitable sharing of benefits from traditional knowledge

2.1.6 Ethnobotany of Wild Food Resources

Humans have relied on nature throughout the ages, for their basic needs such as, production of food stuffs, shelters, means of transportation, clothing, fertilizers, flavors and fragrances, and not the least, medicines. Millions of people in many developing countries do not have enough food to meet their daily requirements and a further more people are deficient in one or more micronutrients. Thus, in most cases rural communities depend on wild resources including wild edible plants to meet their food needs in periods of food crisis. The diversity in wild species offers variety in family diet and contributes to household food security (Food and Agriculture Organization of the United Nations (FAO) 2004).

Term "wild-food" is used to describe all plant resources outside of agricultural areas that are harvested or collected for the purpose of human consumption in forests, savannah, and other bush land areas. Wild-foods are incorporated into the normal livelihood strategies of many rural people, shifting cultivators, continuous croppers, or hunter gatherers (Bell 1995). Indigenous knowledge of wild edible plant is important for sustaining utilization of those plant species (Jasmine et al. 2007).

Of the Earth's half million plant species, only about 3,000 species have been used as agricultural crops and only 150 species have been cultivated on a large scale. However, while development of genetically modified crops may play an important role in achieving enhanced productivity that is essential for human survival, developing new crops by domesticating currently wild edible species offers considerable potential (Heywood and Skoula 1999). Wild edible plants have played an important role in human life since time immemorial. The diversity in wild plant species offers variety in family diet and contributes to household food security. Today, most human plant food is based on rather limited number of crops, but it is clear that in many parts of the world the use of wild plants is not negligible (Cavender 2006; Pieroni et al. 2007).

Wild edible food species are known to be excellent source of nutrients such as minerals, vitamins, and carbohydrates in form of soluble sugars, cellulose, and starch. They are very vital portion of an adequate diet and they serve as food supplement, and an appetizer. Sometimes the nutritional value of traditional wild plants is higher than several known common vegetables and fruits (Orech et al. 2007). There are over 20,000 species of wild edible plants in the world, yet fewer than 20 species now provide 90 % of our food. However, there are hundreds of less well-known edible plants from all around the world which are both delicious and nutritious. There has been renewed or increasing interest in consuming wild food plants (Ladio and Lozada 2004). Evidence indicates that more than 300 million people throughout the contemporary world gain part or all of their livelihood and food from forests (Pimentel et al. 1997).

The utilization of wild plants and animals continues to greatly benefit the society to this day, from processes mostly involving local experimentation through indigenous and local knowledge (Ladio and Lozada 2004; Kristensen and Balslev 2003; Scherrer et al. 2005). Out of these came the direct dependence of numerous resource-poor households on indigenous plant resources in rural areas of most countries as an integral part of their livelihoods (Arnold and Rruiz Perez 2001; Shackleton and Shackleton 2004). Such dependence is predominant in areas where there are easily accessible communal area resources, in conjunction with limited economic options (Dovie et al. 2005).

2.1.7 Socio-Economic Consequence of Wild Food Resources

The world has witnessed growing scientific and commercial interests in wild plants and plant-based products, mainly due to their vast economic potential and widespread cultural acceptability; however, only less than 5 % species have been analyzed as potential medicine, while rests (95 %) of the plants are still there to be analyzed (Shinwari et al. 2009). Today, most human plant food is based on rather limited number of crops, but it is clear that in many parts of the world the use of wild plants is not negligible (Pieroni et al. 2007; Cavender 2006).

Throughout the world, and more especially in developing countries, wild plants make an important contribution to the life of local communities. They play a significant role in a wide range of agricultural systems as a source of wild foods and fuel wood, and they have an important socio-economic role through their use in medicines, dyes, poisons, shelter, fibers, religious and cultural ceremonies (Heywood and Skoula 1999). It has been estimated that 46 % of world's poor live in South Asia (Bhattarai 1998) of which 75 million dwell Himalayas (Dutta and Pant 2003) and the biomass extraction is most widespread pressure on forests where rural people significantly depend for their household and livelihood needs and income generation through the sale of wild harvested materials (Pattanayak et al. 2003). Food production must be actively combined with evaluation, selection, and domestication and greater utilization of under-utilized or wild edible plants that are of local or regional importance to effectively increase nutritional security. By improving the linkage between production and utilization, the biodiversity of traditional crops can be safe-guarded. The knowledge of traditional food plant is important to promote the cereal staples in a more sustainable way. This is so because this know-how will enable for better food security/nutrition at household level (Ohiokpehai 2003).

It is well-established that wild foods contain higher concentrations of minerals, vitamins, carbohydrates, proteins, lipids, dietary fibers, and many bioactive compounds than exotic species. Although there is little information on the yields obtainable from wild food species, it is probable that because they are highly adapted to variable climatic conditions and in particular drought, their food have an extremely important function in times of famine, war and drought-induced nutritional stress. There is global search for means of combating micronutrient deficiencies especially in developing countries of the world.

Pakistan is one of the few places on earth with such a unique biodiversity, comprising different climatic zones with a wide range of plant species having medicinal and nutritional values. There are several hot spot areas of wild edible fruits and vegetables particularly in the Northern Pakistan ranging from Salt range, Kalla Chitta Hills, Tilla Gogian and Murree of Potohar region; Margalla Hills (foot hills of Himalayas) at Islamabad; Moist temperate areas of Ayoubia and Galiat; Siran valley, Kaghan-Naran valley, Palus valley of Kohistan and Babusar pass leading to Chalas and Gilgit in Hazara; Neelam valley and its allied areas in Kashmir; Skardu, Ghanche, Naltar valley, Ferry Meadows, Hunza valley, Yasin and Gupus valley, Astor valley, Khungrab pass leading to Tibet and China, Khaplu and Shigar valley leading to K-2 area (second highest peak in the world), Deosai areas (one of the highest plateau of the world) in the Karakorum ranges of Gilgit-Bultistan are potential places of wild edible fruits and vegetables. In Swat Kalam valley, Mahodhand, Malam Jaba, Chail valley and Madian valley; Bamburit valley of Chitral, Kalakot and Madan valleys of Dir are famous for wild edible fruits and vegetables. One cannot forget the potential of biodiversity of Baluchistan especially the Junipers forests at Ziarat (one of the largest Juniper forests of the world). Baluchistan is also famous for apple, cherries, and grapes along with a vast number of wild edible fruits and vegetables species.

Northern part of Pakistan and epically the Himalayas is predominantly a tribal-dominated area having harsh climate. These tribal communities regularly suffer from severe seasonal hardship, when household food availability and farm employment dries up. As wild edible food species grow there naturally during rainy season, it can generate good source of food and income for the tribal communities. They can be sold in the local market and fetch good prices. Awareness about economic values of wild edibles can be promoted through participation in various events such as food bazaars and exhibition, field day and participatory evaluation, demonstration plots, women's training programs, distribution of seed and flyers, techno-guides, seminars, and poster presentation. It can create a number of job opportunities in the rural and suburban areas and in the complementary fields of business that arise from productions, such as marketing, processing, and transportation. With little efforts these wild edible vegetables can be sold in villages and city markets and to be exported to neighboring cities also. Considering the future growth of big cities and anticipated increase in population of Pakistan particularly and the world in general, production, grading, and packaging of wild edible species in suburban areas and villages are essential to promote for creating rural employment, alleviate poverty, and sustain high quality of food resources.

2.2 Ethnomedicines

2.2.1 History

No one knows when or where plants first began to be used in the treatment of diseases, but the connection between plants and health has existed for thousands of years; evidence of this early association has been found in the grave of a Neanderthal man buried 60,000 years ago. An accidental discovery of some new plant food or juice that eased pain or relieved fever might have been the beginning of folk knowledge, which was passed down for generations and eventually became the foundation of medicine (Levetin and McMahon 1999). The earliest known medical document is a 4,000 years old Sumerian clay tablet that recorded plant remedies for various illnesses. By the time of the ancient Egyptian civilization, a great wealth of information already existed on medicinal plants. This information, along with hundreds of other remedies, was preserved in the "Ebers papyrus" about 3,500 years ago. Ancient China is also a source of information about the early medicinal uses of plants. The "Pun-tsao," a pharmacopoeia published in 1600, contained thousand of herbal cures. In India, herbal medicine dates back several thousands years to the "Rig-Veda". This led to a system of healthcare known as Ayurvedic medicine. Western medicine can be traced back to the Greek physician Hippocrates (460–377 B.C.), known as the father of medicine. He believed that diseases had natural causes and used various herbal remedies in his treatments. Early Romans writings also influenced the development of western medicine, especially the work of Dioscorides (first century A.D.). Although Greek by birth, Dioscorides compiled this information in De Materia Medica, which contained an account of over 600 species of plants with medicinal value. Dioscorides work remained the standard medical reference in most of Europe for the next 1,500 years, since little new knowledge was added during the Middle Ages. Although medical botany was nearly at a standstill in Europe, progress was being made in the Islamic world. Most notables are the early eleventh century Persian, Aveicenna, who wrote a canon of medicine that included new information on herbal medicine.

Some of the most richly illustrated herbals were those written by the four German "Fathers of botany" in the sixteen century: Otto Brunfels, Jerome Bock, Leonhart Fuchs, and Valerius Cordus. During this same period, the Englishman John Gerard published his famous work, "The herbal or general Histories of plants" in 1597. Other English herbals published in the mid seventeen century were John Parkin son's "Theartrum Botanicum". The "Complete herbal" of Nicholas Culpeper was one of the most popular herbals of the day. All the herbals focused on the medicinal uses of plants. The most famous advocate of the doctrine of signatures was early sixteenth century Swiss herbalist Paracelsus. He believed that medicinal uses of plants could easily be ascertained by recognizing distinct "Signatures" visible on plants that corresponded to human anatomy. For example, the red juice of blood wart should be used to treat blood disorders, and lobed appearance of liverworts suggested their use in treating liver complaints. During eighteenth century, a dichotomy in medicine developed between practitioners of herbal medicines and regular physicians. Although herbalism banned in the eighteenth and nineteenth centuries, many of the remedies employed by the herbalists provided effective treatment. Some of these became useful prescriptions, as physicians began experimenting with therapeutic agents. William Withering was the first in the medical field to scientifically investigate a folk remedy. His studies (1775–1785) of foxglove as a treatment of dropsy set the standard for pharmaceutical chemistry.

In nineteenth century, scientist began purifying the active extracts from medicinal plants. In 1806 Friedrich Serturner isolated morphine from the Opium poppy. Although the direct use of plant extracts continued to decrease in the late nineteenth and early twentieth centuries, medicinal plants still contribute significantly to prescription drugs. It is estimated that 25 % prescriptions written in the United States contained plant-derived active ingredients.

Today there is a renewed interest in investigating plants for medically useful compounds, with some of the leading pharmaceutical and research institutes involved in this search (Shinwari et al. 2006). In this land of subcontinent (Indo–-Pakistan), the first record of medicines, where plants held a predominant position, was compiled in Rigveda (4500–1600 B.C.) and Ayurveda (2500–600 B.C.). The system of Ayurveda is still in practice. The newly introduced system that incorporated many of medicines from Ayurved is known as Unani medicine. The origin of unani medicine goes back to "Hippocrates" son of "Heraclides" (460–370 B.C.) With the advent of the British rule in India, both of indigenous systems lost support of the regime, and allopathic system gradually established (Zaman and Khan 1970). Scientific knowledge about the medicinal plants is very important for the Hakims, doctors, and research workers. Seventy to eighty percent of the rural population of the world still relies on herbal medicine as their only healthcare. The long tradition of herbal medicine continues to the present day in China, India, and many countries in Africa and South America. In many villages, in market places, medicinal herbs are sold alongside vegetables and other wares. Practitioners of herbal medicine often undergo a rigorous and extended training to learn the names, uses, and preparation of native plants (Levetin and McMahon 1999).

Pakistan has a diverse range of phytogeographic and climatic conditions which results in varied flora containing numerous plant species. The northern mountains of Pakistan are well-known for their biodiversity as they are situated at the junction of three great Mountain ranges, i.e., Karakorum, Himalaya, and Hindu Kush. These three mountain ranges collectively contain about 25,000 species (about 10 % of world plant species), out of which around 10,000 are economically or medicinally useful. Estimated total flora of Pakistan is comprised of 6,000 species, out of which more than 4,000 plant species grow in mountainous regions of Hindukush and Himalayas of Pakistan (Shinwari et al. 2006; Qureshi et al. 2006; Stewart 1972). Over 75 % of population in Pakistan is cured by means of traditional medicines, prescribed by more than 50,000 traditional herb practitioners and the folk knowledge of plant curing pass down from family to family of herb practitioners and within communities (Shinwari et al. 2009). Majority of the population, especially in villages, is getting healthcare by herbal practitioners (Hakims).

2.2.2 Ethnomedicinal Value of Wild Food Resources

The use of wild plants as a source of medicine is based on the experience of many generations of traditional physicians and herbalists found in the different ethnic societies. Wild edible plants have always been important in the folk traditions of the Mediterranean region (Hadjichambis et al. 2007). However, food and medicinal uses of these plants have been two of the most relevant and consistent reasons for popular plant management, even in cultures that are increasingly losing their close relationship with nature. It is for this reason that ethno-directed research is very useful in the discovery and development of new drug and food resources (Khafagi and Deward 2000; Heinrich and Gibbons 2001). In remote rural settlements where vegetable cultivation is not practiced and market supplies are not organized, local inhabitants depend on indigenous vegetables, both cultivated in kitchen gardens and wild, for enriching the diversity of food. Knowledge of such foods is part of traditional knowledge which is largely transmitted through participation of individuals of households (Misra et al. 2008). It is of outmost importance to obtain data about popular uses of wild edible plants before this knowledge disappears. In many Mediterranean countries these traditions are at risk of disappearing, and hence the crucial need to study such knowledge systems and find innovative ways of infusing them to the future generations (Hadjichambis et al. 2007; Pieroni et al. 2005).

Despite agricultural societies' primary reliance on crop plants, the tradition of eating wild plants has not completely disappeared, because of their nutritional role and health benefits. Wild edible plants are important as dietary supplements, providing trace elements, vitamins, and minerals. However, consumption is determined less by calorie input and more by the pleasure of gathering wild resources, recreating traditional practices, and enjoying characteristic flavors (Pardo-de-Santayana et al. 2007). Both food and medicinal plants have interventional uses. This exists mainly in indigenous and local traditions. Food can be used as medicine and vice versa. However, certain wild edible plants are used because of their assumed health benefits and thus can be called medicinal foods (Etkin 1994).

Food provides not only essential nutrients needed for life but also other bioactive compounds for health promotion and disease prevention. Previous epidemiologic studies have consistently shown that diet plays a crucial role in the prevention of chronic diseases (Temple 2000). Consumption of fruit and vegetables, as well as grains, has been strongly associated with reduced risk of cardiovascular disease, cancer, diabetes, Alzheimer disease, cataracts, and age-related functional decline. Heart disease, cancer, and stroke are the top 3 leading causes of death in the world. This convincing evidence suggests that a change in dietary behavior such as increasing consumption of fruit, vegetables, and grains is a practical strategy for significantly reducing the incidence of chronic diseases (Willett 1995; Espin et al. 2007; Bazzano et al. 2001; Liu 2003). The relationship between food and health becomes increasingly significant as consumers now demand healthy, tasty, and natural functional foods that have been grown in uncontaminated environment.

There is a desire for a wide variety of choice and willingness to pay more for such foods in order to maintain healthy, well-balanced diet (Ercisli 2007).

References

Ahmad, H. (1999). Issues regarding medicinal plants of Pakistan. *Udyana Today, 6*(3), 6–7.

Arnold, J. E. M., & Rruiz Perez, M. (2001). Can non timber forest products match tropical forest conservation and development objectives? *Ecological Economics, 39*, 437–447.

Bazzano, L. A., He, J., & Ogden, L. G. (2001). Legume consumption and risk of coronary heart disease in US men and women: NHANES I epidemiologic follow-up study. *Archives of Internal Medicine, 161*, 2573–2578.

Bell, J. (1995). The hidden harvest. In *Seedling. The Quarterly Newsletter of Genetic Resources Action International*. Retrieved from www.grain.org/publications/

Bhattarai, N. K. (1998). Home herbal remedies of the urban population of Kathmandu valley, Nepal. *Journal of Nepal Pharmaceutical Association, 15*(1–2), 13–27.

Cavender, A. (2006). Folk medicinal uses of plant foods in southern Appalachia, United States. *Journal of Ethnopharmacology, 108*, 74–84.

Dovie, D. B. K., Witkowski, E. T. F., & Shackleton, C. M. (2005). Monetary valuation of livelihoods for understanding the composition and complexity of rural households. *Agriculture and Human Values, 22*, 87–103.

Dutta, A., & Pant, K. (2003). The nutritional status of indigenous people in the Garhwal Himalayas, India. *Mountain Research and Development, 23*(3), 278–283.

Ercisli, S. (2007). Chemical composition of fruits in some rose (Rosa spp.) species. *Food Chemistry, 104*, 1379–1384.

Espin, J. C., Conesa, M. T. G., & Tomás-Barberán, F. A. (2007). Nutraceuticals: Facts and fiction: A review. *Phytochemistry, 68*, 2986–3008.

Etkin, N. (1994). The cull of the wild. In N. L. Etkin (Ed.), *Eating on the wild side: The pharmacologic, ecologic, and social implications of using non cultigens* (pp. 1–21). Tucson: University of Arizona Press.

Food and Agriculture Organization of the United Nations (FAO). (2004). The State of food insecurity in the world. Monitoring the progress towards the world food summit and millennium development goals. Annual report, Rome.

Hadjichambis, A. C. H., Hadjichambi, D. P., Della, D., Giusti, M., Pasquale, D. E. C., Lenzarini, C., Censorii, F., Tejero, M. R. G., Rojas, C. P. S., Gutierrez, J. R. R., Skoula, M., Johnson, C. H., Sarpakia, A., Hmomouchi, M., Jorhi, S., Demerdash, M. E., Zayat, M., & Pioroni, A. (2007). Wild and semi-domesticated food plant consumption in seven circum-Mediterranean areas. *International Journal of Food Science and Nutrition, 59*(5), 383–414.

Hamilton, A. (2002). *Curriculum development in applied ethnobotany*. Proceedings of the workshop on curriculum development in applied ethnobotany, Nathiagali, Pakistan (pp. 91–95).

Heinrich, M., & Gibbons, S. (2001). Ethnopharmacology in drug discovery an analysis of its role and potential contribution. *Journal of Pharmacy and Pharmacology, 53*, 425–432.

Heywood, V., & Skoula, M. (1999). The MEDUSA Network: Conservation and sustainable use of wild plants of the Mediterranean region. In J. Janick (Ed.), *Perspectives on new crops and new uses* (pp. 148–151). Alexandria, VA: ASHS Press. Retrieved from http://www.ethnobiomed.com/content

Jasmine, T. S., Jeeva, S., Febreena, G. L., Mishra, B. P., & Laloo, R. C. (2007). Wild edible plants of Meghalaya, North-east India. *Natural Product Radiance, 6*, 410–426.

Khafagi, I. K., & Deward, A. (2000). The efficiency of random versus ethno-directed research in the evaluation of Sinai medicinal plants for bioactive compounds. *Journal of Ethnopharmacology, 71*, 365–376.

Kristensen, M., & Balslev, H. (2003). Perceptions, use and availability of woody plants among the Gourounsi in Burkina Faso. *Biodiversity and Conservation, 12*, 1715–1739.

Ladio, A. H., & Lozada, M. (2004). Patterns of use and knowledge of wild edible plants in distinct ecological environments: A case study of a Mapuche community from northwestern Patagonia. *Biodiversity and Conservation, 13*(6), 1153–1173.

Levetin, E., & McMahon, K. (1999). *Plants and society* (pp. 384). New York: McGraw-Hill.

Liu, R. H. (2003). Health benefits of fruits and vegetables are from additive and synergistic combination of phytochemicals. *American Journal of Clinical Nutrition, 78*, 517S–520S.

Misra, S., Maikhuri, R. K., Kala, C. P., Rao, K. S., & Saxena, K. G. (2008). Wild leafy vegetables: A study of their subsistence dietetic support to the inhabitants of Nanda Devi Biosphere Reserve, India. *Journal of Ethnobiology and Ethnomedicine, 4*, 15.

Ohiokpehai, O. (2003). Promoting the nutritional goodness of traditional food products. *Pakistan Journal of Nutrition, 2*(4), 267–270.

Orech, F. O., Hansen, J. A., & Friis, H. (2007). Ethnoecology of traditional leafy vegetables of the Luo people of Bondo district, western Kenya. *International Journal of Food Science and Nutrition, 58*(7), 522–530.

Pardo-de-Santayana, M., Tardo, J., Blanco, E., Carvalho, A. M., Lastra, J. J., San Miguel, E., & Morales, R. (2007). Traditional knowledge of wild edible plants used in the northwest of the Iberian Peninsula (Spain and Portugal): A comparative study. *Journal of Ethnobiology and Ethnomedicine, 3*, 27.

Pattanayak, S., Sills, E. O., Mehta, A. D., & Kramer, R. A. (2003). Local use of parks: Uncovering use of household production from forests of Siburet, Indonesia. *Censer Society, 1*(2), 209–222.

Pieroni, A., Houlihan, L., Ansari, N., Hussain, B., & Aslam, S. (2007). Medicinal perceptions of vegetables traditionally consumed by South-Asian migrants living in Bradford, Northern England. *Journal of Ethnopharmacology, 113*, 100–110.

Pieroni, A., Nebel, S., Santoro, R. F., & Heinrich, M. (2005). Food for two seasons: Culinary uses of non-cultivated local vegetables and mushrooms in a south Italian village. *International Journal of Food Science and Nutrition, 56*(4), 245–272.

Pimentel, D., Nair, M. M., Buck, L., Pimentel, M., & Kamil, J. (1997). The value of forests to world food security. *Human Ecology, 25*(1), 91–120.

Qureshi, R. A., Ghufran, M. A., Sultana, K. N., Ashraf, M., & Khan, A. G. (2006). Ethnobotanical studies of medicinal plants of Gilgit district and surrounding areas. *Ethnobotany Research and Applications, 5*, 115–122.

Scherrer, A. M., Motti, R., & Weckerle, C. S. (2005). Traditional plant use in the areas of Monte Vesole and Ascea, Cilento National Park (Campania, Southern Italy). *Journal of Ethnopharmacology, 97*, 129–143.

Shackleton, C., & Shackleton, S. (2004). The Pterocarpus angolensis DC. based woodcraft industry in the Bushbuckridge District, South Africa. In T. Sunderland & O. Ndoye (Eds.), *Forest products, livelihoods and conservation* (pp. 203–228). Centre for International Forest Management: Bogor-Barat, Indonesia.

Shinwari, Z. K., Khan, J., Naz, S., & Hussain, A. (2009). Screening of medicinal plants of Pakistan for their antibacterial activity. *African Journal of Biotechnology, 8*(24), 7082–7086.

Shinwari, Z. K., Rehman, M., Watanabe, T., & Yoshikawa, Y. (2006). *Medicinal and aromatic plants of Pakistan (a pictorial guide)* (pp. 492). Kohat, Pakistan: Kohat University of Science and Technology.

Stewart, R. R. (1972). *Flora of West Pakistan. An annotated catalogue of vascular plants of West Pakistan and Kashmir* (pp. 110–250). Karachi, Pakistan: Fakhri Printing Press.

Temple, N. J. (2000). Antioxidants and disease: More questions than answers. *Nutrition Research, 20*, 449–459.

Willett, W. C. (1995). Diet, nutrition, and avoidable cancer. *Environmental Health Perspectives, 103*(8), 165–170.

Zaman, M. B., & Khan, M. S. (1970). *Hundred drug plants of West Pakistan* (pp. 1–73). Islamabad, Pakistan: Printing Corporation of Pakistan.

Chapter 3
Phytochemicals and Nutraceuticals

3.1 Phytochemistry: Historical Perspective

Phytochemistry deals with chemical processes associated with plant life and the chemical compounds produced by plants. Phytochemistry is mainly concerned with the enormous variety of organic substances that are elaborated and accumulated by plants, their chemical structures, biosynthesis, turnover, metabolism, natural distribution, and biological function (Harborne 1973). A few compounds such as cane-sugar, starch, camphor, and benzoic acid had long been known. Aqueous extraction and evaporation had long been used for the preparation of sugar from sugarcane. The first chemical substance isolated from the plants was benzoic acid in 1560. K. W. Scheele (1742–1786) successfully isolated citric, gallic, malic, oxalic, lactic, tartaric and prussic acids, and glycerol from various herbal sources. In the nineteenth century progress became more rapid. In 1803 narcotine, the first alkaloid was isolated; morphine, strychnine, emetine, and many other followed quickly.

Search for useful medicines from plant kingdom, however, did not really begin until about 1806, when Serturner isolated morphine from the dried latex of *Papaver somniferum* L. (Opium). Between 1813 and 1823 Cheverul elucidated the chemical nature of fats and fixed oil. Later, many more alkaloids and other natural products were extracted from different plant extracts. Coniine was the first alkaloid to have its structure established by Schiff (1870) and to be synthesized by Ladenburgh (1886). Until the middle of the twentieth century the main emphasis in natural product chemistry remained the isolation and structure determination of various compounds. Indeed by this time the attention of natural product chemists was turning to the elucidation of the actual biosynthetic pathways found in the plant. Such studies were made possible by the introduction of new techniques of separation and analysis (Trease and Evans 1983).

A.M. Abbasi et al., *Wild Edible Vegetables of Lesser Himalayas*,
DOI 10.1007/978-3-319-09543-1_3

3.2 Classification and Significance of Phytochemicals

The term "Phytochemicals" refers to a wide variety of non-nutritive plant-based compounds which have protective or disease preventive properties. Phytochemicals are commonly found in fruits, vegetables, beans, and grains. Scientists have identified thousands of phytochemicals, although only a small fraction has been studied closely. Phytochemicals are nonessential nutrients; they are not required by the human body for sustaining life. It is well-known that plants produce these chemicals to protect themselves, but recent research demonstrates that they can also protect humans against various diseases.

There are more than thousand known phytochemicals but following are the major groups of the phytochemicals:

3.2.1 Phenolic Compounds

Phenolic compounds are the most important class of phytochemicals present in plant-based food sources. These compounds, including hydroxyl benzoic acid and their derivatives, flavonoids, flavanones, flavonols, flavones, catechols, anthocyanins, anthraquinones, etc., are the fundamental constituents of the food plants (Aliyu et al. 2008). It has long been known that phenolic compounds are effective antioxidants. Indeed, they have been used for decades as industrial antioxidants. The ability of phenolic compounds to quench free radicals arises because of both their acidity (ability to donate protons) and their delocalized π-electrons (ability to transfer electrons while remaining relatively stable) characteristic of benzene rings (Brown 1995). Perhaps the most well-known physiological phenolic antioxidant is vitamin E. It is an apt antioxidant for membranes because its long aliphatic tail can embed itself in lipid membranes while exposing its phenolic head group to the aqueous environment. A free radical can easily abstract hydrogen from phenol's hydroxyl group, creating a phenolic radical. The phenolic radical stabilizes itself by converting to a quinone (McKay and King 1980). Its conversion to quinine terminates the chain reaction and protects surrounding lipids from oxidation. Phenolic antioxidant therefore acts as a bullet-proof vest. The biological functions of phenolic compounds including antibacterial, anti-inflammatory, antiallergic, hepatoprotective, antithrombotic, antiviral, anticarcinogenic, and vasodilatory actions (Soobrattee et al. 2005) have been accredited to their free radical scavenging and antioxidant activity. There are numerous types of phenolic phytochemicals, but they can be organized into two main groups; polyphenols and flavonoids (Escarpa and González 2000).

3.2.2 Polyphenols

Polyphenols are the most abundant antioxidants in the diet. Their total dietary intake could be as high as ~1 g/day, which is much higher than all other classes of phytochemicals and known dietary antioxidants. For perspective, this is ~10 times higher than the intake of vitamin C and 100 times higher that the intakes of vitamin E and carotenoids (Manach et al. 2004). Their main dietary sources are fruits and plant-derived beverages such as fruit juices, tea, coffee, and red wine. Vegetables, cereals, chocolate, and dry legumes also contribute to the total polyphenol intake. Until the mid-1990s, the most widely studied antioxidants were antioxidant vitamins, carotenoids, and minerals. Research on flavonoids and other polyphenols, their antioxidant properties, and their effects in disease prevention truly began after 1995. Current evidence strongly supports a contribution of polyphenols to the prevention of cardiovascular diseases, cancers, and osteoporosis and suggests a role in the prevention of neurodegenerative diseases and diabetes mellitus. Polyphenols function as efficient free radical scavengers by donating their alcoholic hydrogen or one of their delocalized electrons. These compounds can defend body against cancer by protecting DNA from radical damage and avert cardiovascular disease by preventing oxidation of lipids and cholesterol (Escarpa and González 2000).

3.2.3 Flavonoids

Flavonoids are actually the most plentiful and effective group of phenolic compounds and so are given their own category (Madsen et al. 1996). The abundance of flavonoids in plants might be responsible for their therapeutic effectiveness against wide array of microorganisms (Lovkova et al. 2001). Apples, berries, and other brightly colored fruits and vegetables are important sources of flavonoids. These compounds play dual role for plants. They provide bright colors to the petals which attract pollinating insects, while protect lipids and vital cell components from damaging oxidation (Escarpa and González 2000). Many flavonoids are shown to have antioxidative activity, free-radical scavenging capacity, coronary heart disease prevention, and anticancer activity, while some flavonoids exhibit potential for anti-human immunodeficiency virus functions. Flavonoids are potent free radical scavengers of singlet oxygen and a variety of free radicals which are possibly concerned in DNA damage and tumor promotion (Marchand 2002), because they can donate their alcoholic hydrogen atom to free radicals (Madsen et al. 1996). Teas, broccoli, brussels sprouts, cabbage, cauliflower, grapes, eggplant, red cabbage, and radishes all contain *anthocyanidins—flavonoids* that are thought to act as antioxidants and may protect against some cancers and heart disease. Quercetin, another flavonoid that appears to have antioxidant properties, is found in apples, onions, teas, and red wine.

3.2.4 Flavonols

Flavonols are a subgroup of flavonoids that occur in varying levels in commonly consumed fruits, vegetables, and plant-based beverages. Apple, onion, grape berries, red lettuce, tomatoes, black tea, green tea, red wine, and broccoli are major sources of the flavonols. The key dietary flavonols include Quercetin, Kaempferol, and Myricetin. Flavonols provide numerous health benefits. For example, increased intake of flavonols is associated with reduced risk of cardiovascular diseases. This can be attributed to their antioxidant properties which helps improve endothelial function and reduce platelet activity. Furthermore, antioxidant properties of flavonols also help prevent oxidative damage to cells, lipids, and DNA. The antioxidant properties of the flavonols stem from the presence of aromatic rings of the flavonoid molecules, which allow the donation and acceptance of electrons from free radical species. This helps quench free radicals. In addition, intake of flavonols is associated with reduced risk of cancer and stroke. Furthermore, some flavonols also promote bone health, prevent osteoporosis, and have anti-inflammatory properties. Flavonols also promote healthy brain, as they possess neuroprotective properties. Hence, consumption of food rich in flavonols is associated with long-term health benefits.

3.2.5 Vitamins

Vitamins are organic compounds occurring in natural foods especially in vegetables either as such or as utilizable precursors. Vitamins are needed for maintenance of skin, mucous membranes, bones, teeth and hair, vision, and reproduction. They help body to absorb calcium and phosphorous; needed for bone growth and maintenance. Vitamins are involved in blood clotting, normal functioning of nervous system, and endocrine glands. They are also needed for metabolism of macro molecules (Chatterjea and Shinde 1998). ***Ascorbic acid*** is a six-carbon lactone that is synthesized from glucose in the liver of most mammalian species, but not by humans, non-human primates, and guinea pigs. These species do not have the enzyme gulonolactone oxidase, which is essential for synthesis of the ascorbic acid immediate precursor 2-keto-L-gulonolactone. The DNA encoding for gulonolactone oxidase has undergone substantial mutation, resulting in the absence of a functional enzyme (Nishikimi and Yagi 1996). Consequently, when humans do not ingest vitamin C in their diets, a deficiency state occurs with a wide spectrum of clinical manifestations. Clinical expression of vitamin C deficiency, scurvy, is a lethal condition unless appropriately treated. Thus, humans must ingest vitamin C to survive. Vitamin C is an electron donor and therefore a reducing agent. All known physiological and biochemical actions of vitamin C are due to its action as an electron donor. Ascorbic acid donates two electrons from a double bond between the second and third carbons of the 6-carbon molecule. Vitamin C is called an antioxidant because, by donating its electrons, it prevents other compounds from being oxidized. However, by the very nature of this reaction, vitamin C itself is oxidized in the process (Padayatty et al. 2003).

3.2.6 Phytoestrogens

Phytoestrogens are seemingly estrogen-like substances from plant sources which mimic estrogen, the primary sex hormone in women, and is responsible for the menstrual cycle. Fruits, such as, dates, cherries, and apples and vegetables, such as, carrots and potatoes contain varying amounts of phytoestrogens. They may play a role in the development of and protection against some hormone-dependent cancers such as some types of breast and prostate cancer and menopausal symptoms.

Iso-flavones make up the most common form of phytoestrogens. They have a common diphenolic structure that resembles the structure of the potent synthetic estrogens, diethylstilbestrol and hexestrol. Two of the major iso-flavones found in humans are genistein and daidzein. The iso-flavones are found in foods and supplements such as, soy products, red clover, garbanzo beans, fruits, and vegetables. The consumption of soy-products has many health benefits like protection against breast and prostate cancer. However, they may increase the incidence of epithelial hyperplasia and cause goiter and hyperthyroidism. A recent study has demonstrated that iso-flavones have potent antioxidant properties, comparable to that of the well-known antioxidant vitamin E. The antioxidant powers of iso-flavones can reduce the long-term risk of cancer by preventing free radical damage to DNA. Genistein is the most potent antioxidant among the soy iso-flavones, followed by daidzein.

3.2.7 Carotenoids

These are colorful plant pigments, which give carrots, yams, cantaloupe, butternut squash, and apricots their attractive color. These chemical substances are powerful antioxidants and provide protection against heart diseases and cancer. Bright-orange beta-carotene is the most important carotenoid for adequate vitamin A intake because it yields more vitamin A than alpha- or gamma-carotene. Lycopene, the orange-red pigment found in tomatoes, red peppers, pink grapefruit, and watermelon, is a powerful antioxidant even more potent than beta-carotene. Generally, the deeper color of the fruit or vegetable is an indication of a higher concentration of carotenoids. Carrots, for example, are especially good sources of beta-carotene. Green leafy vegetables such as spinach, asparagus, and broccoli also contain large amounts of carotenoids. The lutein and zeaxanthin are carotenoids found in spinach, kale, and turnip greens that may reduce the risk of some cancers and eye diseases.

3.2.8 Alkaloids

Alkaloids are nitrogenous organic molecules having pharmacological effect on humans and animals. These phytochemicals are the derivatives of pyrrolidine, pyridine, piperidine, quinoline, isoquinoline, indulge purine, etc. (Lovkova et al. 2001). Alkaloids are important constituents of most of the valuable drugs. They are active constituents with high biological activity found in herbs and plants. Probably one of the most famous alkaloids is morphine, for its medical uses and powerful properties. Other less known but also very important alkaloid is serotonin, a powerful neurotransmitter and tryptophan. More than 800 alkaloids have been discovered so far; cocaine, morphine, dopamine, and nicotine are some important and powerful alkaloids.

3.2.9 Glycosides

Glycosides are unique molecules used by plants to safely store potentially dangerous compounds and by animals to eliminate toxins. They consist of a sugar group, called a glycone, chemically attached to a non-carbohydrate compound. The non-carbohydrate is referred to as an aglycone, or a genin. These molecules have a variety of natural uses, such as defence mechanisms, as well as commercial ones, like medications, cleaning agents, and natural sweeteners. Glycosides are divided into cardiac glycosides, anthra glycosides, iridoids, etc. Cardiac glycosides induce strong specific effects on the myocardium and enhance the strength of cardiac contractions (Lovkova et al. 2001; Edeoga and Eriata 2001).

Other important phytochemical substances found in plants are: ***Terpenoids*** including monoterpenes, diterpenes, tetraterpenes, polyterpenes, steroid saponins; ***Polysaccharides*** including pectins, resins, and mucilages; and ***Essential oils*** isolated from different plants (Lovkova et al. 2001). Likewise, ***Indoles*** are found in cabbages, stimulate enzymes that make the estrogen less effective, and could reduce the risk for breast cancer. ***Saponins*** are found in beans and interfere with the replication of cell DNA, thereby preventing the multiplication of cancer cells. ***Capsaicin*** found in hot peppers protects DNA from carcinogens. Another phytochemical ***allicin*** from garlic has anti-bacterial properties. Some phytochemicals bind physically to cell walls, thereby preventing the adhesion of pathogens to human cell walls. ***Proanthocyanidins*** are responsible for the anti-adhesion properties of cranberry. Consumption of cranberries will reduce the risk of urinary tract infections and will improve dental health. Other phytochemicals, which interfere with enzymes, are ***protease*** inhibitors (soy and beans), ***terpenes*** (citrus fruits and cherries). In fact, most foods contain phytochemicals except for some refined foods such as, sugar or alcohol. Some foods, such as whole grains, vegetables, beans, fruits, and herbs, contain many phytochemicals. The easiest way to get more phytochemicals is to consume more fruits and vegetables. For instance, carrot contains more than hundred phytochemicals. The recommended daily intake of the fruits and/or vegetable is at least 5–9 servings. Fruits and vegetables are also rich in minerals, vitamins, and fibers and low in saturated fat.

3.3 Plant-Based Nutrients

A chemical substance used by an organism to live and grow or a substance used in an organism's metabolism which must be taken in from its environment is called nutrient (Whitney et al. 2009). Nutrients are essential to build and repair tissues, regulate body processes and as a source of energy. Organic nutrients include carbohydrates, fats, proteins (or their building blocks, amino acids), and vitamins. Inorganic chemical compounds such as dietary minerals, water, and oxygen are also considered as nutrients (Sizer and Ellie 2007). A nutrient is said to be "essential" if it must be obtained from an external source, either because the organism cannot synthesize it or produces insufficient quantities. Food plants especially the vegetables are important nutritional components, with enriched contents of essential nutrients and energy. Most of the vegetables are rich in essential oils, glycosides, pigments, etc. which stimulates appetite. Large number of species and varieties of food plants/vegetables act as raw materials for preparing a variety of foods. Major components of vegetables are water contents, sugars, proteins, starch, fats, etc. In comparison with the foods of animal origin, vegetable products have a lower food value and a lower heat, but have special importance in human nutrition, with high content of vitamins, minerals, and fibers.

Nutrients are classified into two main groups; macronutrients and micronutrients (Audrey 1994). The chemical substances that humans consume in the largest quantities are called macronutrients, such as, carbon, hydrogen, nitrogen, oxygen, phosphorus, sulphur, etc. These nutrients can be further divided into two groups; primary nutrients (nitrogen, phosphorus, potassium) and secondary nutrients (sulphur, silicon, calcium, magnesium). The chemical elements that humans consume in smaller amount such as, boron, iron, copper, manganese, molybdenum, zinc, and cobalt are called micronutrients. The chemical compounds consumed by humans in the largest quantities and providing the bulk of energy are carbohydrates, proteins, and fats.

3.3.1 Carbohydrates

Carbohydrates are composed of different types of sugars and are classified by their number of sugar units: monosaccharides (such as, glucose and fructose), disaccharides (such as, sucrose and lactose), oligosaccharides, and polysaccharides (such as, starch, glycogen, and cellulose). Many vegetables are good sources of carbohydrates (Butnariu 2007). Average carbohydrates contents of the vegetable are 9–10 %.

Generally, the average 1,800-calorie diet should contain between 210 and 310 g of carbohydrates each day, which is equal to 45–65 % of our daily calories.

3.3.2 Proteins

Proteins are organic compounds consisting of the amino acids joined by peptide bonds. Humans cannot manufacture some of the amino acids (termed as essential amino acids); the diet must supply these. Proteins are broken down through digestion by proteases back into free amino acids. In food, plant proteins include albumins, globulins, glutelins, and peptides, comprising about 2 % of total weight of the vegetables. Humans get 10–35 % calories from proteins and roughly about 50–145 g of proteins are required each day. The energy contents of carbohydrates and proteins are about 4 kcal/g (~16.7 kJ/g).

3.3.3 Lipids

Lipids are composed of fatty acids, fatty acid esters, phospholipids, and glycolipids. Fats or lipids contents are about 0.3 % of total weight of the vegetable. Fats consist of a glycerine molecule with three fatty acids, which may be saturated or unsaturated. Fats are vital in functioning of cell membranes, to insulate body organs against shock, to keep body temperature stable, and to maintain healthy skin and hair. According to Coyle (1995) the energy contents of the fats are about 9 kcal/g (~37.7 kJ/g).

3.3.4 Minerals and Trace Elements

Vegetables are considered as good sources of the minerals and contribute the essential nutrients, which are required by the human body for healthy functioning. These minerals are naturally occurring inorganic solids (O'Donoghue 1990) and are essential constituents for normal metabolic activities of human body tissues, bones, teeth, blood, muscles, hair, and nerve cells. Vitamins cannot be properly assimilated without the correct balance of the minerals (Sonni 2002). The mineral substances are present as salts of organic or inorganic acids or as complex organic combinations; they are in many cases dissolved in cellular juice. Some minerals that are particularly important to the body are calcium, magnesium, potassium, sodium, iron, phosphorus and chloride, etc.; these are needed in larger amounts than others which are required in smaller quantities, are usually less than 0.01 % of the total body weigh, and are called trace minerals/metals. Most of them are abundantly available in fresh vegetables.

Trace metals occur naturally in the environment and become integrated into plants and animals through soil, food, air, and water. Trace metals, such as, chromium, cobalt, copper, manganese, and zinc, occur normally at very low levels in the environment. Living organisms need very small amounts of these trace metals; elevated levels of these metals might be toxic. Concentrations of trace metals in the environment increase when they are released either through natural processes or human activities. Natural processes include breakdown of rocks, spreading of mid-ocean ridges, and volcanic activity. Human activities that release trace metals into the environment include mining, smelting, coal combustion, industrial emissions, automobile emission, and wastewater disposal (Marsden and Rainbow 2004). For the most part, human contributions of trace metals to the environment have been approaching or even exceeding the natural inputs.

Nitrogen is a part of all living cells (about 3 % by weight in human body) and also essential constituent of all proteins, amino acids, nucleic acids, enzymes, and metabolic processes involved in the synthesis and transfer of energy, digestion of food, and growth of foetus during pregnancy. Main source of nitrogen in soil is fertilizer application, although air/water contributes very little nitrogen to the plants (Uchida 2000). Phosphorus is also essential and makes up about 1 % of total body weight, although 85 % of the phosphorus is found in the bones and teeth. Recommended daily intake of phosphorus is 1000 mg/day. It plays an important role in the body's utilization of carbohydrates/fats and in proteins synthesis for the growth, maintenance, and repair of cells and tissues. It also assists in the contraction of muscles, in the functioning of kidneys, in maintaining the regularity of the heartbeat, and in nerve conduction. In a limited supply, phosphorus has the biggest impact on legumes which generally rely on N_2 fixation for nitrogen supply, but the role of phosphorus in the symbiotic process is yet not clear (Robson et al. 1981). In other studies, phosphorus deficiency has been found to decrease nodule number, decrease nodule mass more than shoot growth, and decrease the specific nitrogenous activity of nodules (Jakobsen 1985; Singleton et al. 1985; Israel 1987; Hart 1989; Ribet and Drevon 1995; Vadez et al. 1996; Drevon and Hartwig 1997).

Sulphur is an important mineral element that can be obtained through dietary proteins. Half of the sulphur is found in muscles, skin, and bones. In addition to bonding with proteins, sulphur is also required for the proper structure and biological activity of enzymes. It plays a pivotal role in plant metabolism and development. Cysteine, the first organic-reduced sulphur compound, methionine, and its derivatives contribute not only to life as building blocks in proteins and their activity, but also as precursors for synthesis of glutathione, essential vitamins, sulphur esters, and derivatives (Leustek et al. 2000; Saito et al. 1994; Noctor et al. 2002). In addition, sulphur is also found in the sulpho head group of sulpholipid, a component of plant photosynthetic membranes derived from sulphite (Sanda et al. 2001). Boron is also a vital mineral that is required for normal growth/health; its recommended daily intake is less than 20 mg/day. In plants it helps in the regulation of other nutrients and involve in production of sugar and carbohydrates. In human body many diseases like arthritis, menopausal symptoms, and osteoporosis are naturally managed by boron. It is believed that boron improves the natural ability of human body to absorb calcium and magnesium (Shorrocks 1997; Yan et al. 2006).

3.4 Significance of Essential and Toxic Metals

3.4.1 Calcium

Calcium is one of the most important elements which is found in greater quantities than any other inorganic divalent ion in plants and has been recognized as an essential nutrient for plant growth and in the structure of animal's skeleton (Taylor and Locascio 2004). It is the most abundant metal in the human body and is the main constituent of bones and teeth. It is most commonly found in milk/milk products, vegetables, nuts, and beans. Calcium regulates membrane activity; it assists nerve impulse transfer and hormone release, stabilizes the pH of the body, and is an essential part of conception. It can also regulate the passage of proteins along the secretary pathway (Dolman and Tepikin 2006). Hard water may assist in strengthening bones and teeth because of its high calcium concentration, which may also decrease the risk of heart conditions. It plays a role in neuromuscular excitability, good function of the conducting myocardial system, heart and muscle contractility, intracellular information transmission, and blood coagulability. Osteoporosis and osteomalacia are the most common manifestations of calcium deficiency; a less common but proved disorder attributable to Ca deficiency is hypertension (Kožíšek 2003). In plants, calcium plays an important role in cell division and elongation, movement of cellular organelles such as, the spindle apparatus and secretary vesicles, cell wall development, nitrate uptake and metabolism, enzyme activity, control development process in cytoplasm, and starch metabolism. Calcium is transported in the xylem via an ion exchange mechanism. It attaches to lignin molecules and exchange must occur with calcium or another similar cation. Calcium is not very mobile in the soil, or in plant tissue; therefore, a continuous supply is essential although its deficiency symptoms can be rather vague since the situation often is accompanied by a low soil pH (Lenntech 2010a; Jaleel et al. 2007).

3.4.2 Cadmium

Cadmium has been recognized as an occupational health hazard for many decades. The risks to environmentally exposed populations were emphasized later when it was concluded that the infamous "itai itai" disease was caused by intake of cadmium-polluted rice. It is a well-known toxic metal occurring in the environment naturally and as a pollutant emanating from industrial and agricultural sources.

Food is the main source of cadmium intake in the non-smoking population. The bioavailability, retention, and toxicity are affected by several factors including nutritional status such as low iron status. It is nephrotoxic and can also cause bone damage, either via a direct effect on bone tissue or indirectly as a result of renal dysfunction. Atmospheric deposition of airborne cadmium, mining activities, and the application of cadmium containing fertilizers/sewage sludge on farm land may lead to the contamination of soil and increased cadmium uptake by crops and vegetables grown for human consumption. It is present in virtually all foods; but the concentrations vary to a great extent, depending on type of food and level of environmental contamination. High levels are found in oil seeds, cocoa beans, and in certain wild mushrooms. Food from plants generally contains higher concentrations of cadmium than meat, egg, milk and dairy products, and fish muscle (Järup and Akesson 2009). Among food from plants, cereals such as, rice and wheat, green leafy vegetables, potato, and root vegetables contain higher concentration than other food from plants. Based on estimation of cadmium intake, more than 80 % of the food-cadmium comes from cereals, vegetables, and potatoes. Cadmium in drinking water contributes only to less than a few percent of the total cadmium intake (Olsson et al. 2002). Tobacco smoking is another important source of cadmium exposure (Ilyasova and Schwartz 2005). A possible effect of cadmium on breast cancer risk was assessed in a case–control study; there was a statistically significant increase in risk with increasing cadmium level (McElroy et al. 2006).

3.4.3 Chromium

Chromium (III) occurs naturally in many vegetables, fruits, meats, yeasts, and grains; however, various ways of food preparation and storage may alter the chromium contents of food (Ahalya et al. 2010). It may exist in several chemical forms with oxidation numbers 0–VI, though two of them, trivalent and hexavalent, are stable enough in the environment. Cr (IV) and Cr (V) form only unstable intermediates in reactions of trivalent and hexavalent oxidation states with oxidizing and reducing agents, respectively (Horcsik et al. 2006). Chromium in environment may originate from natural sources, such as, weathering of rock constituents, wet precipitation/dry fallout from the atmosphere, and run-off from the terrestrial systems, while major anthropogenic sources are discharge of wastewater from the metallurgical, electroplating and tanning industries, dying, sanitary landfall leaching, water cooling towers, chrome-bearing sludge and refuse from industrial activity (Kotas and Stasicka 2000).

Chromium is a constituent of complex known as glucose tolerance factor (GTF), which also contains the nicotinic acid and various amino acids. It facilitates the binding of insulin to its cellular receptor, increases the insulin receptor number, and also improves the insulin internalization and sensitivity, which in turn lowers the glucose levels and increases the amino acids and fatty acid uptake after meal. Chromium is also important for growth, synthesis of DNA, RNA, and proper immune system (Manore et al. 2009). Excessive chromium can affect human oxidation–reduction and hydrolysis reactions. It can lead to denaturation of protein, precipitation of nucleic acid, and interfere with normal enzymatic activity. In addition, elevated Cr levels can lead to DNA damage, which may result in cancer-causing gene mutations. Cr (VI) is considered as human carcinogen by the route of inhalation, while ingesting its large amounts can cause stomach upsets and ulcers, convulsions, kidney, and liver damage (Li 2008). In plants high concentrations of chromium are required for chlorosis, necrosis, and a host of other growth abnormalities and anatomical disorders. It is also essential for the regulation of the mineral metabolism, enzyme activity, and other metabolic processes (Samantaray et al. 1998).

3.4.4 *Cobalt*

Cobalt is widely dispersed in the environment and humans may be exposed to it by breathing, drinking water, and eating food along with the dermal adsorption (ATSDR 2000; Hass 2010). It functions as a cofactor in enzyme-catalyzed reactions and is involved in the production of erythropoietin, a hormone that stimulates the formation of erythrocytes. It is stored in the red blood cells and the plasma, as well as in the liver, kidney, spleen, and pancreas. Cobalt is beneficial for humans because it is a part of vitamin B_{12} (hydroxycolalamin), which is essential for human health. Meat, liver, kidney, clams, oysters, and milk are a good source of cobalt, which is used to treat anemia with pregnant women, because it stimulates the production of red blood cells. This property of cobalt was applied in the past as a therapy for anemia (Boeck et al. 2003). The total daily intake of Co is variable, but almost all will pass through the body unabsorbed except absorbed as a part of vitamin B_{12}. However, too high concentrations of Co may damage human health resulting in vomiting, nausea, vision problems, heart problems, and thyroid damage. Increased intake may also cause overproduction of red blood cells, thickened blood, and increased activity in the bone marrow. International Agency for Research on Cancer (IARC) has listed cobalt and cobalt compounds within group 2B (agents which are possibly carcinogenic to humans) (Lenntech 2010b). In plants, Co promotes many growth processes, including stem and coleoptile elongation, leaf disc expansion, curvature of slit stems, opening of hypocotyl hooks, and bud development (Grover and Purves 1976).

3.4.5 *Copper*

Copper is released into the environment by natural sources which are wind-blown dust, decaying vegetation, forest fires, and sea spray, while human activities include mining, metal production, wood production, and phosphate fertilizer production. It plays an important role in energy metabolism; as cofactor enzyme of electron transport chain and required for super oxidase and antioxidant that protect the body against oxidative damage. It also plays an important role for synthesis and maintenance of collagen protein. Enzymes in the body that require the Cu (II) ion as cofactor are cytochrome-c-oxidase, superoxide dismutase, amino oxidase, laccase, plastocyanin, and polyphenol oxidase (Manore et al. 2009).

Copper can be found in many kinds of food, drinking water, and air. Long-term exposure to Cu can cause irritation of the nose, mouth, and eyes and it causes headaches, stomach-aches, dizziness, vomiting, and diarrhea. Very high uptakes may cause liver and kidney damage and even death (Lenntech 2010c). Industrial exposure to its fumes, dusts, or mists may result in metal fume fever with atrophic changes in nasal mucous membranes. The chronic poisoning results in Wilson's disease, characterized by a hepatic cirrhosis, brain damage, demyelination, renal disease, and copper deposition in the cornea (Mohan and Sumitha 2008). World Health Organization (WHO) recommended a maximum acceptable concentration of Cu^{2+} in drinking water of 1.5 mg/L. The excessive intake of Cu leads to severe mucosal irritation, widespread capillary damage, hepatic and renal damage, central nervous problems, followed by depression, gastrointestinal irritation, and possible necrotic changes in the liver and kidney (Sengil et al. 2009). It negatively influences the activity of microorganisms and earthworms due to which the decomposition of organic matter may seriously slow down (Shakya et al. 2006). In plants, Cu is an essential redox-active metal that is involved in many physiological processes in plants. It acts as a structural element in regulatory proteins and participates in photosynthetic electron transport, mitochondrial respiration, oxidative stress responses, cell wall metabolism, and hormone signalling (Marschner 1995; Raven et al. 1999).

3.4.6 Iron

Iron can be found in meat, whole meal products, potatoes, and vegetables. It is an essential part of hemoglobin; the red coloring agent of the blood that transports oxygen. Chronic inhalation of excessive concentrations of iron oxide fumes or dusts may result in development of a benign pneumoconiosis, called siderosis (Adebiyi et al. 2008), and may enhance the risk of lung cancer (Atolaiye et al. 2009). A more common problem for humans is iron deficiency, which leads to anemia. Recommended average daily intake of iron is 7 mg in man and 11 mg in woman; normal diet generally provides the required amounts. The total amount of iron in the human body is approximately 4 g, of which 70 % is present in red blood cells (Saljoughian 2007). Iron is essential for the production of hemoglobin, myoglobin, several essential enzymes, and involved in DNA synthesis. The essential enzymes that require Fe (II) as cofactor in body are pyruvate oxidase, mitochondrial cytochrome, ribonucleotide reductase, tyrosine and praline hydrolase, monoamine oxidase, glucose-6-phosphate (Manore et al. 2009). Normal brain functions are iron-dependent. It is also involved in the formation of red blood cells and helps body to fight against diseases (Wang and Chen 1990; Leung et al. 1999). When iron exceeds the required amount, it is stored in the liver. The bone marrow contains high amounts of iron, because it produces hemoglobin (Nadadur et al. 2008). Iron is also essential for the formation of chlorophyll in plants (Gray and Smith 2005).

3.4.7 Potassium

Potassium is essential for plant growth/reproduction and is often referred to as "the regulator" since it is involved in more than 60 different enzymes in plants. It helps plants to resist drought and excessive temperatures. It aids plants in the production of starch, controls root growth, and regulates the opening and closing of pores in plant cells, which is important for efficient water use (Blatt 2004). Adequate amounts of potassium may promote the growth; increase the shelf life of fruits; increase the stem length; enhance the green color; and increase the size and quality of grains/vegetables. The consequences of low potassium levels are apparent in a variety of symptoms: restricted growth; reduced flowering; and lower yields. In human body potassium plays an important role in the physical fluid system and it assists nerve functions. Inhalation of potassium dust or mists can irritate the eyes, nose/throat/lungs with sneezing, coughing, and sore throat. Higher exposures may cause a build-up of fluid in the lungs. The total amount of potassium in the human body lies somewhere between 110 and 140 g and mainly depends upon muscle mass, while the normal potassium level in human blood ranges from 3.5 to 5.2 mmol/L (Blatt 2004). The muscles contain most potassium after red blood cells and brain tissue. It protects the heart/arteries and may even prevent cardiovascular disease. Potassium shortages are relatively rare, but may lead to depression, muscle weakness, heart rhythm disorder, and confusion. These effects ultimately determine the morbidity and mortality related to this condition. Potassium is taken up by the plants in larger amounts than any other mineral element except nitrogen and calcium. Main sources of potassium for plants are soil minerals, organic materials, and fertilizer (Uchida 2000).

3.4.8 Magnesium

Magnesium is a dietary mineral for humans, one of the microelements that are responsible for membrane function, nerve-stimulant transmission, muscle contraction, protein construction, and DNA replication. The human body contains about 25 g of magnesium, of which 60 % is present in the bones and 40 % is present in muscles and other tissue. Daily intake is approximately 250–350 mg, while the minimum requirement is at least 200 mg, but the body deals very effectively with this element. There is no evidence of systemic poisoning associated with it, although persistent over-indulgence may lead to muscle weakness, lethargy, and confusion. Magnesium and calcium often perform the same functions within human body and are generally antagonistic (Hemat 2009). Magnesium is needed for more than 300 biochemical reactions in humans. It helps maintain normal muscle and nerve function, keeps heart rhythm steady, supports a healthy immune system, and keeps bones strong. Magnesium also helps regulate blood sugar levels, promotes normal blood pressure, and is known to be involved in energy metabolism and protein synthesis. There is an increased interest in the role of magnesium in preventing and managing disorders such as hypertension, cardiovascular disease, and diabetes (Chillemi and Chillemi 2007). Early signs of magnesium deficiency include loss of appetite, nausea, vomiting, fatigue, and weakness. Severe magnesium deficiency can result in low levels of calcium in the blood or hypocalcemia. Magnesium deficiency is also associated with low levels of potassium in the blood (Morley and Thomas 2007). In plants, magnesium is an essential constituent of chlorophyll, an activator of more than 300 enzymes and is involved in regulation of ions, transport as well as cations balance in plants (Hailes et al. 1997; Reuter and Robinson 1997). Its deficiency occurs particularly in plants growing in highly acidic soils with low cation-exchange capacity (Tan et al. 1991; Aitken et al. 1999).

3.4.9 Manganese

Manganese is found in small amounts in human body where it is a component of manganese super-oxide dismutase enzyme and acts as a powerful antioxidant that seeks out the free radicals in the human body and neutralizes these damaging particles. It is one of the toxic essential trace elements, which means that it is necessary for humans to survive, but at very high concentrations it is also toxic. Daily recommended intake of manganese is 1.8–2.3 mg. Some of its health benefits include healthy bone structure, bone metabolism, formation of connective tissues, absorption of calcium, proper functioning of the thyroid gland and sex hormones, regulation of blood sugar level, metabolism of fats/carbohydrates, and acting as co-enzyme to assist metabolic activity. Its metallo-enzymes include arginase, glutamine synthetase, phosphoenolpyruvate decarboxylase and manganese super-oxide dismutase (Otten et al. 2006). Exposure to high concentrations of manganese can cause harm to the nervous system; a disorder similar to Parkinson's disease can result (Finley et al. 2003). It is found in the environment either as inorganic manganese compounds or as organic manganese compounds; former are contributed by processes involved in steel production, batteries, ceramics, combustion in motor vehicles, and coal-burning industrial plants while the latter are used in some pesticides, fertilizers, and in a gasoline additive called methyl cyclopentadienyl manganese tricarbonyl (MMT) (ATSDR 2000). Manganese has been shown to be deposited in certain regions of the brain, and exposure to high concentrations in occupational studies was associated with permanent damage, with symptoms of impaired neurological and neuromuscular control, mental and emotional disturbances, muscle stiffness, lack of coordination, tremors, difficulties with breathing or swallowing, and other neuromuscular problems (USEPA 2010). Higher concentrations are found in foodstuffs, such as, grains, rice, soya beans, eggs, nuts, olive oil, green beans, and oysters (Mohan and Sreelakshmi 2008). Deficiency of manganese can also cause health effects such as fatness, glucose intolerance, blood clotting, skin problems, lowered cholesterol levels, skeleton disorders, birth defects, changes of hair color, and neurological symptoms. In plants, it plays a direct role in photosynthesis, accelerates seed germination and maturity, increases the validity of P and Ca, supports movement of iron in plant, and acts as catalysts in oxidation and reduction reaction within the plant tissue (Lenntech 2010d).

3.4.10 Sodium

Sodium is a component of many foodstuffs and it is necessary for humans to maintain the balance of the physical fluids system. It is also required for nerve and muscle functioning. Too much sodium can damage the kidneys and increase the chances of high blood pressure. Although it is essential, but controversy surrounds the amount required (Permyakov 2009). Contact of sodium with water, including perspiration, causes the formation of sodium hydroxide fumes, which are highly irritating to skin, eyes, nose, and throat. This may cause sneezing and coughing (Sittig 1991). Very severe exposures may result in difficult breathing, coughing, and chemical bronchitis. Contact to the skin may cause itching, tingling, thermal, and caustic burns and permanent damage. It regulates extra-cellular fluids, acid–base balance, and membrane potential, partially together with potassium. Sodium shortages may lead to dehydration, convulsion, muscle paralysis, decreased growth, and general numbness. Generally, humans require about 300 mg sodium chloride per day to warrant a balanced sodium level (Hemat 2009). Sodium is involved in the regeneration of phosphoenolpyruvate and can also replace potassium regulation in stomatal opening in plants (Lehr 2006).

3.4.11 Lead

Lead has been mined, smelted, refined, and used for hundreds of years. Although it is naturally occurring, significant amounts are released into the environment from anthropogenic sources. It has been used as an additive in paints and gasoline, leaded pipes, solders, crystals, and ceramics. Higher concentrations are observed near mining/smelting sites, industrial/municipal facilities, and in areas adjacent to highways. Lead released into the environment makes its way into the air, soil, and water. The Pb in fuels contributes to air pollution, especially in urban areas (Paz-Alberto et al. 2007). Its salts enter the environment through the exhausts of automobiles.

The larger particles drop to the ground immediately and pollute the soils or surface water. This lead-cycle caused by human production is much more extended than the natural lead-cycle. Food, such as, fruits, vegetables, meats, grains, seafood, soft drinks, and wine may contain significant amounts of lead. It can also enter the drinking water through corrosion of pipes (Mongillo and Linda 2000). The harmful effects of the lead included disruption of the biosynthesis of hemoglobin and anemia, a rise in blood pressure, kidney damage, miscarriages/subtle abortions, disruption of nervous systems, brain damage, declined fertility of men through sperm damage, diminished learning abilities of children, behavioral disruptions of children, such as, aggression, impulsive behavior, and hyperactivity. Lead can enter a fetus through the placenta of the mother, thus can cause serious damage to the nervous system and the brains of unborn children (Jarup 2003).

3.4.12 Zinc

Zinc is essential for human health; it is required for structure and function of more than 300 enzymes in human body. Its deficiency decreases the resistance against bacterial, viral, fungal, and parasitic pathogens. Enzymes that require the Zn (II) ion as cofactor for their proper functions are iodate dehydrogenase, carbonic anhydrase, malate dehydrogenase, carboxypeptidase, alkaline phosphate, and alcohol dehydrogenase (Calder et al. 2002). Too little availability of Zn may result in a loss of appetite, decreased sense of taste and smell, slow wound healing, and skin sores. Its shortage can even cause birth defect and sexual immaturity. Although humans can handle proportionally large concentrations of zinc, too much concentration can still cause eminent health problems, such as stomach cramps, skin irritations, vomiting, nausea, and anemia. Very high Zn levels can damage the pancreas and disturb the protein metabolism and cause arteriosclerosis (Jeejeebhoy 2009). Higher Zn application appears to protect people from Cd poisoning. It may also decrease Pb absorption. Elevated levels of Zn can be a danger to unborn and newborn children who are exposed to it through blood or milk of their mothers (Lenntech 2010e).

Water may be polluted by Zn due to the presence of its large quantities in wastewater of industrial plants. One of the consequences is that rivers are depositing zinc-polluted sludge on their banks. Plants often have a zinc uptake that their systems cannot handle, due to which on zinc-rich soils only a limited number of plants have a chance of survival. It can also interrupt the activity in soils, as it negatively influences the activity of microorganisms and earthworms (Simeonov and Sargsyan 2008). It is commonly found in red meat and poultry, beans, nuts, seafood, whole grains, fortified breakfast cereals, and dairy products (Fraga 2005). In plants zinc plays significant role in many metabolic processes (Rout and Das 2003; Aravind and Prasad 2003, 2004a, b, 2005a, b; Vallee and Auld 1990). It is also involved in chlorophyll formation by taking part in the regulation of concentrations of nutrients in cytoplasm and increases the biosynthesis of chlorophyll and carotenoids (Aravind and Prasad 2005a).

3.5 Nutraceuticals in Plants

The term "nutraceutical" is a combination of two words "nutrition" and "pharmaceutical" and was first coined by Dr. Stephen L. DeFelice. Nutraceuticals are the substances that may be considered as food or part of food and provide health benefits like the prevention and treatment of various diseases. Such substances may range from isolated nutrients, dietary supplements, herbal products, specific diets, genetically modified foods, and processed products, such as, cereals, soups, and beverages. Some examples of the neutraceuticals are dietary fiber, polyunsaturated fatty acids, proteins, peptides, amino acids, keto acids, minerals, vitamins, and antioxidants (Andlauer and Furst 2002; Temple 2000). A nutraceutical is demonstrated to have a physiological benefit or provide protection against chronic diseases.

Plants provide not only essential nutrients needed for life, but also some bioactive compounds for health promotion and disease prevention. Previous epidemiologic studies have consistently shown that diet plays a crucial role in the prevention of chronic diseases (Kruger and Mann 2003). Consumption of fruit and vegetables has been strongly associated with reduced risk of cardiovascular disease, cancer, diabetes, Alzheimer disease, cataracts, and age-related functional decline. Heart disease, cancer, and stroke are the top three leading causes of death in the world. This convincing evidence suggests that a change in dietary behavior such as increasing consumption of fruit, vegetables, and grains is a practical strategy for significantly reducing the incidence of chronic diseases (Willett 1995; Espin et al. 2007; Bazzano et al. 2001; Liu 2003). The relationship between food and health becomes increasingly significant as consumers now demand healthy, tasty, and natural functional foods that have been grown in uncontaminated environment. There is a desire for a wide variety of choice and willingness to pay more for such foods in order to maintain health through well-balanced diet (Ercisli 2007).

3.6 Free Radicals

Radicals are formed as necessary intermediates in a variety of normal biochemical reactions, but when generated in excess or not appropriately controlled, they can wreak havoc on a broad range of macromolecules. A prominent feature of radicals is that they have extremely high chemical reactivity which explains not only their normal biological activities but how they inflict damage on cells. By definition a free radical is any atom with at least one unpaired electron in the outermost shell and is capable of independent existence. A free radical is easily formed when a covalent bond between entities is broken and one electron remains with each newly formed atom. Free radicals are highly reactive due to the presence of unpaired electron (Karlsson 1997).

3.6.1 Formation of Reactive Oxygen Species

There are many types of radicals but those of most concern in biological systems are derived from oxygen and known collectively as *reactive oxygen species* (ROS). These oxygen-derived radicals are generated constantly as part of normal aerobic life. They are formed in mitochondria as oxygen is reduced along the electron transport chain. ROS are also formed as necessary intermediates in a variety of enzyme reactions. Examples of situations in which oxygen radicals are overproduced in cells include:

- ***White blood cells*** such as neutrophils specialize in producing oxygen radicals, which are used in host defence to kill invading pathogens.
- ***Cells exposed to abnormal environments*** such as hypoxia or hyperoxia generate abundant and often damaging ROS. A number of drugs have oxidizing effects on cells and lead to production of oxygen radicals.
- ***Ionizing radiation*** is well-known to generate oxygen radicals within biological systems. Interestingly, the damaging effects of radiation are higher in well-oxygenated tissues than in tissues deficient in oxygen.

3.6.2 Types and Structures of Free Radicals

Any free radical involving oxygen can be referred to as ROS. Oxygen-centered free radicals contain two unpaired electrons in the outer shell. When free radicals steal an electron from a surrounding compound or molecule, a new free radical is formed in its place. In turn, the newly formed radical then looks to return to its ground state by stealing electrons with anti-parallel spins from cellular structures or molecules. Thus, the chain reaction continues and can be thousand of events long (Goldfarb 1999). Following are the most common ROS:

3.6.2.1 Superoxide Anion

Superoxide anion (O_2^-) is the most widely produced free radical. It is a negatively charged mono-radical resulting from the mono-valent reduction of molecular oxygen. Superoxide anions are formed when O_2 acquires an additional electron, leaving the molecule with only one unpaired electron. Such type of radical is formed within the mitochondria. The rate of formation depends on the amount of oxygen flowing through the mitochondria at any given time. This radical is essentially produced enzymatically by: NADPH oxidases (phagocytosis), Mit Cyt oxidase (cell respiration), Liver Cyt P_{450} (oxid metabolism of Xenobiotics) and Xanthine oxidase (ishemic reperfusion).

In the presence of protons, its dis-mutation leads to $O_2 + H_2O_2$

$$2O_2 \xrightarrow[SOD]{2H^+} H_2O_2 + O_2$$

3.6.2.2 Peroxide (Hydrogen Peroxide)

Hydrogen peroxide is produced in vivo by many reactions and can be converted to the highly damaging hydroxyl radical or be catalyzed and excreted harmlessly as water. Glutathione peroxidase is essential for the conversion of glutathione to oxidized glutathione, during which H_2O_2 is converted to water (Alessio and Blasi 1997).

3.6.2.3 Hydroxyl Radical

Hydroxyl radicals are short-lived but the most damaging radicals within the body. This type of free radical can be formed from O_2^- and H_2O_2 via the Harber–Weiss reaction. The interaction of copper or iron and H_2O_2 also produce OH^- as first observed by Fenton (Halliwell and Gutteridge 1985). This radical is several times more reactive than O_2^- and leads easily to the formation of new free radicals. Hydroxyl radical is also produced from H_2O_2 within tissues in the presence of Ferrous ions and leads to OH^- and ·OH, according to Fenton reaction.

$$X\text{-}Fe^{2+} + H_2O_2 \rightarrow X\text{-}Fe^{3+} + OH^- + \cdot OH$$

3.6.2.4 Singlet Oxygen

Singlet oxygen ($1O_2$) is not a free radical but can be formed during radical reactions and also cause further reactions. Singlet oxygen violates Hund's rule of electron filling in that it has eight outer electrons existing in pairs leaving one orbital of the same energy level empty. When oxygen is energetically excited, one of the electrons can jump to empty orbital creating unpaired electrons. Singlet oxygen can then transfer the energy to a new molecule and act as a catalyst for free radical formation. The molecule can also interact with other molecules leading to the formation of a new free radical (Karlsson 1997).

3.6.3 Biological Effects of Free Radicals

Free radicals play an important role in a number of biological processes. Many of these are necessary for life, such as, the intracellular killing of bacteria by phagocytic cells, in certain cell signalling processes called redox signalling (Pacher et al. 2007). For example, addition of superoxide or hydrogen peroxide to a variety of cultured cells leads to an increased rate of DNA replication and cell proliferation. In other words, these radicals function as mitogens.

Despite their beneficial activities, ROS clearly can be toxic to cells. The radicals possess an unpaired electron which makes them highly reactive and thereby able to damage all macromolecules, including lipids, proteins, and nucleic acids. The superoxide and hydroxyl radical derived from molecular oxygen under reducing conditions can participate in unwanted side reactions resulting in cell damage. Excessive amounts of these free radicals can lead to cell injury and death which may contribute to many diseases such as cancer, stroke, myocardial infarction, diabetes, and other disorders. Many forms of cancer are thought to be the result of reactions between free radicals and DNA, potentially resulting in mutations that can adversely affect the cell cycle and lead to malignancy (Mukherjee et al. 2004). Some of the symptoms of aging such as atherosclerosis are also attributed to free radical-induced oxidation of many of the chemicals in body. In addition, free radicals contribute to alcohol-induced liver damage, perhaps more than alcohol itself. Free radicals may also be involved in Parkinson's disease, senile and drug-induced deafness, schizophrenia, and Alzheimer's (Floyd 1999). The classic free-radical syndrome, the iron-storage disease hemochromatosis, is typically associated with a constellation of free radical-related symptoms including movement disorder, psychosis, skin pigmentary, melanin abnormalities, deafness, arthritis, and diabetes mellitus. The free radical theory of aging proposes that free radicals underlie the aging process itself. Similarly, the process of mitohormesis suggests that repeated exposure to free radicals may extend life span.

One of the best known toxic effects of oxygen radicals is damage to cellular membranes (plasma, mitochondrial, and endomembrane systems), which is initiated by a process known as *lipid peroxidation*. A common target for peroxidation is unsaturated fatty acids present in membrane phospholipids. A peroxidation reaction involving a fatty acid is depicted in Fig. 3.1.

Reactions involving radicals occur in chain reactions. Note in Fig. 3.1, that hydrogen is abstracted from the fatty acid by hydroxyl radical, leaving a carbon-cantered radical as part of the fatty acid. That radical then reacts with oxygen to yield the peroxy radical, which can then react with other fatty acids or proteins. Peroxidation of membrane lipids can have numerous effects, including:

- Increased membrane rigidity
- Decreased activity of membrane-bound enzymes (e.g., Sodium pumps)
- Altered activity of membrane receptors.
- Altered permeability

In addition to effects on phospholipids, radicals can also directly attack membrane proteins and induce lipid–lipid, lipid–protein, and protein–protein crosslinking, all of which obviously have effects on membrane function.

Because free radicals are necessary for life, the body has a number of mechanisms to minimize free radical-induced damage and to repair damage that occurs, such as the enzymes superoxide dismutase (SOD), catalase, glutathione peroxidase, and glutathione reductase. In addition, antioxidants play a key role in these defence mechanisms. These are often the three vitamins, vitamin A, vitamin C, and vitamin E and polyphenol antioxidants. Further, there is good evidence bilirubin and uric

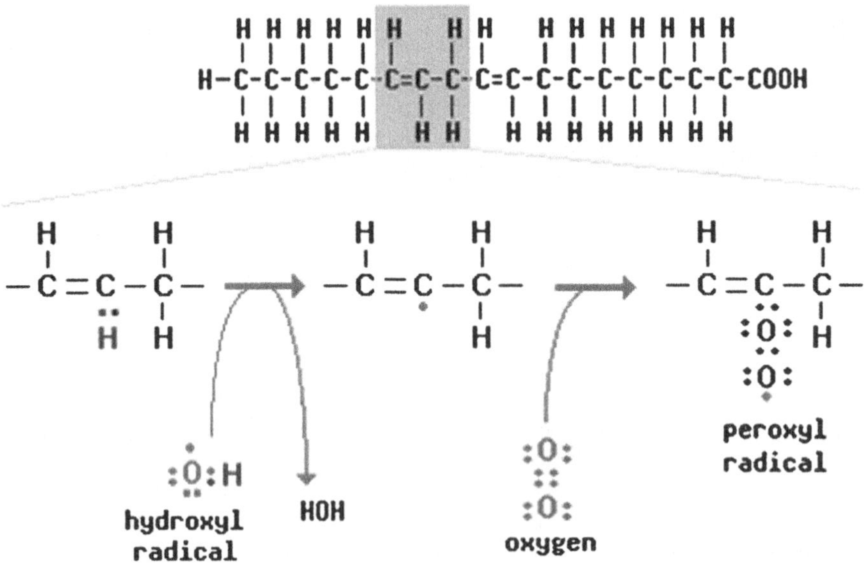

Fig. 3.1 Lipid peroxidation by free radicals

acid can act as antioxidants to help neutralize certain free radicals. Bilirubin comes from the breakdown of red blood cells' contents, while uric acid is a breakdown product of purines.

3.7 Antioxidants

As the name implies, antioxidants are the substances capable of counteracting the oxidative damage to the living tissues. Recent years have witnessed increased interest in the role of free radical oxidative damage in human diseases and aging. Free radical oxidative stress has a probable role in the pathogenesis of variety of human diseases, thus promoting the use of agents that can supplement the natural antioxidant defence. Various natural antioxidant enzymes, vitamins, and even synthetic agents with antioxidant enzymes have a potential role in various conditions like diabetes mellitus, inflammatory diseases, reperfusion injury, atherosclerosis, cancer, and aging as per the recent studies. Much interest has been shown on the anti-aging effect of these antioxidants and extensive research is underway for agents that can simulate natural SOD, glutathione peroxidase, glutathione reductase, and glutathione S transferase, etc.

Antioxidants may be nutrients (vitamins and minerals), as well as enzymes (proteins that assist in chemical reactions). They are found in fruits and vegetables with bright distinctive colors such as red cherries, orange carrots, purple blueberries, etc. Some antioxidants, such as phytochemicals, are also found in beverages such as wine and tea. Antioxidants play a critical role in preventing the development of chronic diseases. Their molecules can safely interact with free radicals and terminate the chain reaction before the vital molecules are damaged. Although there are several enzyme systems within the body that scavenge free radicals, the principle micronutrient (vitamin) antioxidants are vitamin E, beta-carotene, lycopene, and vitamin C. Additionally, selenium, a trace metal that is required for proper function of one of the body's antioxidant enzyme systems, is sometimes included in this category. It is mostly found in fish, shellfish, red meat, grains, eggs, chicken, and garlic. This metal plays significant role to control cardiovascular diseases, cancer, thyroid hormone problems, and immunity. Commonly antioxidants are classified into two groups; non-enzymatic antioxidants and enzymatic antioxidants.

3.7.1 Non-enzymatic Antioxidants

3.7.1.1 Vitamin A and Carotenoids

Vitamin A and carotenoids are commonly provided through food. Beta-carotene is well known to reduce the damage caused by the free radicals. Foods high in carotenoids include mostly the vegetables and fruits (tomatoes, carrots, spinach, brussels sprouts, sweet potatoes, winter squash, kale, collards, cantaloupe, peaches, apples, and broccoli). The chemical constituents of these foods may be effective allies against the cancer.

3.7.1.2 Vitamin C

Vitamin C, perhaps, is the best-known antioxidant and it offers a wide variety of health benefits, including protecting from infection and damage to body cells, helping produce collagen (the connective tissue that holds bones and muscles together), protecting the human body from bruising by keeping capillary walls and blood vessels firm, and helping in the absorption of other nutrients. To take advantage of these benefits, intake of foods rich in vitamin C such as citrus fruits (oranges, grapefruits, and tangerines), strawberries, tomatoes, green peppers, leafy green vegetables, broccoli, and potatoes is highly recommended.

3.7.1.3 Vitamin E

Vitamin E is generally found in nuts/seeds, peanut butter, salad dressing, margarine, wheat germ, whole-grain products, green leafy vegetables, vegetable oil, and liver oil. It helps to protect human body from cell damage that can lead to cancer, heart disease, and cataracts. Vitamin E works with other antioxidants like vitamin C to offer protection from various chronic diseases.

3.7.1.4 Polyphenols and Flavonoids

These phytochemicals are available in soy, red wine, purple and red grapes, pomegranates, cranberries, and tea. These are potent water-soluble antioxidants/ free radical scavengers, which prevent oxidative cell damage and exhibit strong anti-cancer activity.

3.7.1.5 Lycopene

It is found in tomatoes, pink grapefruit, and watermelon. It may be effective against cancers of the mouth, pharynx, esophagus, stomach, colon, and rectum.

3.7.1.6 Lutein

It is present in dark green vegetables such as kale, kiwi, broccoli, brussels sprouts, and spinach. It may help to reduce the risk of macular degeneration.

3.7.2 Enzymatic Antioxidants

3.7.2.1 Superoxide Dismutases

These are enzymes that catalyze the conversion of two superoxides into hydrogen peroxide and oxygen. The benefit here is that hydrogen peroxide is substantially less toxic than superoxide. SOD accelerates this detoxifying reaction roughly 10,000-fold over the non-catalyzed reaction. These SODs are metal-containing enzymes that depend on bound manganese, copper, or zinc for their antioxidant activity. In mammals, the manganese-containing enzyme is most abundant in mitochondria, while the zinc or copper forms predominant in cytoplasm. Interestingly, SODs are inducible enzymes—exposure of bacteria or vertebrate cells to higher concentrations of oxygen results in rapid increases in the concentration of SOD.

3.7.2.2 Catalase

It is found in peroxisomes in eucaryotic cells. It degrades hydrogen peroxide to water and oxygen, and hence finishes the detoxification reaction started by SOD.

3.7.2.3 Glutathione Peroxidase

It is a group of enzymes, the most abundant of which contain selenium. These enyzmes, like catalase, degrade hydrogen peroxide. They also reduce organic peroxides to alcohols, providing another route for eliminating toxic oxidants. In addition to these enzymes, glutathione transferase, ceruloplasmin, hemoxygenase, and possibly several other enzymes may participate in enzymatic control of oxygen radicals and their products.

Antioxidants work to protect lipids from peroxidation by the radicals. They are effective because of their capability of giving up their own electrons to free radicals. When a free radical gains the electron from an antioxidant, it cannot attack the cell and the chain reaction of oxidation is broken (Dekker et al. 1996). After donating an electron an antioxidant becomes a free radical by definition. Antioxidants in this state are not harmful because they have the ability to accommodate the change in electrons without becoming reactive. The human body has an elaborate antioxidant defence system. There are two lines of antioxidant defence within the cell. The first line, found in the fat-soluble cellular membrane, consists of vitamin E, beta-carotene, and coenzyme Q (Kaczmarski et al. 1999). Of these, vitamin E is considered the most potent chain breaking antioxidant within the membrane of the cell. Inside the cell water-soluble antioxidant scavengers are present. These include vitamin C, glutathione peroxidase, SOD, and catalase (Dekker et al. 1996). Some compounds contribute to antioxidant defence by chelating transition metals and preventing them from catalyzing the production of free radicals in the cell. Particularly important is the ability to sequester iron, which is the function of iron-binding proteins such as transferrin and ferritin (Imlay 2003).

3.8 Pro-Oxidant Activities

Pro-oxidants are the chemical substances that induce oxidative stress, either by generating ROS or by inhibiting antioxidant systems. Transition metals like cadmium, copper, and iron and hydrogen peroxide are some common pro-antioxidants (Han et al. 2005; Duarte and Lunec 2005). Antioxidants that are reducing agents can also act as pro-oxidants. For example, vitamin C has antioxidant activity when it reduces oxidizing substances such as hydrogen peroxide (Zingg and Azzi 2004); however, it will also reduce metal ions that generate free radicals through the Fenton reaction (Sen et al. 2006).

$$2\,Fe^{3+} + \text{Ascorbate} \rightarrow 2\,Fe^{2+} + \text{Dehydroascorbate}$$
$$2\,Fe^{2+} + 2\,H_2O_2 \rightarrow 2\,Fe^{3+} + 2\,OH + 2\,OH^-$$

The relative importance of the antioxidant and pro-oxidant activities of antioxidants is an area of current research, but vitamin C, which exerts its effects as a vitamin by oxidizing polypeptides, appears to have a mostly antioxidant action in the human body (Valko et al. 2005). However, less data are available for other dietary antioxidants, such as vitamin E, (Schneider 2005), or the polyphenols (Ristow and Zarse 2010). Likewise, the pathogenesis of diseases involving hyperuricemia likely involves uric acid's direct and indirect pro-oxidant properties. That is, paradoxically, agents which are normally considered antioxidants can act as conditional pro-oxidants and actually increase oxidative stress. Besides ascorbate, medically important conditional pro-oxidants include uric acid and sulfhydryl amino acids such as homocysteine. Typically, it involves some transition metal such as copper or iron as catalyst.

3.9 Contamination of the Vegetables

Contaminants may enter in the vegetables by several pathways; by uptake with soil-pore water, by diffusion from soil or air, by deposition of soil or airborne particles, or by direct application. Persistent polar and non-volatile contaminants have highest potential for accumulation from soil and the concentration in leaves may be several hundred times higher than in the soil. However, in most of the cases the accumulation of contaminants in vegetables is much lower. Lipophilic contaminants are mainly transported to leaves by attached soil particles or from air. Volatile contaminants have low potential for accumulation because they quickly escape to air. Uptake of organic contaminants into the vegetables may lead to human health risks, but it may also be used to delineate subsurface plumes and monitor natural attenuation (Trapp 2011).

3.9.1 Organic Pollutants

Most common contaminants in vegetables are persistent organic pollutants which tend to bio-accumulate in the food chain due to their long chemical half life and their lipophilic characteristics. They may cause different adverse health effects in human and animals, such as, endocrine disruption, reproductive, neurobehavioral, and developmental disturbances (Jones and Voogt 1999).

3.9.2 Antibiotics

Antibiotics have been widely used to prevent/treat the diseases and promote the growth. Antibiotics cannot be absorbed or metabolized absolutely in vivo, and 30–90 % of antibiotics are excreted into the environment (Heberer 2002; Bound and Voulvoulis 2004). Some antibiotics are still biologically active, after they enter into the environment, even after wastewater treatment (Zhou et al. 2006). Antibiotics are usually accumulated with the form of organic manure in soil. Furthermore, antibiotics can be taken up by various plants, crops, and soil animals on farmland. The contamination of antibiotics is an urgent problem in organic vegetables because of food security crises (Smukler et al. 2008; Siderer et al. 2005).

3.9.3 *Nitrate Ion*

Nitrate ion is another contaminant most widely found in the vegetables. Nitrate ion is ubiquitous in the environment arising from source such as fertilizer and release from decaying vegetation. It is water-soluble and is readily taken up by plants from soil and is also available to animals. It is present in some water supplies but usually only at low level. Nitrate ion can be converted into nitrite ion which interacts with hemoglobin causing the blood to be less efficient in transporting oxygen. This condition is known as methemoglobinemia. It also interacts with secondary and tertiary ammine to form N-nitroso compounds which are potent carcinogens (Bonell 1995).

3.9.4 *Heavy Metals*

Intake of heavy metal-contaminated vegetables may pose serious risk to the human health so the metal contamination of the vegetables should not be underestimated as these foodstuffs are chief components of human diet. The metal contamination of the food items is one of the most important aspects of food quality assurance (Marshall 2004; Radwan and Salama 2006; Wang et al. 2005; Khan et al. 2008b). International and national regulations on food quality have lowered the maximum permissible levels of toxic metals in the food items due to an increased awareness of the risk these metals pose to food chain contamination (Radwan and Salama 2006). Mining, industrial processing, pesticide, chemical fertilizer, and automobile exhaust are the main sources of the metal contamination in the environment leading to the contamination in vegetable (Graneroa and Domingo 2002; Lee et al. 2005). These metals may accumulate to toxic levels which can lead to impairment in the quality of human life (Wang et al. 2005). The main threats to human health from the metals are associated with exposure to Pb, Cd, and Hg. Their exposure may pose adverse health effects, including kidney damage, memory deterioration, behavioral disturbances, and possibly also bone effects and fractures (World Health Organization (WHO) 1992; Jarup 2003). Among other metals, Cr, Cu, and Zn showed non-carcinogenic hazardous effects to human health when exposures exceed the tolerable reference dose (US Environmental Protection Agency (USEPA) 2000). Dietary intake is the main route of exposure to heavy metals for most people, although inhalation can play an important role in very contaminated sites. Some researchers have assessed the risk of heavy metals from consuming food grown on soils irrigated with sewage and food chain transfer (Wang et al. 2005; Khan et al. 2008a).

The uptake and bioaccumulation of toxic metals in vegetables are influenced by a number of factors such as climate, atmospheric depositions, the concentrations of metals in soil, the nature of soil on which the vegetables are grown, and the degree of maturity of the plants at the time of harvest (Scott et al. 1996; Voutsa et al. 1996; Agrawal 2003). Elevated levels of heavy metals in vegetables are reported from the areas having long-term uses of treated or untreated wastewater (Sinha et al. 2004, 2005; Sharma et al. 2009). Other anthropogenic sources of the metals include the addition of manures, sewage sludge, fertilizers, and pesticides, which may affect the uptake of metals by modifying the physicochemical properties of the soil such as, pH, organic matter, and soil texture/structure. These metals can be harmful due to their potential to accumulate in different body parts of the human beings. Even in low concentrations, they showed adverse health effects because of non-biodegradable and persistent nature (Duruibe et al. 2007; Ikeda et al. 2000). Food chain translocation of the metals is one of the consequences of soil contaminated with toxic metals (Nasreddine and Parent-Massin 2002; Turkdogan et al. 2003). Ingestion of heavy metals can seriously cause depletion of some essential nutrients in the body, which in turn causes a decrease in immunological defences, intrauterine growth retardation, psychosocial dysfunctions, disabilities associated with malnutrition, and a high prevalence of various types of cancers (Turkdogan et al. 2003; Iyengar and Nair 2000).

References

Adebiyi, A. P., Adebiyi, A. O., Ogawa, T., & Muramoto, K. (2008). Purification and characterization of antioxidative peptides from unfractionated rice bran protein hydrolysates. *International Journal of Food Science and Technology, 43*, 35–43.

Agrawal, M. (2003). *Enhancing food chain integrity: Quality assurance mechanisms for air pollution impacts on fruit and vegetable system.* Final technical report II submitted to Department of International Development, London.

Ahalya, N., Kanamadi, R. D., & Ramachandra, T. V. (2010). Removal of hexavalent chromium using coffee husk. *International Journal of Environment and Pollution, 43*(1), 106–116.

Aitken, R. L., Dickson, T., Hailes, K. J., & Moody, P. W. (1999). Response of field-grown maize to applied magnesium in acidic soils in northeastern Australia. *Australian Journal of Agricultural Research, 50*, 191–198.

Alessio, H. M., & Blasi, E. R. (1997). Physical activity as a natural antioxidant booster and its effect on a healthy lifestyle. *Research Quarterly for Exercise and Sport, 68*(4), 292–302.

Aliyu, A. B., Musa, A. M., Oshanimi, J. A., Ibrahim, H. A., & Oyewale, A. O. (2008). Phytochemical analysis and mineral elements composition of some medicinal plants of Northern Nigeria. *Nigerian Journal of Pharmaceutical Sciences, 7*(1), 119–125.

Andlauer, W., & Furst, P. (2002). Nutraceuticals: A piece of history, present status and outlook. *Food Research International, 35*, 171–176.

Aravind, P., & Prasad, M. N. V. (2003). Zinc alleviates cadmium induced toxicity in *Ceratophyllum demersum*, a fresh water macrophyte. *Plant Physiology and Biochemistry, 41*, 391–397.

Aravind, P., & Prasad, M. N. V. (2004a). Carbonic anhydrase impairment in cadmium-treated *Ceratophyllum demersum* L. (free floating freshwater macrophyte): Toxicity reversal by zinc. *Journal of Analytical Atomic Spectrometry, 19*, 52–57.

Aravind, P., & Prasad, M. N. V. (2004b). Zinc protects chloroplasts and associated photochemical functions in cadmium exposed *Ceratophyllum demersum* L., a freshwater macrophyte. *Plant Science, 166*, 1321–1327.

Aravind, P., & Prasad, M. N. V. (2005a). Modulation of cadmium-induced oxidative stress in *Ceratophyllum demersum* by zinc involves ascorbate-glutathione cycle and glutathione metabolism. *Plant Physiology and Biochemistry, 43*, 107–116.

Aravind, P., & Prasad, M. N. V. (2005b). Cadmium–Zinc interactions in hydroponic system using *Ceratophyllum demersum*: Adaptive plant ecophysiology, biochemistry and molecular toxicology. *Brazilian Journal of Plant Physiology, 17*, 3–20.

Atolaiye, B. O., Babalola, J. O., Adebayo, M. A., & Aremu, M. O. (2009). Equilibrium modeling and pH dependence of the adsorption capacity of *Vitex doniana* leaf for metal ions in aqueous solutions. *African Journal of Biotechnology, 8*(3), 507–514.

ATSDR. (2000). *Toxicological profile for manganese* (pp. 336–338). Atlanta, GA: Agency for Toxic Substances and Diseases Registry, US Department of Health and Human Services.

Audrey, H. E. (1994). *Foods & nutrition encyclopedia* (Vol. 1). Boca Raton, FL: CRC Press. 2415 pp.

Bazzano, L. A., He, J., & Ogden, L. G. (2001). Legume consumption and risk of coronary heart disease in US men and women: NHANES I epidemiologic follow-up study. *Archive of Internal Medicine, 161*, 2573–2578.

Blatt, M. R. (2004). *Membrane transport in plants*. Oxford, England: Blackwell Publishing. 372 pp.

Boeck, B. M., Volders, K. M., & Lison, D. (2003). Cobalt and antimony: Genotoxicity and carcinogenicity. *Mutation Research, 533*, 135–152.

Bonell, A. E. (1995). Nitrate concentrations in vegetables. In: *Proceedings of the International Workshop on Health Aspects of Nitrates and its Metabolites (Particularly Nitrite), Bilthoven, Netherlands, 8–10 November 1994* (pp. 11–20). Strasbourg: Council of Europe Press.

Bound, J. P., & Voulvoulis, N. (2004). Pharmaceuticals in the aquatic environment: A comparison of risk assessment strategies. *Chemosphere, 56*, 1143–1155.

Brown, R. H. (1995). Free radicals, programmed death and muscular dystrophy. *Current Opinion in Neurology, 8*, 373–378.

Butnariu, M. (2007). *Noţiuni teoretice şi practice de biochimie vegetală* (pp. 95). Timişoara: Editura Mirton.

Calder, P. C., Field, C. J., & Gill, H. S. (2002). *Nutrition and immune function*. Oxford, England: CABI Publishing.

Chatterjea, M. N., & Shinde, R. (1998). *Text book of medical biochemistry* (p. 173). New York: Jaypee Brothers Medical.

Chillemi, S., & Chillemi, M. (2007). *Australian tea tree oil first aid handbook: 101 plus ways to use tea tree oil*.

Coyle, E. F. (1995). Substrate utilization during exercise in active people. *American Journal of Clinical Nutrition, 61*(4 Suppl.), 968S–979S.

Dekker, J. C., Van Doornen, L. J. P., & Kemper, H. C. G. (1996). The role of antioxidant vitamins and enzymes in the prevention of exercise include muscle damage. *Sports Medicine, 24*, 213–238.

Dolman, N. J., & Tepikin, A. V. (2006). Calcium gradients and the Golgi. *Cell Calcium, 40*, 505–512.

Drevon, J. J., & Hartwig, U. A. (1997). Phosphorus deficiency increases the argon-induced decline of nodule nitrogenase activity in soybean and alfalfa. *Planta, 201*, 463–469.

Duarte, T. L., & Lunec, J. (2005). Review: When is an antioxidant not an antioxidant? A review of novel actions and reactions of vitamin C. *Free Radical Research, 39*(7), 671–686.

Duruibe, J. O., Ogwuegbu, M. D. C., & Egwurugwu, J. N. (2007). Heavy metal pollution and human biotoxic effects. *International Journal of Physical Sciences, 2*, 112–118.
Edeoga, H. O., & Eriata, D. O. (2001). Alkaloid, tannin and saponin contents of some Nigeria medicinal plants. *Journal of Medicinal and Aromatic Plant Sciences, 23*, 344–349.
Ercisli, S. (2007). Chemical composition of fruits in some rose (Rosa spp.) species. *Food Chemistry, 104*, 1379–1384.
Escarpa, A., & González, M. C. (2000). Approach to the content of total extractable phenolic compounds from different food samples by comparison of chromatographic and spectrophotometric methods. *Analytica Chimica Acta, 427*, 119–127.
Espin, J. C., Conesa, M. T. G., & Tomás-Barberán, F. A. (2007). Nutraceuticals: Facts and fiction: A Review. *Phytochemistry, 68*, 2986–3008.
Finley, J. W., Penland, J. G., Pettit, R. E., & Davis, C. D. (2003). Dietary manganese intake and type of lipid do not affect clinical or neuropsychological measures in healthy young women. *Journal of Nutrition, 133*, 2849–2856.
Floyd, R. A. (1999). Antioxidants, oxidative stress, and degenerative neurological disorders. *Proceeding of the Society for Experimental Biology and Medicine, 222*, 236–245.
Fraga, C. G. (2005). Relevance, essentiality and toxicity of trace elements in human health. *Molecular Aspects of Medicine, 26*, 235–244.
Goldfarb, A. H. (1999). Nutritional antioxidants as therapeutic and preventive modalities in exercise-induced muscle damage. *Canadian Journal of Applied Physiology, 24*(3), 249–266.
Graneroa, S., & Domingo, J. L. (2002). Levels of metals in soils of Alcala de Henares, Spain: Human health risks. *Environment International, 28*, 159–164.
Gray, E. J., & Smith, D. L. (2005). Intracellular and extracellular PGPR: Commonalities and distinctions in the plant-bacterium signaling processes. *Soil Biology and Biochemistry, 37*, 395–412.
Grover, S., & Purves, W. K. (1976). Cobalt and plant development. *Plant Physiology, 57*, 886–889.
Hailes, K. J., Aitken, R. L., & Menzies, N. W. (1997). Magnesium in tropical and subtropical soils from north–eastern Australia. II. Response by glasshouse-grown maize to applied magnesium. *Australian Journal of Soil Research, 35*, 629–641.
Halliwell, B., & Gutteridge, J. M. (1985). The importance of free radicals and catalytic metal ions in human diseases. *Molecular Aspects of Medicine, 8*(2), 89–193.
Han, S. G., Kim, Y., Kashon, M. L., Pack, D. L., Castranova, V., & Vallyathan, V. (2005). Correlates of oxidative stress and free-radical activity in serum from asymptomatic shipyard welders. *American Journal of Respiratory and Critical Care Medicine, 172*(12), 1541–1548.
Harborne, J. B. (1973). *Phytochemicals methods* (pp. 49–188). London: Chapman & Hall.
Hart, A. L. (1989). Distribution of phosphorus in nodulated white clover plants. *Journal of Plant Nutrition, 12*, 159–171.
Hass, E. M. (2010). *Cobalt*. Retrieved from http://www.healthy.net/scr/article.aspx?ID=2051
Heberer, T. (2002). Occurrence, fate, and removal of pharmaceutical residues in the aquatic environment: A review of recent research data. *Toxicology Letters, 131*, 5–17.
Hemat, R. A. S. (2009). *Water, Urotext*. ISBN: 1-903737-12-5.
Horcsik, Z., Oláh, V., Valgo, A., Mészáros, I., Simon, L., & Lakatos, G. (2006). Effect of chromium (VI) on growth, element and photosynthetic pigment composition of *Chlorella pyrenoidosa*. *Acta Biologica Szegediensis, 50*, 19–23.
Ikeda, M., Zhang, Z. W., Shimbo, S., Watanabe, T., Nakatsuka, H., Moon, C. S., Matsuda, I. N., & Higashikawa, K. (2000). Urban population exposure to lead and cadmium in east and south–east Asia. *Science of the Total Environment, 249*, 373–384.
Ilyasova, D., & Schwartz, G. G. (2005). Cadmium and renal cancer. *Toxicology and Applied Pharmacology, 207*, 179–186.
Imlay, J. A. (2003). Pathways of oxidative damage. *Annual Review of Microbiology, 57*, 395–418.
Israel, D. W. (1987). Investigation of the role of phosphorus in symbiotic dinitrogen fixation. *Plant Physiology, 84*, 835–840.

Iyengar, V., & Nair, P. (2000). Global outlook on nutrition and the environment: Meeting the challenges of the next millennium. *Science of the Total Environment, 249*, 331–346.

Jakobsen, I. (1985). The role of phosphorus in nitrogen fixation by young pea plants (*Pisum sativum*). *Physiologia Plantarum, 64*, 190–196.

Jaleel, C. A., Manivannan, P., Lakshmanan, G. M. A., Sridharan, R., & Panneerselvam, R. (2007). NaCl as a physiological modulator of proline metabolism and antioxidant potential in *Phyllanthus amarus*. *Comptes Rendus Biologies, 330*, 806–813.

Jarup, L. (2003). Hazards of heavy metal contamination. *British Medical Bulletin, 68*, 167–182.

Järup, L., & Akesson, A. (2009). Current status of cadmium as an environmental health problem. *Toxicology and Applied Pharmacology, 238*(3), 201–208.

Jeejeebhoy, K. (2009). Zinc: An essential trace element for parenteral nutrition. *Gastroenterology, 137*, S7–S12.

Jones, C. K., & Voogt, D. P. (1999). Persistent organic pollutants (POPs): State of the science. *Environmental Pollution, 100*, 209–221.

Kaczmarski, M., Wojicicki, J., Samochowiee, L., Dutkiewicz, T., & Sych, Z. (1999). The influence of exogenous antioxidants and physical exercise on some parameters associated with production and removal of free radicals. *Pharmazie, 54*, 303–306.

Karlsson, J. (1997). Introduction to nutraology and radical formation. In: *Antioxidants and exercise* (pp. 1–143). Champaign, IL: Human Kinetics Press.

Khan, S., Cao, Q., Zheng, Y. M., Huang, Y. Z., & Zhu, Y. G. (2008a). Health risk of heavy metals in contaminated soils and food crops irrigated with waste water in Beijing, China. *Environmental Pollution, 152*, 686–692.

Khan, M. S., Zaidi, A., Oves, M., & Wani, P. A. (2008b). Heavy metal toxicity to legumes. In E. B. Samuel & C. W. William (Eds.), *Heavy metal pollution* (pp. 197–225). Hauppauge, NY: Nova Science.

Kotas, J., & Stasicka, Z. (2000). Chromium occurrence in the environment and methods of its speciation. *Environmental Pollution, 107*, 263–283.

Kožíšek, F. (2003). Health significance of drinking water calcium and magnesium. *Environmental Research Section, 1*(84), 219–227.

Kruger, C. L., & Mann, S. W. (2003). Safety evaluation of functional ingredients. *Food and Chemical Toxicology, 41*, 793–805.

Lee, J., Chon, H., & Kim, K. (2005). Human risk assessment of As, Cd, Cu and Zn in the abandoned metal mine site. *Environmental Geochemistry and Health, 27*, 185–191.

Lehr, J. J. (2006). Sodium as a plant nutrient. *Journal of the Science of Food and Agriculture, 4*, 449–496.

Lenntech. (2010a). *Cadmium—Cd*. Retrieved March 11, from http://www.lenntech.com/periodic/elements/cd.htm

Lenntech. (2010b). *Cobalt—Co*. http://www.lenntech.com/periodic/elements/co.htm

Lenntech. (2010c). *Copper—Cu*. Retrieved from http://www.lenntech.com/periodic/elements/cu.htm

Lenntech. (2010d). *Manganese—Mn*. Retrieved from www.lenntech.com/periodic/elements/mn.htm

Lenntech. (2010e). *Health effects of zinc*. Retrieved from www.lenntech.com/periodic/elements/zn.htm

Leung, P. L., Huang, H. M., Sun, D. Z., & Zhu, M. G. (1999). Hair concentrations of calcium, iron, and zinc in pregnant women and effects of supplementation. *Biological Trace Element Research, 69*, 269–282.

Leustek, T., Martin, M. N., Bick, J. A., & Davies, J. P. (2000). Pathways and regulation of sulfur metabolism revealed through molecular and genetic studies. *Annual Reviews of Plant Physiology and Plant Molecular Biology, 51*, 141–165.

Li, X. L. (2008). *Study on the effect of growth, absorption and accumulation of lead and chromium in tea plants* (pp. 40–41). Sichuan, China: Sichuan Agriculture University.

Liu, R. H. (2003). Health benefits of fruits and vegetables are from additive and synergistic combination of phytochemicals. *American Journal of Clinical Nutrition, 78*, 517S–520S.

Lovkova, M. Y., Buzuk, G. N., Sokolova, S. M., & Kliment'eva, N. I. (2001). Chemical features of medicinal plants (review). *Applied Biochemistry and Microbiology, 37*(3), 229–237.

Madsen, H. L., Nielsen, B. R., Bertelsen, G., & Skibsted, L. H. (1996). Screening of antioxidative activity of spices. A comparison between assays based on ESR spin trapping and electrochemical measurement of oxygen consumption. *Food Chemistry, 57*, 331–337.

Manach, C., Scalbert, A., Morand, C., & Jimenez, L. (2004). Polyphenols: Food sources and bioavailability. *American Journal of Clinical Nutrition, 79*, 727–747.

Manore, M.M., Meyer, N.L., Thompson, J.L. (2009). Sport Nutrition for Health and Performance, 2nd Ed. Human Kinetics Publishers, Champaign, IL, 2009, 543 pages. ISBN: 978-0-7360-5296-5.

Marchand, L. L. (2002). Cancer preventive effects of flavonoids—A review. *Biomedicine and Pharmacotherapy, 56*, 296–301.

Marschner, H. (1995). *Mineral nutrition of higher plants* (2nd ed.). London: Academic.

Marsden, I. D., & Rainbow, P. S. (2004). Does the accumulation of trace metals in crustaceans affect their ecology-the amphipod example? *Journal of Experimental Marine Biology and Ecology, 300*, 373–408.

Marshall, F. (2004). Enhancing food chain integrity: Quality assurance mechanism for air pollution impacts on fruits and vegetables systems. *Crop Post Harvest Program.* Retrieved February 27, 2012, from http://www.sussex.ac.uk/spru/1-4-7-1-11-1.html

McElroy, J. A., Shafer, M. M., Trentham-Dietz, A., Hampton, J. M., & Newcomb, P. A. (2006). Cadmium exposure and breast cancer risk. *Journal of the National Cancer Institute, 98*, 869–873.

McKay, B. P., & King, M. M. (1980). Vitamin E: Its role as a biological free radical scavenger and its relationship to the microsomal mixed-function oxidase system. In T. Machlin (Ed.), *Vitamin E* (pp. 289–317). New York: Marcel Dekker.

Mohan, S., & Sreelakshmi, G. (2008). Fixed bed column study for heavy metal removal using phosphate treated rice husk. *Journal of Hazardous Materials, 153*, 75–82.

Mohan, S., & Sumitha, K. (2008). Removal of copper (II) adsorption using *Casuarina equisetifolia* bark. *Environmental Engineering and Science, 25*(4), 89–97.

Mongillo, J. F., & Linda, Z. W. (2000). *Encyclopedia of environmental science* (p. 104). Rochester, NY: University of Rochester Press. ISBN 978-1-57356-147-1.

Morley, J. E., & Thomas, D. R. (2007). *Geriatric nutrition.* Boca Raton, FL: CRC Press.

Mukherjee, P., Abate, L. E., & Seyfried, T. N. (2004). Antiangiogenic and proapoptotic effects of dietary restriction on experimental mouse and human brain tumors. *Clinical Cancer Research, 10*, 5622–5629.

Nadadur, S. S., Srirama, K., & Mudipalli, A. (2008). Iron transport & homeostasis mechanisms: Their role in health & disease (a review). *Indian Journal of Medical Research, 128*, 533–544.

Nasreddine, L., & Parent-Massin, D. (2002). Food contamination by metals and pesticides in the European Union. Should we worry? *Toxicology Letters, 127*, 29–41.

Nishikimi, M., & Yagi, K. (1996). Biochemistry and molecular biology of ascorbic acid biosynthesis. *Subcellular Biochemistry, 25*, 17–39.

Noctor, G., Gomez, L., Vanacker, H., & Foyer, C. H. (2002). Interactions between biosynthesis, compartmentation and transport in the control of glutathione homeostasis and signaling. *Journal of Experimental Botany, 53*, 1283–1304.

O'Donoghue, M. (1990). *Rocks and minerals.* New York: Gallery Books. ISBN 10: 0831769645/ ISBN 13: 9780831769642.

Olsson, I. M., Bensryd, I., Lundh, T., Ottosson, H., Skerfving, S., & Oskarsson, A. (2002). Cadmium in blood and urine—Impact of sex, age, dietary intake, iron status, and former smoking association of renal effects. *Environmental Health Perspectives, 110*, 1185–1190.

Otten, J. J., Hellwig, J. P., & Meyers, L. D. (2006). *DRI, dietary reference intakes: The essential guide to nutrient requirements.* Washington, DC: National Academies Press.

Pacher, P., Beckman, J. S., & Liaudet, L. (2007). Nitric oxide and peroxynitrite in health and disease. *Physiological Reviews, 87*, 315–424.

Padayatty, S. J., Karz, A., Wang, Y., Eck, P., Kwon, O., Lee, J. H., Chen, S., Corpe, C. H., Dutra, S., & Levine, M. (2003). Vitamin C as an antioxidant: Evaluation of its role in disease prevention. *Journal of the American College of Nutrition, 22*, 18–35.

Paz-Alberto, A. M., Sigua, G. C., Bellrose, G. B., & Prudente, J. A. (2007). Phytoextraction of lead contaminated soil using vetiver grass (*Vetiveria zizanioides* L.), Cogongrass (*Imperta cylindrica* L.) and Carboagrass (*Paspalum conjugatum* L.). *Environmental Science and Pollution Research, 14*(7), 505–509.

Permyakov, E. A. (2009). Sodium-binding and potassium-binding proteins. In *Metalloproteomics*. Hoboken, NJ: Wiley. doi: 10.1002/9780470447758.ch14.

Radwan, A. M., & Salama, K. A. (2006). Market basket survey for some heavy metals in Egyptian fruits and vegetables. *Food and Chemical Toxicology, 44*, 1273–1278.

Raven, J. A., Evans, M. C. W., & Korb, R. E. (1999). The role of trace metals in photosynthetic electron transport in O_2-evolving organisms. *Photosynthesis Research, 60*, 111–150.

Reuter, D. J., & Robinson, J. B. (1997). *Plant analysis: An interpretation manual* (2nd ed.). Melbourne, Australia: CSIRO.

Ribet, J., & Drevon, J. J. (1995). Phosphorus deficiency increases the acetylene-induced decline in nitrogenase activity in soybean (*Glycine max* (L.) Merr). *Journal of Experimental Botany, 46*, 1479–1486.

Ristow, M., & Zarse, K. (2010). How increased oxidative stress promotes longevity and metabolic health: The concept of mitochondrial hormesis (mitohormesis). *Experimental Gerontology, 45* (6), 410–418.

Robson, A. D., Hara, G. W., & Abbott, L. K. (1981). Involvement of phosphorus in nitrogen fixation by subterranean clover (*Trifolium subterraneum* L.). *Australian Journal of Plant Physiology, 8*, 427–436.

Rout, G. R., & Das, P. (2003). Effect of metal toxicity on plant growth and metabolism. *Agronomie, 23*, 3–11.

Saito, K., Kurosawa, M., Tatsuguchi, K., Takagi, Y., & Murakoshi, I. (1994). Modulation of cysteine biosynthesis in chloroplasts of transgenic tobacco over expressing cysteine synthase O-acetylserine (thiol) lyase. *Plant Physiology, 106*, 887–895.

Saljoughian, M. (2007). Iron Deficiency Anemia: A Closer Look. US Pharmacists, 32(8), HS26–HS37.

Samantaray, S., Rout, G. R., & Das, P. (1998). Role of chromium on plant growth and metabolism. *Acta Physiologiae Plantarum, 20*(2), 201–212.

Sanda, S., Leustek, T., Theisen, M. J., Garavito, R. M., & Benning, C. (2001). Recombinant arabidopsis SQD1 converts UDP-glucose and sulfite to the sulfolipid head group precursor UDP-sulfoquinovose in vitro. *Journal of Biology and Chemistry, 276*, 3941–3946.

Schneider, C. (2005). Chemistry and biology of vitamin E. *Molecular Nutrition and Food Research, 49*(1), 7–30.

Scott, D., Keoghan, J. M., & Allen, B. E. (1996). Native and low input grasses: A New Zealand high country perspective. *New Zealand Journal of Agricultural Research, 39*, 499–512.

Sen, C. K., Khanna, S., & Roy, S. (2006). Tocotrienols: Vitamin E beyond tocopherols. *Life Sciences, 78*(18), 2088–2098.

Sengil, I., Ozacar, M., & Turkmenler, H. (2009). Kinetic and isotherm studies of Cu (II) biosorption onto valonia tannin resin. *Journal of Hazardous Material, 162*, 1046–1052.

Shakya, B., Rai, S. K., Singh, A., & Shrestha, A. (2006). Intestinal parasitosis among the elderly people in Kathmandu Valley. *Nepal Medical College Journal, 8*, 243–247.

Sharma, K. R., Agrawal, M., & Marshall, M. F. (2009). Heavy metals in vegetables collected from production and market sites of a tropical urban area of India. *Food and Chemical Toxicology, 47*, 583–591.

Shorrocks, V. (1997). The occurrence and correction of boron deficiency. *Plant and Soil, 193*, 121–148.

Siderer, Y., Maquet, A., & Anklam, E. (2005). Need for research to support consumer confidence in the growing organic food market. *Trends in Food Science and Technology, 16*, 332–343.

Simeonov, L., & Sargsyan, V. (2008). *Soil chemical pollution, risk assessment, remediation and security*. Dordrecht, The Netherlands: Springer.

Singleton, P. W., AbdelMagid, H. M., & Tavares, J. W. (1985). Effect of phosphorus on the effectiveness of strains of *Rhizobium japonicum*. *Soil Science Society of America Journal, 49*, 613–616.

Sinha, S., Gupta, A. K., Bhatt, K., Pandey, K., Rai, U. N., & Singh, K. P. (2005). Distribution of metals in the edible plants grown at Jajmau, Kanpur (India) receiving treated tannery wastewater: Relation with physiochemical properties of the soil. *Environmental Monitoring and Assessment, 115*, 1–22.

Sinha, K. P., Mohon, D., Sinha, S., & Dalwani, R. (2004). Impact assessment of treated/untreated wastewater toxicants discharge by sewage treatment plants on health, agricultural and environmental quality in waste water disposal area. *Chemosphere, 55*, 227–255.

Sittig, M. (1991). *Handbook of toxic and hazardous chemicals and carcinogens* (3rd ed.). Park Ridge, NJ: Noyes.

Sizer, F. S., & Ellie, W. N. (2007). *Nutrition: Concepts and controversies*. Monterey, CA: Brooks/Cole. 848 pp. ISBN-10: 0495390658, ISBN-13: 9780495390657.

Smukler, S. M., Jackson, L. E., Murphree, L., Yokota, R., Koike, S. T., & Smith, R. F. (2008). Transition to large-scale organic vegetable production in the Salinas Valley, California. *Agriculture Ecosystems and Environment, 126*, 168–188.

Sonni, A. (2002). *Importance of minerals and trace minerals in human nutrition*. http://www.mgwater.com/impor.shtml

Soobrattee, M. A., Neergheen, V. S., Luximon-Ramma, A., Aruoma, O. I., & Bahorun, T. (2005). Phenolics as potential antioxidant therapeutic agents: Mechanism and actions. *Mutation Research, 579*, 200–213.

Tan, K. Z., Keltjens, W. G., & Findenegg, G. R. (1991). Role of magnesium in combination with liming in alleviating acid soil stress with the aluminum-sensitive sorghum genotype Cv323. *Plant and Soil, 136*, 65–71.

Taylor, M. D., & Locascio, S. J. (2004). Blossom-end rot: A calcium deficiency. *Journal of Plant Nutrition, 27*, 123–139.

Temple, N. J. (2000). Antioxidants and disease: More questions than answers. *Nutrition Research, 20*, 449–459.

Trapp, S. (2011). Uptake of organic contaminants from soil into vegetables and fruits'. In F. A. Swartjes (Ed.), *Dealing with contaminated sites: From theory towards practical application* (Vol. 9, pp. 369–408). Dordrecht, The Netherlands: Springer.

Trease, G. E., & Evans, W. C. (1983). *Pharmacognosy* (12th ed., pp. 309–707). London: Oxford University Press.

Turkdogan, M. K., Kilicel, F., Kara, K., Tuncer, I., & Uygan, I. (2003). Heavy metals in soil, vegetables and fruits in the endemic upper gastrointestinal cancer region of Turkey. *Environmental Toxicology and Pharmacology, 13*, 175–179.

Uchida, R. S. (2000). Recommended plant tissue nutrient levels for some vegetable, fruit, and ornamental foliage and flowering plants in Hawaii. In J. A. Silva & R. S. Uchida (Eds.), *Plant nutrient management in Hawaii's soils* (pp. 57–65). Honolulu, HI: College of Tropical Agriculture and Human Resources, University of Hawaii.

US Environmental Protection Agency (USEPA). (2000). *Risk based concentration table*. Washington DC: United States Environmental Protection Agency.

USEPA. (2010). *Information on the toxic effects of various chemicals and Groups of Chemicals Unites States Environmental Protection Agency*. Retrieved February 6, 2012, from http://www.epa.gov/R5Super/ecology/html/toxprofiles.htm#

Vadez, V., Rodier, F., Payre, H., & Drevon, J. J. (1996). Nodule permeability to O_2 and nitrogenase-linked respiration in bean genotypes varying in the tolerance of N_2 fixation to P deficiency. *Plant Physiology and Biochemistry, 34*, 871–878.

Valko, M., Morris, H., & Cronin, M. T. (2005). Metals, toxicity and oxidative stress. *Current Medicinal Chemistry, 12*(10), 1161–1208.

Vallee, B. L., & Auld, D. S. (1990). Zinc coordination, function, and structure of zinc enzymes and other proteins. *Biochemistry, 29*, 5647–5659.

Voutsa, D., Grimanis, A., & Samara, C. (1996). Trace elements in vegetables grown in industrial areas in relation to soil and air particulate matter. *Environmental Pollution, 94*, 325–335.

Wang, J., & Chen, C. (1990). Biosorption of heavy metals by *Saccharomyces cerevisiae. Biotechnology Advances, 24*, 427–451.

Wang, X. L., Sato, T., Xing, B. S., & Tao, S. (2005). Health risks of heavy metals to the general public in Tianjin, China via consumption of vegetables and fish. *Science of the Total Environment, 350*, 28–37.

Whitney, E. N., & Rolfes, S. R. (2007). Understanding Nutrition, 11th Ed. Wadsworth Publishing, Belmont, CA, USA. ISBN-10: 0495116696

Willett, W. C. (1995). Diet, nutrition, and avoidable cancer. *Environmental Health Perspectives, 103*(8), 165–170.

World Health Organization (WHO). (1992). *Environmental health criteria* [Monograph] (Vol. 134, pp. 280). Geneva, Switzerland: Cadmium International Programme on Chemical Safety (IPCS).

Yan, X., Wu, P., Ling, H., Xu, G., Xu, F., & Zhang, Q. (2006). Plant nutriomics in China: An overview. *Annals of Botany, 98*, 473–482.

Zhou, J. S., Pillidge, C. J., Gopal, P. K., & Gill, H. S. (2006). Antibiotic susceptibility profiles of new probiotic Lactobacillus and Bifidobacterium strains. *International Journal of Food Microbiology, 98*, 211–217.

Zingg, J., & Azzi, A. (2004). Non-antioxidant activities of vitamin E. *Current Medicinal Chemistry, 11*(9), 1113–1133.

Chapter 4
Ethnobotanical Aspects of Wild Edible Vegetables

4.1 Ethnobotanical Methods

4.1.1 Research Design

Ethnobotany is a multidisciplinary approach that integrates techniques from biology, ecology, anthropology, ethnology, linguistics, economy, and other fields. Ethnobotanists not only work in primary forests, they are also involved in the management of broad range of vegetation ranging from home-gardens to mature forests, where valuable plants are found. Ethnobotanical research focuses on the traditional knowledge of the indigenous people and is mostly conducted in rural areas. However, some studies, such as descriptions of animals or plants sold in urban market places, are carried out in urban areas. Ethnobotanists deal with hypothetical questions about the association among humans and their environment and the answers contribute to rural development, poverty elevation, healthcare, and protection (Martin 1995).

Identification of goals or objectives is essential prior to selection of any approach which suits the research interests, the research budget, and the time schedule to complete the task. Ethnobotanical and ethnoecological investigations are tedious, may be costly and time consuming as several field surveys are necessary to collect field data and samples in proposed ethnobotanical research area. Usually the most successful studies are those which span over several seasons and continue for a number of years. Long-term projects allow working with local inhabitants to record ethnobotanical knowledge in ritual occasions and seasonal farming activities. The following activities are necessary to conduct an ethnobotanical study;

- Collaboration with government officials, scientists, and local people.
- Researchers must obtain permission from local representatives or community members before conducting any research, especially when conducting research in indigenous communities.

A.M. Abbasi et al., *Wild Edible Vegetables of Lesser Himalayas*,
DOI 10.1007/978-3-319-09543-1_4

- It is important that researchers should clearly and honestly discuss their goals, methodology, and consequences of the research to local communities.
- Intra-cultural variation and specialization must be taken into account when selecting informants.
- Local informants and communities should compensate with a fair amount of money and/or through presents and services.
- The intellectual property rights regarding the ecological knowledge of the indigenous people must be taken into consideration.
- Researchers must take into consideration and respect the national legal frames about biodiversity prospective before designing an ethnobotanical project. Furthermore, they should handle according to the Code of Ethics of the Society of Ethnobiology (International Society of Ethnobiology 1998).

4.1.2 Field Equipments

The equipments mentioned (by Martin 1995; Alexiades and Sheldon 1996; Cotton 1996) used in fieldwork during ethnobotanical survey are:

- Research diary
- A topographic and high resolution map of the study area or aerial photographs of the whole study area
- Relevant field guides and reference books for on-the-spot identification of unknown plant species
- Field note book, questionnaires
- Global positioning system (GPS)/altimeter/compass
- Audio recorder
- Digital camera for field photography
- Plant presser, labels, pencils, newspaper, cardboard, blotters, etc.
- Gloves, scissors, cutter, dagger, and plant clippers to harvest the plant
- List of names and contact details (address, telephone number) of all our respondents in the study area and relevant experts
- A large plastic sheet or rain coat to cover ourselves and all our equipments from expected rain
- Empty plastic and paper bags of different sizes, wire twists or other means to seal bags and boxes, and a permanent pen to store and label seeds and other artifacts
- Water, snacks, insect repellent, suntan lotion, and a hat

4.1.3 Mapping

A map enclosed the topographic allocation of the plant species, home gardens, households, and local infrastructure of a particular area (Townsley 1996). Existing maps, GPS records taken during the survey, and aerial photographs represent the basic topographic information. The latitude, longitude, and altitude of the collected plant specimens should be recorded using GPS. As a starting point for discussion with some individuals and groups of local people, rough maps of the area can be drawn up. These maps provide

1. A physical focus for discussions.
2. Something concrete for people to refer especially when they are talking about local circumstances, changes in conditions or particular matters. This also allows them to illustrate their views about environment and what is relevant in that environment for them.
3. Help to complete limited information provided by existing maps, which are often incomplete, and to get a better understanding of key local particularities.
4. A basic sketch map for use to fill in thematic information, as historical changes in settlement and land use, social and economic particularities among the local inhabitants or land cover.

4.1.4 Collecting and Identifying Plants

Plant collection is one of the most basic criteria in ethnobotany. The collected plant species are used as voucher specimens, which are permanent records of the plants collected from a certain area, place, or community. They allow taxonomists to identify the family, genus, and species of the collected specimen (Martin 1995).

- The selection of samples should be based on the representativeness of the plant species from a certain area, place, or community.
- Plants specimen should include flowers, fruits, or both in order to make their identification easier.
- Plant specimens should be pressed in the field whenever possible.
- A good herbarium specimen consists of a dried, pressed section of a plant containing well-preserved vegetative and reproductive structures (Alexiades 1996). The identification follows at the affiliated Herbarium. The authenticity of identification is based on the quality of the collected specimen.

4.1.5 Ethnobotanical Data Collection

During field survey ethnobotanical data collection is based on interviews or questionnaires.

4.1.5.1 Interviews

Interview with the local people from a certain area or community is one of the basic techniques used to collect ethnobotanical data of that particular area. The setting of the interview and the types of questions are predetermined. Furthermore, the questions asked determine the degree of control of the interviews. Some basic types of interviews mentioned (by Martin 1995; Alexiades 1996) are as under:

- Informal interview
  The informal interview has no structure; the researcher simply makes notes during or after casual conversations.
- Unstructured interview
  This type of interview has the appearance of a casual conversation, but the people involved in the conservation know that it is an interview. It develops within the framework established by the researcher.
- Semi-structured interview
  In this type of interview the fieldworker has a list of questions and topics that need to be covered. However, this list is only a guide which allows the researchers to be flexible. As the discussion goes no, new lines of inquiry arise naturally while some of the prepared questions will fall to the wayside.
- Structured interview
  This interview bases on fixed questions and is mainly used for the evaluation of local knowledge. The structured interviews should be reserved for later stages of the ethnobotanical study, when a high degree of confidence is established between researchers and local inhabitants of a particular area.

4.1.5.2 Questionnaires

A questionnaire is a formalized set of questions for obtaining information from local inhabitants of a particular area. The prime objective is to interpret the researcher's information into a set of specific questions that local people are ready and able to answer. A questionnaire is an important source of collecting primary quantitative data. It enables researchers to collect quantitative data in a uniform way so that the data becomes reliable and logical for analysis. A questionnaire ensures standardization and comparability of the data across interviewers. It facilitates data recording and processing in an efficient manner within due time. There are no scientific principles to guarantee a most favorable or perfect questionnaire. The ideas, ability, and knowledge of the researcher play a major role to design a good questionnaire (Martin and Polivka 1995).

A sample of questionnaire used to collect ethnobotanical data is as under:

Questionnaire

Informants' approval for the participation in the study
I________________________________(name of informant) hereby give my full permission and awareness to participate in this study and declare that to the best of my knowledge the information that I have provided are true, accurate, and complete.
Date...................................... (Signature/Thumb impression of informant)

Informants' details

Name...
Gender..
Age...
Ethnic group...
Occupation..
Education..
Locality/Residence...
Data about wild edible medicinal plant and its use
Plant (Local name)..
Habit (Tree/Herb/Shrub/Climber)..
Habitat...
Flowering/fruiting period...
Plant part used...
Collection season...
Collected by...
Method of crude drug preparation..
Method of collection and storage...
Mode of administration..
Dosage...
Name of disease(s) treated...
Side effects (if any)..
Precautions..
Is it used to cure live stock diseases..
Cooking recipe...
Which part is preferred..
At which stage it is collected...
Other ingredients used...
Taste of cooked vegetable..
Availability in natural resources (easy/difficulty/very difficult)..........................
Conservation status (abundant, common, rare, etc.)...
Conservation needs..
Threats to this plant...
Conservation efforts made by Government and local residents..........................
Other uses (if any)..
Is it sold in local markets...

If sold, price/kg...
Remarks:
Botanical name of plant...
Family name...
Identified by...

Signature of Researcher

Beside interviews and questionnaires some other techniques like free listing and ranking are also used to collect indigenous knowledge of local informants.

Free-Listing

Informants are asked to list items in a given period of time. They could be asked, for example, to name as many medicinal plants as they can in 5 min.

Ranking

Ranking helps to highlight local preferences for different concerns. People are encouraged to divide sets of items or activities into categories and rank them according to different criteria. For example, informants who provide a list of plant used as construction materials could be asked which one is the best or used more often and why.

4.1.6 Market Survey

Most cities in a region have markets with sections where plant products like fruits, vegetables, and medicinal plants are sold. Moreover, such products are sold in informal markets, mainly in rural areas and/or along road sides. These places are important sources of information for conservation, rural development, and resource management programs. Direct observations, interviews, and surveys of traders, vendors, and consumers can be used to obtain qualitative and quantitative data concerning socio-economic, ecological, and cultural aspects of wild plant products and horticulture products. For instance, market survey allows the identification of overexploited wild plant species or plant products with a high potential for horticultural or industrial development (Alexiades 1996; Cunningham 2001).

4.1.7 Analytical Tools for the Evaluation of Traditional Knowledge

Through the use of analytical tools, participants show their cultural preferences and empirical ecological knowledge by ranking objects or sorting them based on criteria such as similarity or relative quality (Martin 1995; Alexiades 1996; Hamilton et al. 2003). The scientific severity of ethnobotanical research has increased significantly in the past two decades due to the adoption of quantitative methods (Phillips 1996). Generally, ethnobotanists have documented and responded to the need for research based upon hallmarks of the scientific method, including testable hypotheses, reproducible methods, and statistical measures of variation. Relative cultural importance (RCI) indices are important analytical tools used to measure cultural preference of plant species in a particular area.

4.1.7.1 Relative Cultural Importance Indices

The application of RCI indices in ethnobotany began during the late 1980s. RCI indices create numerical scales or values per plant taxon, which are based on social sciences and ecology (Alexiades and Sheldon 1996; Kvist et al. 1996; Lykke et al. 2004; Martin 2004; Phillips and Gentry 1993a, b; Phillips et al. 1994; Prance et al. 1987; Reyes-García et al. 2006; Turner 1988).

The percentage of plants used by Panare indigenous informants in Venezuela determined by Boom (1990) was a starting point for quantitative inter-cultural comparisons of plant knowledge.

- In order to recognize that not all uses are equal Prance et al. (1987) applied weighted indices of 1.0 for "important" uses and 0.5 for "minor" uses. This approach was aimed at capturing relative degrees of "importance," but did not address informant variation.
- Publication of Phillips and Gentry (1993a, b) on RCI "use values" was a watershed event in quantitative ethnobotany. They evaluated variation among informants based upon use citation frequencies, considering each as a statistical "event."

RCI indices are considered as key research tools in ethnobotany because of their application and keen interest among researchers. Therefore, ethnobotanists are well-informed about RCI methodologies and strive to develop or maintain competency in their application.

Researchers interested in RCI indices are presently confronted by a disorderly array of RCI models in the literature. With the aim of justifying this situation, RCI indices are compiled within four broad categories including ethnographic methods, use categorization, sampling, and statistical analysis.

- Ethnographic data
  Although the numbers and statistics used in quantitative ethnobotany may look inspiring, but without ethnographic data these are worthless. Various ethnographic methods have been used successfully to collect data agreeable to RCI analyses (Martin 2004; Bernard 2002). More than one method is often necessary to address research questions and environments. An essential activity associated with ethnobotanical interviews is the collection of plant voucher specimens.

Vernacular names and biological species names differ and local names change over time. Scientific names also change due to taxonomic revision. The use of herbarium specimens ensures that future researchers and others can verify results and make useful comparisons of the collected specimens (Alexiades 1996).

Some of the practical ethnographic methods including interviews, questioners, free listing, and ranking are mentioned previously in ethnobotanical data collection section.

- Use categories

  Many RCI indices pool the specific uses cited by informants into "use-categories" The number of categories and sub-categories is potentially endless, but some commonly utilized titles include: food, medicine, fodder, fuel, construction, agricultural tool handles, fencing, hedging, furniture, and other. Uniform categorization helps in compilation, comparison, and well-organized presentation of data sets. Plants are frequently cited for uses which are slightly different (e.g., wood for house beams and wood for fuel). Such plants would receive overstated RCI scores if data are not broken down into use categories.

  In recent years, informant-defined "folk" use categories are presented more often (McClatchey et al. 2006). Such categories give another dimension to the data and improve reproducibility by reducing researcher bias. However, the characteristic nature of folk categories reduces options for comparison with other studies. A solution would be to include and compare results with both researcher and informant-defined categories.

- RCI Sample Size

  During ethnobotanical survey, informant sample size appears to be either subjective or based upon limitations of field time. According to rule more than 35 independent, random samples are required for strong and parametric statistics (Balick 1996; Lozada et al. 2006). Accumulation curves (a.k.a. collector's curves) and richness estimators are used to describe the total sampling effort (species found per unit of sampling effort) and to estimate expected species richness if sampling were to continue indefinitely (Colwell 2005).

  In order to predict the total number of useful species based upon interview samples and to estimate the total number of distinct uses for each species, ethnobotanists could use estimation tools if informant sample size is sufficient (Balick 1996; Lozada et al. 2006). Improvement in ethnobotanical estimation methods would facilitate greater comparability between studies such as maybe accomplished with free software (Colwell 2005; Chao and Shen 2003). Additional work is needed to understand how closely ethnobotanical knowledge might conform to the assumptions of these models.

- Statistical Analysis

  Random selection of participants and sufficient sample size is required for statistical approach to assess the knowledge in a study community (Martin 2004). Participants should be interviewed in isolation from others in the community to satisfy the requirement of statistical independence. In other cases, specialized knowledge of a few "key informants" (Martin 2004) or elders is sought and low sample size will likely preclude robust statistical analysis.

Data with sufficient sampling will be agreeable to statistical analysis and the testing of hypotheses after completion of interviews and calculation of RCI indices. Species are often ranked by RCI values and any two species can be compared for significant differences (Phillips and Gentry 1993a). RCI values are also helpful to compare and test hypotheses concerning the "importance" of vegetation zones, plant families, or growth forms (Kvist et al. 1996; Phillips et al. 1994; Albuquerque et al. 2005; Chazdon 1999; Hammer and Harper 2006; Johnston 1998).

Statistics are sometimes applied to RCI data to reveal predictive relationships between plant characteristics and how they are used (Hoft et al. 1999). The factors of plant frequency, density, stem diameter, growth rate, and growth form with regression and analysis of variance (ANOVA) were examined (Phillips and Gentry 1993b). They found significant, predictable use-value patterns for all tested plant characters. These authors also introduced the Family Use Value (FUV) to distinguish plant families that have more uses than would be expected by random chance.

Multivariate techniques such as cluster analysis (CA), principle components analysis (PCA), and discriminate function analysis are used to determine complex relationships among variables. Free internet software is available for a wide variety of statistical tests. PAST is a fairly comprehensive suite of statistical tools (Hammer and Harper 2006). Moreover, in Metasig tests the null hypothesis patterns of plant use can be explained by random processes (Estabrook 2003).

- RCI indices by category
  RCI indices are designed to transform the complex, multidimensional concept of "importance" into uniform and comparable numerical scales or values. The number of people interviewed can range from one to hundreds, depending upon the research design and the cultural limitations of fieldwork. Plant use citation data from ethnographic plant interviews is applied to RCI formulas to derive values.

According to Phillips (1996) RCI indices are divided into three major categories, including:

- Uses totaled
- Subjective allocation
- Informant consensus

While Kvist et al. (1996) gave different names to the same categories, including

- Researcher-tally (use totaled)
- Researcher-score (subjective allocation)
- Informant-tally (informant consensus)

The term "tally" was used for methods that indiscriminately count every use cited, and "score" for methods that sort uses into pre-determined hierarchical categories (Kvist et al. 1996).

"Informant score" is a fourth tally method that refers to consensus-based RCI values using entirely informant-generated scores.

1. Uses Totaled/Researcher Tally
   The uses totaled methods are among the earliest in quantitative ethnobotany (Boom 1990; Paz et al. 1995). Citations of use (and non-use) are recorded for all plant species within a limited area or encountered during general walks and interviews with community informants. The uses are recorded and may be assigned to use categories.

$$\mathrm{UT} = \sum \mathrm{Uses}_{\mathrm{Species}(i)}$$

   A simple sum of all known uses for each species. The uses can be categorized by utility, taxon, or vegetation type.

   The number of uses are summed and ranked. To remove the bias associated with many similar uses for a taxon the index may be category-limited. In this case a score of 1 is entered for each use category with at least one cited use, but the total number of specific uses is ignored.

   The uses totaled method does not distinguish relative degrees of importance for different uses; the most "important" taxon is simply that with the most use citations. The percent of useful plants and a breakdown of plant uses within specific use categories are provided. Because the method requires the least amount of data collection (a list of species and associated uses), less field time is required than with other methods. In fact, the uses totaled method could be based only upon literature review.

   In terms of statistical relevance and hypothesis testing, the uses totaled method is the least effective. It lacks an explicit method and "importance" scores are sensitive to sampling intensity. Intra-cultural variability cannot be assessed because data is not recorded per-respondent or informant. Furthermore, this method ignores the dynamics of cultural importance, such as distinctions between current and historical use, frequency of use, and relative degrees (rankings) of importance, Subjective Allocation/Researcher Score (Prance et al. 1987; Turner 1988; Piñedo-Vasquez et al. 1990; Stoffle et al. 1990).
2. *Subjective Allocation* (Researcher-Score)
   This group of methods adds the allocation of a score or rank to the same set of data that would be obtained by the "uses totaled" method. The researcher distinguishes between major and minor uses by assigning a weighted score in each use category for each taxon ideally based upon substantial knowledge and experience

   (a) Use Value
   Using a similar approach, Prance et al. (1987) and Turner (1988) determined the percent of useful plants per hectare within pre-established use categories for different cultural groups in Amazonia.

$$\mathrm{UV}_s = \sum_{i}^{n} \mathrm{Value}_{\mathrm{use\ \ category}(i)}$$

The species use value is a sum of researcher generated score for each of its uses. “Major” uses are scoring 1 while “Minor” uses are scored 0.5. Uses refer to use categories (construction or food), not specific uses.

(b) Index of Cultural Significance

The Index of Cultural Significance (ICS), an anthropological approach presented by (Turner 1988) and modified by (Stoffle et al. 1990), calculates importance through researcher-determined weighted ranking of multiple factors. Turner assigned scores on a five-point scale to the variables of quality and intensity of use and assigned a score of 2, 1, or 0.5 for the exclusivity or preference of use.

$$\mathrm{ICS} = \sum_{i=1}^{n} (q \times i \times e)$$

For each species scores of all uses cited from 1 to n are added together. The score of each use is determined from the multiplied score derived from three ordinary scales of significance, q = quality of use [critical resource (5) to little noticed (0)], i = intensity of use [high (5), low (0)], e = exclusivity of use [subsituation available? (2)-(1)-(0.5)].

(c) Cultural significance index

To reduce the subjectivity of above method Stoffle et al. (1990) revised the ICS with a two-point scale for the variables of species management (2 = managed, 1 = not managed), preferred (2 = preferred for a given use, 1 = not the preferred species for a given use), and use frequency (2 = species effectively used for a given use, 1 = species rarely cited for a given use). They also incorporated a consensus method called a correction factor to reduce the sensitivity of this method to sampling intensity. The ethnographic, qualitative approach of the CSI method requires considerable experience and rapport with a cultural group for meaningful results.

$$\mathrm{CSI} = \sum_{i=1}^{n} (i \times e \times c) \times \mathrm{CF}$$

It is designed to combine elements from former indices to reduce subjectivity. i = species management [non-managed (1) or managed (2)], e = exclusivity of use [preferred by at least one informant (2), not mentioned as preferred (1)], c = contemporary usage [contemporary (2) or not (1)].

Subjective allocation methods can save time in the field and provide a more refined data set than the “uses totaled” method. However, these methods introduce researcher bias because degrees of importance and categories are based solely upon researcher assessment. Furthermore, as with other methods in this category, informant responses are not independently recorded, thus eliminating the opportunity for analysis of informant variability.

3. *Informant consensus* (Informant Tally)
   Informant consensus methods mentioned by Phillips and Gentry (1993a, b) require considerably more data to be collected than those previously discussed. Each plant citation is recorded separately and referred to as an "event" and the same plant and same informant may participate in many "events." Initial data collection is simply a count and use citations are not ranked. The use citations are summed for each informant and divided by the total number of "events." The final species use values are calculated as the sum of the species use values for each informant divided by the total number of informants interviewed about a given species. The informant consensus method has had more influence over the past 13 years than any other RCI index. Many researchers have applied or adapted the Phillips and Gentry approach.
   However, some limitations of the use value method have been noted:
   - It does not distinguish degrees of importance and analyzes only the average number of cited uses. Thus, a rarely used plant with two cited uses would be more "important" than a very popular plant with only one use (Kvist et al. 1996).
   - An open-ended tally (no maximum) artificially inflates use values for plants with multiple single-category uses (Kvist et al. 1996).
   - It does not distinguish between cited and observed uses. "The results say more about the structure of people's knowledge than they do about the importance of plants per se" (Wong 2000).
4. Informant Consensus/Informant Score
   Since the earliest efforts at quantitative ethnobotany, researchers have acknowledged that cultural informants are best able to determine RCI. According to (Turner 1988), "Ideally, evaluations of cultural significance should be done by native peoples themselves living within a traditional culture." Four methods that emphasize the judgment of informants rather than researchers in classifying the importance of uses are given below.
   (a) *The Informant Score Method* is similar to the Use values of Phillips and Gentry but with an informant-generated score similar to the score assigned (by Kvist et al. 1996; Lykke et al. 2004; Turner 1988).
   (b) *The Choice Value Method* considers cohorts of substitute products to measure relative preference for specific uses (Kremen et al. 1998).
   (c) *The Importance Value* measures the proportion of informants who regard a species as most important (Byg and Balslev 2001).
   (d) *Rapid Informant Rank* asks informants to list and rank the ten most important species harvested from the forest over the past 10 years (Lawrence et al. 2005).

Some other consensus methods include:

(a) Fidelity level (FL) and Ranked order priority (ROP)
This method was introduced to calculate a ratio between the number of informants who cite the use of a species for the same major purpose and the total number of informants who mentioned any use for the species (Friedman et al. 1986).

$$\mathrm{FL} = \frac{I_\mathrm{p}}{I_\mathrm{u}} \times 100$$

$$\mathrm{ROP} = \mathrm{FL} \times \mathrm{RPL}$$

The FL value quantifies the importance of a species for a given purpose. I_p = number of informants who cited the species for a particular use, I_u = total number of informants that mentioned the plant for any use.
Relative popularity level (RPL) is a number between 0 and 1.

(b) Species use value for one informant (Phillips and Gentry 1993a)

$$\mathrm{UV}_{is} = \left(\sum U_{is}\right)/(n_{is})$$

U_{is} = number of uses mentioned for species s by informant i and n_{is} = number of events in which informant i cites a use for species s.

(c) Species use value for one species across all informants (Phillips and Gentry 1993a)

$$\mathrm{UV}_{s} = \left(\sum \mathrm{UV}_{is}\right)/(n_{i})$$

n_i = total number of informants interviewed for species s. Sum the informant use values for a species and divide by the total number of informants.

(d) Family use value (Phillips and Gentry 1993a)

$$\mathrm{FUV} = \sum \mathrm{UV}_s/(n_s)$$

n_s = total number of species within a given family. Sum the use values for all the species within a given family and divide by n_s.

(e) Cultural, Practical, and Economic Value
Consensus methods are used to distinguish between actual and potential or past uses and also takes economic valuation into consideration. The overall score is a composite value (Reyes-García et al. 2006).
Cultural Value Index

$$\mathrm{CV_e} = \mathrm{UC_e} \times \mathrm{Ic_e} \times \sum \mathrm{IUc_e}$$

Where e = ethno-species (voucher specimens did not match up with herbarium botanical species so referred to as ethnospecies). Uc_e = number of uses reported (through free listing) for an ethno-species divided by the total number of use categories. Ic_e = Number of participants who listed a species as useful divided by total number of participants. $\sum IUc_e$ = Number of participants who mentioned each use (category) for the ethno-species divided by the total number of participants.

Practical Value Index

$$PV_e = Up_e \times Ip_e \times DUp_e$$

Up_e = number of uses reported for an ethno-species divided by the total number of use categories. Ip_e = Number of times an ethno-species was brought to the household for use divided by the total number of participants in scan observations. DU_{pe} = an assignment of duration of use for each item brought to the household. In this case "scan observation" methodology is employed to determine the species "practical value." Randomly selected subjects are asked about the plants they brought home within a 24 h period.

Economic Value Index

$$EV_e = Oe_e \times Pe_e$$

Oe_e = number of times an ethno-species was brought to a household. Pe_e = Price of the ethno-species based on market price or time taken to obtain the species multiplied by the average daily wage.

4.1.8 *Ecological Methods for Ethnobotanists*

The ecological methods allow ethnobotanists to analyze the ecological context within people interact with plants. These methods go beyond the traditional ethnobotanical plant collection, plant identification and documentation while take into account the fact that things happen when people use plants (Hall and Bawa 1993; Peters 1996).

1. Quantitative assessment of species density
   This method measures the number of individuals per unit area, informing the researcher about how much a given plant resource is available for exploitation and where the greatest abundance of this resource is located. It offers the possibility of estimating the long-term sustainability of plant resource exploitation too. The study requires systematic or random plots or transects, which will vary in size depending on the selected plant species.

2. Growth yield studies
   The basic objective of a yield study is to provide an estimate of the quantity of resource (fruits, seeds, latexes, resins, gums, stems, leaves, roots, barks, etc.) produced by a given species growing in a particular area. As it is very difficult to monitor all the individuals of a selected species, a subsample of plants must be selected. The selection should be stratified by two main variables: diameter and condition. Using the results from a plot survey as a guide, the fieldworker should randomly select individuals from different size classes and habitats. Ideally, the number of sample trees selected from each size class should be the same in each site condition or forest type.

4.1.8.1 Other tools for the Evaluation of Local Ecological Knowledge

Other important tools for the evaluation of local ecological knowledge are:

Triad tests (the researcher gives the informants three plant specimens, plant pictures, or other items and they have to choose one that "does not fit" or choose the two "that seem to fit the best").

Pile sorting (participants divide objects into a number of groups according to the overall similarity of items).

24 h recall (participants have to list, for instance, the name of the plant products they have consumed during the last 24 h).

4.2 Ethnobotany of Wild Edible Vegetables

4.2.1 Amaranthus hybridus L.

Amaranthus cruentusv L.
Amaranthus frumentaceus Buch.Ham

English Name: Slim amaranth/Pigweed
Local Name: Ghinar

Fig. 4.1 *Amaranthus hybridus* L

Family Name: Amaranthacea

Description:
An annual herbaceous plant of 1–6 ft height. Stem branched, angular, glabrous having short or long multicellular hairs. Leaves alternate, long petiole, dull green, rough, hairy, lamina broadly lanceolate to ovate with wavy margins, gradually narrowed with subacute tip. Flowers small, yellowish green, terminal spikes formed of cymose clusters. Seeds black, shining (Fig. 4.1).
Flowering: May–August.
Status and habitat: Common on moist and waste places along cultivated fields.
Part used: Whole plant
Distribution:
World: Tropical and subtropical regions. ***Pakistan***: Murree, Hazara, Kaghan, Swat, Rawalpindi, Kashmir, Sind, Punjab, and Baluchistan.
Ethnobotanical uses: Fresh leaves are cooked in water and used as culinary vegetable (sag). Aerial parts are also used as fodder for cattle.
Medicinal uses: Dried seeds of *Amaranthus hybridus* and dried fruits of *Ficus carica*, are grinded together along with sugar. This powder is taken orally to cure eye vision problem. The leaves are emollient, also used as antidote for snake and scorpion bite, in the treatment of intestinal bleeding, diarrhea, and excessive menstruation.

4.2.2 *Amaranthus spinosus L.*

English Name: Spinay amaranth/Prickly amaranth
Local Name: Ghinar

Fig. 4.2 *Amaranthus spinosus* L.

Family Name: Amaranthacea

Description:
Annual, erect or slightly decumbent, much-branched, and bushy herb. Stem stout, branched, reddish, angular, and glabrous with long hairs. Leaves glabrous, long petiolated, lamina ovate to rhomboid-ovate, elliptic, or lanceolate, leaf-axial bearing a pair of fine slenderical and compressed long spines. Flowers green, in the lower part of the plant in axillary clusters. Seeds compressed, black, and shining (Fig. 4.2).
Flowering: May–August.
Status and habitat: Common along road sides and waste places.
Part used: Leaves and seeds
Distribution:
World: Originated from America, consider as cosmopolitan weed in the warmer and temperate regions of the world.
Pakistan: Murree, Haripur, Abbottabad, Balakot, Manshera, Kahouta, Margalla Hills, Islamabad, Muzafarabad, Sind, Punjab, and Baluchistan.
Ethnobotanical uses:
Fresh leaves are cooked in water and used as vegetable (sag). Aerial parts are also used as fodder for cattle.
Medicinal uses:
Dried seeds of *Amaranthus spinosus* and dried fruits of *Ficus carica* are grinded together along with sugar. This powder is taken orally to cure eye vision problem. Plant is used as laxative, diuretic, stomachic, antipyretic, improves the appetite, biliousness, blood diseases, diabetes, burning sensation, leprosy, bronchitis, piles, internal bleeding, menstruation, and leucorrhoea.

4.2.3 Amaranthus viridis L.

Amaranthus gracilis Desf.

English Name: Slender amaranth, Green Amaranth
Local Name: Ghinari

Fig. 4.3 *Amaranthus viridis* L.

Family Name: Amaranthacea

Description:
Annual, erect, or more rarely ascending herb. Stem reddish brown, rough, cylindrical, glabrous, sparingly branched. Leaves green, multifoliate, glabrous-glutinous; lamina deltoid-ovate to rhomboid oblong, long petiolated; leaf insertion cauline to ramal; base broad and narrow to the apex; entire margin. Flowers green, slender, axillary or terminal, bracteates, often peniculate spikes (Fig. 4.3).
Flowering: May–August.
Status and habitat: Common and widely distributed on damp, wet, and open places.
Part used: Leaves and seeds
Distribution:
World: Throughout the tropical and subtropical regions.
Pakistan: Murree, Hazara, Kaghan, Swat, Rawalpindi, Islamabad, Kashmir, Sind, Punjab, and Baluchistan.
Ethnobotanical uses:
Fresh leaves are cooked in water and used as vegetable (sag). Aerial parts are also used as fodder for cattle.
Medicinal uses:
Dried seeds of *Amaranthus viridis* and dried fruits of *Ficus carica* are grinded together along with sugar. This powder is taken orally to cure eye vision problem. Leaves are emollient, cooling, laxative, stomachic, appetizer, and antipyretic; used in burning sensation, leprosy, bronchitis, piles, leucorrhoea, constipation, used in scorpion sting and snake bites.

4.2.4 *Digera muricata* (L.) Mart.

Digera alternifolia (L.) Aschers
Achyranthes muricata L.

English Name: Wild rhubrab
Local Name: Tundulla

Fig. 4.4 *Digera muricata* (L.) Mart

Family Name: Amaranthacea

Description:
Annual, simple with ascending branches herb. Stem glabrous with pale ridges. Branches hairy, pale ridges. Leaf blade linear to ovate, glabrous; petiole slender in the lower leaves, shortening in the upper leaves. Flowers glabrous, white tinged with pink to red, usually becoming greenish white in fruit, axillary racemes. Fruit subglobose, slightly compressed, keeled along each side (Fig. 4.4).
Flowering: July–September
Status and habitat: Common as a weed of Maize fields and waste places.
Part used: Leaves
Distribution:
World: Widespread in southern Asia from tropical Arabia and the Yemen to Afghanistan, India, Ceylon, Malaysia, Indonesia, tropical Africa, and Madagascar.
Pakistan: Murree, Haripur, Abbottabad, Balakot, Manshera, Kahouta, Margalla Hills, Islamabad, Rawalpindi, Muzafarabad, Sindh, Punjab.
Ethnobotanical uses:
Fresh leaves are cooked in water and used as culinary vegetable (sag). Aerial parts are also used as fodder for cattle.
Medicinal uses: Fresh leaves are boiled in water and paste is taken orally to cure constipation. This plant is also effective in urinary disorders and for secondary infertility.

4.2.5 Dryopteris ramosa (Hope) C. Chr.

Nephrodium ramosum C. Hope

English Name: Oak fern
Local Name: Gunji

Fig. 4.5 *Dryopteris ramosa* (Hope) C. Chr.

Family Name: Aspidiaceae

Description: A terrestrial herb with adventitious roots, creeping rhizome with brownish hairs. Leaves green, macrophyllous, compound, pinnate or bipinnate, originate from upper surface of the rhizome. Young leaves circinate, covered with brownish hairs. Sporangia brown, on lower leaf surface (Fig. 4.5).
Spore period: December–March
Status and habitat: Commonly found on shady, cold, and moist places.
Part used: Young leaves
Distribution:
World: Widespread in Europe and Africa except arctic zone and South America.
Pakistan: Murree, Gilyat, Abbottabad, Naran, Shogran.
Ethnobotanical uses: Young circinate leaves are cooked in water and used as culinary vegetable
Medicinal uses: Young leaves cooked as vegetables are taken orally to cure gastric ulcer and constipation.

4.2.6 *Bidens bipinnata L.*

Cosmos bipinnatus Cav.

English Name: Spanish needles
Local Name: Siryala

Fig. 4.6 *Bidens bipinnata* L.

Family Name: Asteraceae

Description:
Annual herb. Stem erect, slender, stiff, four-angled with glabrous, green spreading branches. Leaves opposite, dentate or serrate, foliate, pinnately 3–5-foliolate, terminal leaflet larger than lateral leaflets, without stipules, dark green. Flowers five-numerous, bisexual, yellow (Fig. 4.6).
Flowering period: October–April
Status and habitat: Commonly found in grasses, filed borders.
Part used: Young leaves
Distribution:
World: A cosmopolitan weed found in Asia, Africa, Europe, and America. ***Pakistan***: Murree, Hazara, Punjab, Salt range, Baluchistan, Gilgit, Kashmir.
Ethnobotanical uses:
Young leaves are cooked in water and used as culinary vegetable
Medicinal uses:
Fresh leaves are crushed and mixed in water, this extract is applied topically to cure leprosy and skin cuts. Leaves chewed to cure sore throat and juice is used against earache.

4.2.7 Digera muricata (L.) Mart.

Cichorium intybus L.
English Name: Chicory
Local Name: Kashni

Fig. 4.7 *Cichorium intybus* L.

Family Name: Asteraceae

Description:
A perennial suberect or spreading herb up to 2 ft tall. Stem erect, branched, herbaceous, green. Leaves simple, green, toothed, lanceolate, deeply divided, crowded at the base, spirally arranged on stem. Flowers bright blue, blue purple or pinkish, terminal or axillary. Pappus absent or scaly. Fruit brownish to black (Fig. 4.7).
Flowering period: June–September
Status and habitat: Common weed on waste places, cultivated lands, and grassy fields.
Part used: Young leaves
Distribution:
World: Europe, West Asia, Afghanistan, Iran, India, and Nepal.
Pakistan: Balouchistan, Waziristan, Kurram, Peshawar, Swat, Hazara, Astor, Gilgit, Baltistan, Murree, Rawalpindi, Islamabad, and Kashmir.
Ethnobotanical uses:
Young leaves are cooked in diluted milk (Lusii) and used as culinary vegetable.
Medicinal uses:
Fresh leaves are boiled in water and this decoction is taken orally to cure fever, gas trouble, and body swelling. Leaves juice is used to cure cancer of the uterus, as tonic and blood purifier, while paste is applied topically to heal wounds.

4.2.8 *Launaea procumbens (Roxb).*

Prenanthes procumbens Roxb.

English Name: Bold-leaf launaeae
Local Name: Makhna

Fig. 4.8 *Launaea procumbens* (Roxb).

Family Name: Asteraceae

Description:
Perennial herb branched from base. Stem branched, smooth, glabrous, leafless, or with few leaves. Rosette leaves stipulate, dentate, pinnatly lobed, tapering in to narrow base, margin white cartilaginous, denticulate. Flowers yellowish white (Fig. 4.8).
Flowering period: November–December
Status and habitat: Common on waste, moist, and cold places.
Part used: Young leaves
Distribution:
World: Afghanistan, China, India, Kashmir, Kazakhstan, Myanmar, Nepal, Tajikistan, Turkmenistan, Uzbekistan; SW Asia.
Pakistan: Balochistan, Waziristan, Kurram, Peshawar, Swat, Hazara, Astor, Gilgit, Baltistan, Murree, Rawalpindi, Islamabad, and Kashmir.
Ethnobotanical uses:
Young leaves are cooked in diluted milk (Lusii) and used as culinary vegetable.
Medicinal uses:
Fresh leaves are grinded along with sugar and extract is taken orally to cure painful micturation. Plant is also found as coolant, diuretic, demulcent, and is effective against allergic infection.

4.2.9 Sonchus asper L.

Sonchus nymanii Tineo & Guss

English Name: Spiny sow thistle
Local Name: Dodhal

Fig. 4.9 *Sonchus asper* L.

Family Name: Asteraceae

Description:
Annual herb with erect, glabrous, soft, herbaceous, hollow, unbranched stem emits a milky sap when broken. Leaves variable, obovate; basal leaves rosette; upper alternate, cauline, blades of mid cauline oblong-lanceolate, elliptical, glabrous, dark green, spiny, dentate, cute apex. Pappus caduceus, white. Flowers terminal, head consists of 30–100 ligulate, yellow ray florets (Fig. 4.9).
Flowering period: May–October
Status and habitat: Common on waste places.
Part used: Young leaves
Distribution:
World: Afghanistan, Bhutan, China, India, Japan, Kashmir, Kazakhstan, Korea, Kyrgyzstan, Nepal, Russia, Tajikistan, Thailand, Vietnam, Africa, Australia, New Guinea, North and South America, and New Zealand.
Pakistan: Balouchistan, Sind, Hazara, Waziristan, Kurram, Peshawar, Swat, Gilgit, Murree, Rawalpindi, Islamabad, and Kashmir.
Ethnobotanical uses:
Young leaves are cooked in diluted milk (Lusii) and used as culinary vegetable.
Medicinal uses:
Leaves are boiled in water and decoction is taken orally against fever and constipation. Leaf paste is applied topically as a poultice on wounds and boils.

4.2.10 *Sonchus oleraceus* L.

Sonchus angustissimus H.Lindb.

English Name: Sow thistle/Milky thistle
Local Name: Dodhal

Fig. 4.10 *Sonchus oleraceous* L.

Family Name: Asteraceae

Description:
Annual herb up to 1–2 m high. Stem erect, branched, exudes milky latex. Basal leaves lanceolate upper lanceolate, lobed, with pointed ends clasping basal lobes. Flowers many, small yellow florets in head (Fig. 4.10).
Flowering period: February–May
Status and habitat: Common weed on waste places, and crop fields.
Part used: Young leaves
Distribution:
World: Native of Europe, Asia, and Africa, cosmopolitan in distribution.
Pakistan: Balouchistan, Sind, Hazara, Waziristan, Kurram, Peshawar, Swat, Gilgit, Murree, Rawalpindi, Islamabad, and Kashmir.
Ethnobotanical uses:
Young leaves are cooked in diluted milk (Lusii) and used as culinary vegetable.
Medicinal uses:
Fresh leaves are boiled in water and decoction is taken orally to treat body weakness and constipation. This plant is also used to cure diarrhoea, menstrual disorders, fever, and inflammation.

4.2.11 *Taraxacum officinale* L.

Taraxacum depressum Baratte

English Name: Cankerwort/Milk witch
Local Name: Sheshe haund

Fig. 4.11 *Taraxacum officinale* L.

Family Name: Asteraceae

Description:
A small, glabrous, perennial herb with rosette of basal leaves. Stem underground, long, simple or branched rhizome, exudes milky latex. Leaves in rosette from rhizome, bright green, pinnately lobed, lyrate. Flowers yellow, ligulate, on scapigerous heads. Fruit achene, greenish brown, surrounded by pappus (Fig. 4.11).
Flowering period: February–May
Status and habitat: Common weed on waste places along cultivated lands.
Part used: Young leaves
Distribution:
Cosmopolitan in distribution found throughout the world and Pakistan.
Ethnobotanical uses:
Young leaves are cooked in diluted milk (Lusii) and used as culinary vegetable. Used as fodder for cattle.
Medicinal uses:
Fresh rhizomes are boiled in water and decoction is taken orally against jaundice and as diuretic. The plant is used to cure liver, gallbladder, kidney, and joint problems. It is also considered a blood purifier and is used for eczema and cancer.

4.2.12 *Bombax malabaracum DC.*

Bombax aculeatum L.
Bombax ceba L.

English Name: Silk cotton tree
Local Name: Dug sumbal

Fig. 4.12 *Bombax malabaracum* DC.

Family Name: Bombacaceae

Description:
A large tree up to 20 m tall. Stem erect, branched, woody, gray bark armed with prickles. Leaves compound, 5–7 leaflets, green, entire, elliptic to lanceolate, long stalked. Flowers large, red fleshy, cup-shaped, appearing on branches before leaves. Fruit woody, capsule, dehiscing by five valves (Fig. 4.12).

Flowering period: March–October

Status and habitat: Rare plant grows on waste dry places.

Part used: Aerial parts

Distribution:

World: India, Bhutan, and South China.

Pakistan: Sub-Himalayan tract, Rawalpindi, Islamabad, Hazara, Muzfarabad.

Ethnobotanical uses:
Young flowering buds are cooked as vegetables. Wood is used in furniture and as fuel, also cultivated as ornamental plant. Branches are used for fencing.

Medicinal uses:
Fresh bark is crushed and paste is applied topically to cure skin eruptions, pimples, wounds, and joint pain. Powdered flowers along with honey are used for menorrhagia. Root bark extract is used as a tonic in case of sexual debility. Root powder mixed with sucrose and milk is taken to avoid impotency.

4.2.13 *Capsella bursa-pastoris (L.) Medic.*

Capsella thracicus (Velen.) Stoj. & Stef.

English Name: Shepherd's purse
Local Name: Saag

Fig. 4.13 *Capsella bursa-pastoris* L.

Family Name: Brassicaceae

Description:
Annual or biennial herb up to 45 cm tall. Stem erect, glabrous, or hairy with simple or branched hairs. Basal leaves rosulate, lyrate to entire, shortly stalked; cauline leaves smaller, sessile. Racemes many flowered, white (Fig. 4.13).
Flowering period: March–June
Status and habitat: Common weed of waste places.
Part used: Aerial parts
Distribution:
World: Cosmopolitan found in cold areas.
Pakistan: Throughout the country.
Ethnobotanical uses:
Young leaves are cooked in water as vegetables. Aerial parts are used as fodder for cattle.
Medicinal uses:
Fresh leaves are boiled in water and decoction is taken orally to cure menstrual disorders. Plant is effective against both internal and external bleeding and diarrhea.

4.2.14 *Nasturtium officinale R. Br.*

Nasturtium microphyllum Boenn. ex Rchb.

English Name:
Watercress
Local Name: Saag

Fig. 4.14 *Nasturtium officinale* R.Br.

Family Name: Brassicaceae

Description:
A small perennial, rhizomatous herb up to 2 ft long. Roots adventitious, below at nodes, floating, glabrous. Leaves lyrate, pinnate; lower stalked, upper sessile, 5–9 leaflets; terminal leaflets cordate, lateral entire or toothed. Racemes 10–25 flowered, white. Fruit siliqua, with many seeds (Fig. 4.14).
Flowering period: April–July
Status and habitat: Commonly found in water and marshy places.
Part used: Leaves
Distribution:
World: Wildly distributed in Europe and Asia.
Pakistan: Chitral, Hazara, Kashmir, Islamabad, Murree, Margalla Hills. Punjab, Peshawar, Balochistan, Sind.
Ethnobotanical uses:
Young leaves are cooked in water and used as vegetables.
Medicinal uses:
Fresh leaves are cooked in water and paste is taken orally to cure constipation. This plant is also used as depurative, expectorant, purgative, stimulant, and stomachic. The freshly extracted juice is used internally and externally in the treatment of chest and kidney complaints, chronic irritations and inflammations of the skin, and as hair tonic.

4.2.15 *Bauhinia variegata* L.

Phanera variegata Benth.

English Name: Camel's foot tree/Orchid tree
Local Name: Kalyar/Kichnar

Fig. 4.15 *Bauhinia variegata* L.

Family Name: Caesalpiniodeae

Description:
A medium-sized tree with dark brown, smooth bark. Leaves petiolate, lamina long, as broad as, with a medium cleft, lobe obtuse, base deeply heart shaped. Inflorescence raceme. Flowers pinkish to purplish white, terminal, bracteate. Pods long, hard, flat, dehiscent 10–15 seeded (Fig. 4.15).
Flowering period: February–April
Status and habitat: Found in rough areas, along the road sides, gardens, and forests.
Part used: Whole plant
Distribution:
World: India, Nepal, Burma, China; widely cultivated in tropics.
Pakistan: Hazara, Kashmir, Islamabad, Margalla Hills, Salt range, Peshawar, Karachi.
Ethnobotanical uses:
Flowering buds are cooked with meat, while flowers are cooked in diluted milk and used as vegetables. Leaves are used as fodder, wood is used as fuel, in construction, making tool handles; plant is also cultivated for ornamental purpose.
Medicinal uses:
Leaves and flowers are crushed and paste is given to cattle against diarrhoea. Roots are carminative, antidote to snake bite; bark used as tonic, astringent, to cure skin infections. Juice of flowers is used against diarrhoea and dysentery.

4.2.16 *Bauhinia variegata var. candida Voigt*

Bauhinia alboflava de Wit.

English Name: Mountain ebony
Local Name: Safed Kalyaar

Fig. 4.16 *Bauhinia variegata* var. *candida* Voigt

Family Name: Cesalpiniodee

Description:
A medium-sized deciduous tree with dark brown, smooth bark. Leaves petiolate, lamina long, as broad as, with a medium cleft, lobe obtuse, base deeply heart-shaped. Inflorescence raceme. Flowers white petals, terminal, bracteate, steminiods absent. Pods long, hard, flat, dehiscent 10–15 seeded (Fig. 4.16).
Flowering period: February–March
Status and habitat: Cultivated in gardens, also found along the road sides.
Part used: Whole plant
Distribution:
World: China; widely cultivated in tropical forest.
Pakistan: Hazara, Islamabad, Karachi.
Ethnobotanical uses:
Flowering buds are cooked with meat, while flowers are cooked in diluted milk and used as vegetables. Leaves are used as fodder, wood is used as fuel, in construction, making tool handles; plant is also cultivated for ornamental purpose.
Medicinal uses:
Leaves and flowers are crushed and paste is given to cattle against diarrhoea.

4.2.17 Bauhinia variegata L. var. variegata

Phanera variegata (L.) Bentham

English Name: Chinensis Candolle
Local Name: Jamni Kalyaar

Fig. 4.17 *Bauhinia variegata* var. *variegata*

Family Name: Caesalpiniodeae

Description:
A medium-sized deciduous tree with dark brown, smooth bark. Leaves petiolate, lamina long, as broad as, with a medium cleft, lobe obtuse, base deeply heart-shaped. Inflorescence raceme. Flowers whitish petals with color spots, terminal, bracteate, steminiods 1–5. Pods long, hard, flat, dehiscent 10–15 seeded (Fig. 4.17).
Flowering period: February–March
Status and habitat: Cultivated in gardens, also found along the road sides and waste places.
Part used: Whole plant
Distribution:
World: Cambodia, Laos, Myanmar, Thailand, Vietnam; widely cultivated in the tropics and subtropics.
Pakistan: Hazara, Islamabad.
Ethnobotanical uses:
Flowering buds are cooked with meat, while flowers are cooked in diluted milk and used as vegetables. Leaves are used as fodder, wood is used as fuel, in construction, making tool handles; plant is also cultivated for ornamental purpose.
Medicinal uses:
Leaves and flowers are crushed and paste is give to cattle against diarrhoea.

4.2.18 *Silene conoidea* L.

Conosilene conoidea Fourr.
Cucubalus conoideus Lam.

English Name: Conoid catchfly
Local Name: Doda

Fig. 4.18 *Silene conoidea* L.

Family Name: Caryophyllaceae

Description:
Annual herb with erect, branched or simple, densely glandular-pubescent stem. Leaves linear, lanceolate, base clasping at the nodes, sessile, glandular-pubescent. Bracts similar to leaves, but smaller in size. Inflorescence paniculate, with 3–5 pink to red flowers. Seeds brown (Fig. 4.18).
Flowering period: March–April
Status and habitat: Common weed of waste places and crop filed.
Part used: Leaves
Distribution:
World: France, Italy, Turkey, the Eastern Mediterranean, the Caucasus, Iran, Central Asia, the Himalaya, Africa, and North America. ***Pakistan***: Hazara, Kashmir, Islamabad, Margalla Hills, Salt range, Peshawar, Taxila, Balochistan.
Ethnobotanical uses: Young leaves and unripe fruits are cooked in water as well as in diluted milk as vegetable. Aerial parts are also used as fodder for cattle.
Medicinal uses: Fresh leaves are crushed and paste is applied topically to cure skin infections. The plant is said to be emollient and is used as a fumigant. The juice of the plant is used in the treatment of ophthalmia.

4.2.19 *Stellaria media (L.) Vill*

Stellaria cupaniana (Jord. & Fourr.) Nyman

English Name: Chickweed
Local Name: Makhni

Fig. 4.19 *Stellaria media* (L.) Cyr.

Family Name: Caryophyllaceae

Description:
Annual herb, with prostrate to ascending stems. Stems leafy, hairy on one side. Leaves elliptic to ovate, sessile or petiolate, acute to subacute glabrous with hairy margin, hairy on the nerves. Flowers white in panicles (Fig. 4.19).
Flowering period: April–August
Status and habitat: Common weed of cold waste places.
Part used: Aerial parts
Distribution:
World: Cosmopolitan in distribution, found throughout the world.
Pakistan: Hazara, Kashmir, Swat, Islamabad, Margalla Hills, Taxila, Baluchistan, Gilgit.
Ethnobotanical uses:
Young leaves are cooked in water as vegetable. Aerial parts are also used as fodder for cattle.
Medicinal uses:
Fresh leaves are crushed and paste is applied topically to treat swelling joints, bone fracture. Leaves are boiled in water and decoction is taken orally against constipation.

4.2.20 *Chenopodium album L.*

Chenopodium acuminatum (Willd.) Kuntze

English Name: Pigweed
Local Name: Bathu

Fig. 4.20 *Chenopodium album* L.

Family Name: Chenopodiaceae

Description:
Annual usually erect, variously branched herb. Stem yellowish to green, sometimes reddish. Lower and medium leaves petiolate, variously trullate, ovate to lanceolate, clearly longer than broad, base narrowly to broadly cuneate, margins irregularly serrate to entire, often somewhat three-lobed; uppermost leaves lanceolate, usually entire. Inflorescence a cymosely branched panicle, mostly terminal. Seeds horizontal, black, somewhat ovate (Fig. 4.20).
Flowering period: January–September
Status and habitat: Common weed on waste places.
Part used: Aerial parts
Distribution:
World: Cosmopolitan, common in subtropical to temperate zones.
Pakistan: Hazara, Kashmir, Swat, Chitral, Murree, Islamabad, Margalla Hills, Taxila, Baluchistan, Gilgit, Karachi.
Ethnobotanical uses:
Young leaves are cooked in diluted milk as vegetable. Aerial parts are also used as fodder for cattle.
Medicinal uses:
Fresh leaves are cooked as vegetable and eaten raw to cure constipation and intestinal worms. Juice of fresh leaves is taken orally to treat jaundice and urinary disorders.

4.2.21 Commelina benghalensis L.

English Name: Day flower/Dew flower
Local Name: Naria

Fig. 4.21 *Commelina benghalensis* L.

Family Name: Commelinaceae

Description:
An erect to prostrate perennial herb, dichotomously branched from base, glabrous to pubescent. Leaves ovate to oblong, subsessile to shortly petioled, entire, acute to subround at apex. Upper cymes 2–3-flowered, lower 1–2-flowered. Aerial flowers blue, in short cymes, maturing earlier than the subterranean flowers (Fig. 4.21).
Flowering period: June–September
Status and habitat: Common weed in crop fields and waste places.
Part used: Aerial parts
Distribution:
World: Tropical and subtropical Asia and Africa.
Pakistan: Hazara, Kashmir, Swat, Murree, Islamabad, Margalla Hills, Taxila, Baluchistan.
Ethnobotanical uses:
Young leaves are cooked in water as vegetable. Aerial parts are also used as fodder for cattle.
Medicinal uses:
Dried roots are grinded and powder is taken orally to cure epilepsy, whereas fresh roots are boiled in water and decoction is used to treat stomach disorders. This plant is also used to cure infected wounds, for treating sterility and as mucilage for treating burns, sore eyes, and sore throats.

4.2.22 *Evolvulus alsinoides* L.

Evolvulus adscendens (House) Ooststr.

English Name: Dwarf morning glory
Local Name: Kalowa

Fig. 4.22 *Evolvulus alsinoides* L.

Family Name: Convolvulaceae

Description:
Annual Herbs. Stem herbaceous, ascending, pubescent. Leaves oblong, elliptic-oblong or lanceolate, subsessile, acute to obtuse basally and apically, pubescent. Flowers blush white on peduncles. Fruit capsular, glabrous. Seeds black, smooth (Fig. 4.22).
Flowering period: February–October
Status and habitat: Common weed in crop fields and waste places.
Part used: Aerial parts
Distribution:
World: America, Africa, Iran, India, China, Ceylon, and Malaysia.
Pakistan: Hazara, Kashmir, Swat, Islamabad, Margalla Hills, Taxila.
Ethnobotanical uses:
Young leaves are cooked in water as vegetable. Aerial parts are also used as fodder for cattle.
Medicinal uses:
Leaves are crushed along with sugar and water, extract is taken orally to cure indigestion and constipation. This plant is also used as brain tonic, to cure fever, cough, and cold.

4.2.23 *Dioscorea deltoidea* Wall. ex, Griseb

Dioscorea orbiculata Prain & Burkill

English Name: Yam
Local Name: Jungli kachaloo

Fig. 4.23 *Dioscorea deltoidea* Wall. ex, Kunth

Family Name: Dioscoreaceae

Description:
An annual tubers climber. Stems glabrous, twining clockwise. Leaves alternate, simple, ovate, often cordate, the basal lobes rounded, acuminate, glabrous. Male spikes solitary, axillary, simple or sometimes branched, slender; flowers small distant clusters, female spikes solitary, few-flowered. Fruits capsule, obovate or obcordate. Seeds winged round (Fig. 4.23).
Flowering period: May–July
Status and habitat: Rare climber on *Punica granatum* and other shrubs.
Part used: Tubers, leaves
Distribution:
World: Afghanistan, China, India and throughout the Himalayas.
Pakistan: Hazara, Kashmir, Swat, Murree, Margalla Hills.
Ethnobotanical uses:
Young leaves are cooked in water and used as vegetable. Aerial parts are also used as fodder for cattle.
Medicinal uses:
Paste of fresh tubers is applied topically and taken orally against intestinal worms and to kill lice. Roots are crushed and paste is taken orally against constipation.

4.2.24 *Lamium amplexicaule* L.

Lamium albiflorum D.M.Moore

English Name: Henbit dead nettle
Local Name: Phumbra

Fig. 4.24 *Lamium amplexicaule* L.

Family Name: Lamiaceae

Description:
Annual, slender herb. Stems erect, ascending or prostrate branched from the base, almost glabrous, purplish below or not, with few leaves. Leaves broadly ovate to reniform crenate to lobed, obtuse, rounded or cordate at base, pubescent, influences verticillasters. Flowers purple-pink, upper lip pubescent to vinous, straight; lower lip reflexed, shorter than upper (Fig. 4.24).
Flowering period: December–April
Status and habitat: Common on waste places and crop filed margins.
Part used: Aerial parts
Distribution:
World: Throughout Europe and Asia, introduced weed in South Africa.
Pakistan: Chitral, Hazara, Kashmir, Swat, Murree, Margalla Hills. Quetta, Biltistan, Punjab.
Ethnobotanical uses:
Young leaves are cooked in water as vegetable. Aerial parts are also used as fodder for cattle.
Medicinal uses:
Fresh leaves are crushed and paste is applied topically on swelling joints. This plant is found antirheumatic, diaphoretic, excitant, febrifuge, laxative, and stimulant.

4.2.25 *Origanum vulgare subsp. Hirtum (Link) Letsw.*

Origanum albidum Bellynck

English Name: Wild marjoram/Oregan
Local Name: Poodni

Fig. 4.25 *Origanum vulgare* subsp. *Hirtum* L.

Family Name: Lamiaceae

Description:
A perennial rhizomatous aromatic herb with fibrous roots. Stem branched from base, thinly to densely pilose with spreading hairs, or glabrous, leafy, purplish, or green. Leaves simple, entire, ovate to narrow elliptic, gland-dotted, apex acute or obtuse, with scattered hairs or glabrous, young leaves usually in leaf axils; Inflorescence branched; verticillasters two-flowered. Flowers purple or white (Fig. 4.25).
Flowering period: June–October
Status and habitat: Common on waste grassy, slopes.
Part used: Leaves
Distribution:
World: S. Europe, Mediterranean countries, Asia, along the Himalayan area to China and Taiwan. ***Pakistan***: Abbottabad, Haripur, Mansehra, Naran, Sugran, Kashmir, Swat, Murree, Margalla Hills.
Ethnobotanical uses: Young leaves are cooked along with other vegetables as flavoring agent. Aerial parts are also grazed by cattle.
Medicinal uses: Fresh leaves are chewed to cure toothache and mouth gums. This plant is used to treat asthma, cough, cold, rheumatism, stomach and urinary disorders, sore throats, to heal wounds and against insects bites.

4.2.26 *Tulipa stellata var. clusiana Hk. f.*

Tulipa chrysantha A.D.Hall

English Name: Benghal day flower/Wild tulip
Local Name: Kakar moona

Fig. 4.26 *Tulipa stellata var. clusiana* Hk. f.

Family Name: Liliaceae

Description:
Annual herb with bulbs lie a little below the soil, ovoid or globose. Stem underground, glabrous. Leaves 3–6, alternate, sparse, or crowded, linear, lanceolate, acute-acuminate, margin often wavy. Flowers solitary terminal, variable in color, flowers white with pink outside or with pink or yellow base. Seeds compressed brown (Fig. 4.26).
Flowering period: March–May
Status and habitat: Common on waste places and in crop fields.
Part used: Leaves, flowers, bulbs
Distribution:
World: Iran, Afghanistan, India, naturalized in Southern Europe.
Pakistan: Hazara, Kashmir, Swat, Murree, Margalla Hills, Sind, Baluchistan, Punjab.
Ethnobotanical uses:
Young leaves and flowers are cooked in diluted milk along with other vegetables. Aerial parts are also grazed by cattle. Flowers are also used for decoration while bulbs are edible.
Medicinal uses:
Fresh bulbs are peeled off and eaten raw as cardiac stimulant.

4.2.27 Malva parviflora L

Althaea microcarpa Alef.
Althaea parviflora (L.) Alef.

English Name: Least mallow
Local Name: Sonchal

Fig. 4.27 *Malva parviflora* L

Family Name: Malvaceae

Description:
Annual, prostrate or procumbent herb, up to 6–10 cm tall. Stem: green, rough, cylindrical, pubescent. Leaves: green, glabrous, pubescent and glistening, orbiculate, petiolate, insertion ramal, cordate base, rounded apex and ciliate margin. Flowers: Purplish white, bracteates, axillary, symmetrical (Fig. 4.27).
Flowering period: May–June
Status and habitat: Common on waste places along cultivated fields.
Part used: Leaves
Distribution:
World: Europe, North Africa, South and West Asia. ***Pakistan***: Hazara, Kashmir, Rawalpindi, Swat, Murree, Margalla Hills, Sind, Baluchistan, Punjab.
Ethnobotanical uses: Young leaves cooked in water as vegetables. Aerial parts are also used as fodder for cattle.
Medicinal uses: Fresh leaves are boiled in water and decoction is taken orally to cure constipation, cough, and fever. The whole plant is emollient, used as a poultice on swellings, running sores and boils, seeds are used in the treatment of coughs and ulcers in the bladder. A decoction of the roots or leaves is used to remove dandruff and to soften the hair.

4.2.28 *Ficus carica* L.

Ficus caprificus Risso

English Name: Fig
Local Name: Trakani
Phagwar

Fig. 4.28 *Ficus carica* L

Family Name: Moraceae

Description:
A small deciduous tree, with spreading branches. Bark smooth, grey or dull white, young twigs glabrous or softly hairy. Leaves with glabrous grooved petiole; lamina broadly ovate to nearly orbicular, divided obscurely palmatifid to mostly palmatipartite, dentate margin. Hypanthodia axillary solitary or paired, borne on peduncles, subsessile to sessile, Figs usually pyriform-obovoid, glabrous or shortly hispid, yellowish to brownish violet (Fig. 4.28).
Flowering period: April–August
Status and habitat: Common on waste places along cultivated lands.
Part used: Whole plant
Distribution:
World: Europe, India, Afghanistan; Russia, Iran, Middle East, N. Africa and Europe; introduced in cultivation elsewhere.
Pakistan: Hazara, Kashmir, Rawalpindi, Swat, Murree, Margalla Hills, Punjab.
Ethnobotanical uses:
Young leaves are cooked in diluted milk, whereas unripe figs are cooked in water and used as vegetables. Leaves are used as fodder, wood as fuel, in making tool handles, sheltering. Ripened figs are eaten raw as fruit.
Medicinal uses:
Powder of dried *Ficus carica* figs and dried seeds of *Amaranthus viridis* is taken orally with water or milk to cure eye vision problem. Milky latex is applied topically to cure skin infections, warts. This plant is used to cure gastric problems, inflammation, and cancer.

4.2.29 *Ficus palmata Forssak.*

Ficus virgata Browicz.

English Name: Fig
Local Name: Phagwar

Fig. 4.29 *Ficus palmata* Forssak

Family Name: Moraceae

Description:
A small tree, up to 10 m tall. Bark smooth, brownish-grey, young twigs densely hairy. Leaves softly tomentose petiole; broadly ovate to suborbicular or orbicular, mostly undivided, dentate to serrate, acute apex. Hypanthodia solitary or sometimes paired, axillary, subglobose to pear-shaped. Male flowers: numerous in the upper half, pedicellate. Female flowers: basal, numerous. Figs constricted or gradually narrowed at base, yellow or purple, hairy (Fig. 4.29).
Flowering period: May–November
Status and habitat: Common on waste places along cultivated lands.
Part used: Whole plant
Distribution:
World: Nepal, India, Afghanistan, Iran, Arabian Peninsula, Somalia, Sudan, Ethiopia, and Egypt.
Pakistan: Hazara, Kashmir, Rawalpindi, Swat, Gilgit, Skardu, Murree, Margalla Hills, Punjab.
Ethnobotanical uses:
Young leaves cooked in diluted milk, whereas unripe figs are cooked in water as vegetables. Leaves are used as fodder, wood is used as fuel, in tool handles, thatching, sheltering. Ripened figs are eaten raw as fruit. Plant is cultivated as shade tree.
Medicinal uses:
Fresh leaves are boiled in the milk of goat and taken orally to cure bowel complaints, whereas milky latex is applied topically on warts and small tumors. Fruits are used against constipation and in the diseases of the lungs and the bladder.

4.2.30 *Oxalis corniculata L*

Oxalis albicans (Kunth) Lourteig.

English Name: Creeping wood sorrel
Local Name: Gandora

Fig. 4.30 *Oxalis corniculata* L.

Family Name: Oxalidaceae

Description:
An annual herb about 30–70 cm long creeping, pubescent herb with roots at nodes. Leaves cauline, trifoliolate, entire, leaflets obcordate, tomentose. Flowers yellow, solitary axillary umbels. Fruit capsule, brown, transversely ribbed (Fig. 4.30).
Flowering period: March–October
Status and habitat: Common on waste places.
Part used: Aerial parts, leaves
Distribution: A cosmopolitan weed found throughout the world and Pakistan.
Ethnobotanical uses:
Young leaves cooked in diluted milk as flavoring agent along with other vegetables. Aerial parts are also grazed by cattle.
Medicinal uses:
Paste of fresh leaves is applied topically on worms and scorpion sting. The plant is used in the treatment of influenza, fever, urinary tract infections, diarrhea, injuries, sprains, and poisonous snake bites. The juice of the plant, mixed with butter, is applied to muscular swellings, boils, and pimples.

4.2.31 Lathyrus aphaca L.

English Name: Yellow pea
Local Name: Paratha

Fig. 4.31 *Lathyrus aphaca* L.

Family Name: Papilionoideae

Description:
The plant is a weak trailing annual herb. Stem erect, glabrous. Leaves are paripinnate, rachis ending in a tendril, large foliaceous, hastate, broadly ovate, paired stipules. Flowers are lemon yellow and in racemes (Fig. 4.31).
Flowering period: February–April
Status and habitat: Common on waste places along cultivated lands.
Part used: Aerial parts, leaves
Distribution:
World: North Africa, Europe, West and Central Asia, India and Nepal.
Pakistan: Hazara, Kashmir, Rawalpindi, Swat, Gilgit, Murree, Margalla Hills, Punjab, Baluchistan.
Ethnobotanical uses:
Young leaves cooked in water as vegetable. Aerial parts are also used as fodder.
Medicinal uses:
Dried seeds are grinded; powder is mixed in tobacco and used for soothing effect and as narcotic.

4.2.32 *Medicago polymorpha* L.

Medicago denticulata (Willd.) Kerguélen

English Name: California bur clover/Toothed bur clover
Local Name: Sinjii

Fig. 4.32 Medicago polymorpha

Family Name: Papilionoideae

Description:
Annual herb up to 30–46 cm long. Stem spreading, erect, herbaceous, yellowish green, rough, cylindrical, hairy. Leaves green, compound, glabrous-pubescent, petiolate and obovate to cuneate, ramal insertion, narrow base, obtuse-truncate apex, laciniate stipules. Base; narrow. Flowers yellow, 2–8 flower peduncle receme, bracteates, terminal, irregular (Fig. 4.32).
Flowering period: March–May
Status and habitat: Common on waste places along cultivated lands.
Part used: Aerial parts, leaves
Distribution:
World: Throughout the world expect tropical regions and deserts.
Pakistan: Hazara, Kashmir, Rawalpindi, Swat, Murree, Margalla Hills, Punjab.
Ethnobotanical uses:
Young leaves cooked in water as vegetable. Aerial parts are also used as fodder.
Medicinal uses:
Fresh leaves are cooked in water and taken orally against constipation and indigestion.

4.2.33 Melilotus alba Ledeb.

Melilotus albus Medik

English Name:
White sweat Clover
Local Name:
Safed Sinjii

Fig. 4.33 *Melilotus alba* Desr.

Family Name: Papilionoideae

Description:
Annual or biennial 4–8 in. high herb. Stem light green, rounded or furrowed on all sides, glabrous. Leaves alternate, trifoliate, petiolate, and stipulate; leaflets greyish, hairless, oblong or ovate, dentate along the upper margin, middle leaflet petiolated, while lateral are sessile. Flowers narrow recemes, white, tubular at the base, while broader towards the outer edges (Fig. 4.33).
Flowering period: May–July
Status and habitat: Rare on waste places.
Part used: Aerial parts, leaves
Distribution:
World: North America, Eurasia, the Mediterranean region from central Europe to Tibet. ***Pakistan***: Hazara, Naran, Shogran, Kashmir, Swat.
Ethnobotanical uses:
Young leaves cooked in water as vegetable. Aerial parts are used as fodder.
Medicinal uses:
Fresh leaves are crushed and paste is applied topically on swelling and inflammatory joints.

4.2.34 Melilotus indicus (L.) All.

Melilotus permixtus (Jord.) Rouy

English Name: Yellow Melilot
Local Name: Sinjii

Fig. 4.34 *Melilotus indicus* (L.) All

Family Name: Papilionoideae

Description:
Annual, erect, slightly pubescent to glabrescent herb. Stem, herbaceous, erect or ascending, rough, cylindrical, pubescent, branching from base. Leaves, trifoliate, glabrous, pubescent, petiolate, stipulate, leaflets; obovate, cuneate to narrowly oblong, ramal insertion, narrow connate base, obtuse-retuse to truncate apex, dentate margin. Flowers yellow, recemes slender, brecteate, terminal, irregular (Fig. 4.34).
Flowering period: March–May
Status and habitat: Common on waste places along crop fields.
Part used: Aerial parts, leaves
Distribution:
World: throughout the world expect or the tropical regions and desert.
Pakistan: Hazara, Rawalpindi, Murree, Kashmir, Swat, Punjab, and Sind.
Ethnobotanical uses:
Young leaves cooked in water as vegetable. Aerial parts are used as fodder.
Medicinal uses:
Fresh leaves are crushed and paste is applied topically on swelling joints.
The seeds are used in the treatment of bowel complaints and diarrhea, leaves are used externally as a poultice on swellings.

4.2.35 *Vicia faba L*

Vicia equine Steud

English Name: Broad Bean/Horse Bean
Local Name: Sarphound

Fig. 4.35 *Vicia faba* L.

Family Name: Papilionoideae

Description:
An erect or climbing annual, 60–120 cm tall herb. Leaflets 2–6, oval to elliptic, obtuse, mucronate, glabrous, stipulate. Flowers 1–6, axillary, white with dark violet wings. Fruit pods, pubescent. Seeds, ovoid-oblong (Fig. 4.35).
Flowering period: July–September
Status and habitat: Common on waste places along crop fields.
Part used: Aerial parts, leaves
Distribution:
World: Native to central Asia and the Eastern Mediterranean reigns, Australia, Europe, China, and Ethiopia.
Pakistan: Hazara, Rawalpindi, Margalla Hills, Murree, Kashmir, Swat, Punjab, Sind.
Ethnobotanical uses:
Young leaves cooked in water as vegetable. Aerial parts are used as fodder.
Medicinal uses:
Fresh leaves are boiled in water and decoction is taken orally to cure kidney stones and eye infection.

4.2.36 *Vicia sativa* L.

Vicia angustifolia (L.) Wahlb.

English Name: Spring vetch
Local Name: Auli Pali

Fig. 4.36 *Vicia sativa* L.

Family Name: Papilionoideae

Description:
An erect or climbing, pubescent to subglahrous, decumbent, annual herb. Leaf pinnately compound, petiolate, linear to lanceolate to oblong or obovate, acute, obtuse or emarginate, thinly pubescent or glabrescent; stipules semisagittate, dentate, branched tendrils. Flowers pale pink, purplish violet. Fruit narrowly oblong, pubescent becoming glabrous when matures (Fig. 4.36).
Flowering period: July–August
Status and habitat: Common on waste places along cultivated fields.
Part used: Aerial parts, leaves
Distribution:
World: Cosmopolitan in distribution, India, China, Russia, Europe.
Pakistan: Throughout the country.
Ethnobotanical uses:
Young leaves cooked in water as vegetable. Aerial parts are used as fodder.
Medicinal uses:
Paste of crushed leaves is applied topically against scorpion sting.

4.2.37 *Plantago lanceolata* L.

Plantago lanuginosa Arcang.

English Name: Ribwort plantain
Local Name: Bhatti

Fig. 4.37 *Plantago lanceoplata* L.

Family Name: Plantaginaceaea

Description:
A perennial, up to 30 cm tall herb. Leaves membranous, narrow lanceolate to narrow elliptic, glabrous slightly pilose or attenuated, acute, at the base narrowed into a narrow petiole, sparsely covered with white hairs. Spikes dense, cylindrical, subglobose to globose, Bracts broad ovate, narrowly caudate, carinate. Sepals, glabrous, anterior sepals connate, obovate, bilobed. Corolla lobes narrow ovate to ovate (Fig. 4.37).
Flowering period: April–July
Status and habitat: Common on waste places along cultivated fields.
Part used: Leaves, seed husk
Distribution: ***World***: Europe, North Africa, south Asia; introduced all over the world. ***Pakistan***: Hazara, Kashmir, Margalla Hills, Murree, Baluchistan, Sind.
Ethnobotanical uses:
Young leaves cooked in diluted milk as vegetable. Aerial parts are used as fodder. **Medicinal uses**: Fresh leaves are crushed and paste is applied topically on sore joints. Seed husk along with sugar (Gur) is mixed in water and taken orally to cure jaundice, internal body inflammation, and constipation.

4.2.38 *Persicaria amplexicaulis (D.Don) Ronse Decr*

Bistorta amplexicaulis (D. Don) Greene.

English Name: Fleece flower/Mountain fleece
Local Name: Saag

Fig. 4.38 *Bistorta amplexicaulis* (D. Don)

Family Name: Polygonaceae

Description:
An erect, 30–65 cm tall, branched or simple, glabrous, perennial, long rhizomatous herb. Leaves broadly lancoelate-ovate, serrate, acuminate, ciliate on midrib and margins, cordate or amplexicaule at base, petiolate. Ochrea lanceolate, tubular, acuminate, with two to three long acuminate lobes. Inflorescence terminal, simple many flowered. Flowers deep purplish red, pedicelate, lanceolate, cartilaginous, with long aristate apex and entire margin. Tepals lanceolate to ovate, obtuse, entire, unequal. Nuts ovate, dark brown to black, glabrous, shining (Fig. 4.38).
Flowering period: June–September
Status and habitat: Common on cold, shady places.
Part used: Leaves
Distribution: ***World***: India, Afghanistan, and China.
Pakistan: Hazara, Kashmir, Murree, Naran, Kaghan, Swat, Kalam, Chitral, Ziarat.
Ethnobotanical uses: Young leaves cooked in diluted milk as vegetable. Aerial parts are used as fodder.
Medicinal uses: Fresh leaves are boiled in water along with sugar and decoction is taken orally to treat fever, joint pain, and flue.

4.2.39 *Polygonum aviculare* L.

Polygonum aequale (Lindm.) Asch. & Graebn.

English Name: Knotweed, Pig weed
Local Name: Trubra

Fig. 4.39 *Polygonum aviculare* L.

Family Name: Polygonaceae

Description:
A suberect-erect, ascending, glabrous, branched, annual herb. Leaves heterophyllous, larger on main branches, smaller on the lateral and upper branches, elliptic, lanceolate or ovate, acute, entire, dotted. Ochrea bifid, silvery, membranous lacerate. Inflorescence solitary, axillary. Flower pedicilate, white, Tepals 5, elliptic-lanceolate or ovate, obtuse-acute, entire. Nuts ovate, trigonous, black, shining, striate (Fig. 4.39).
Flowering period: March–September
Status and habitat: Common on waste places.
Part used: Leaves, aerial parts
Distribution: ***World***: Widely distributed in temperate and subtropical regions.
Pakistan: Throughout the country.
Ethnobotanical uses: Young leaves cooked in diluted milk as vegetable. Aerial parts are used as fodder.
Medicinal uses: Fresh leaves are boiled in water and decoction is taken orally to cure diarrhoea and dysentery. The whole plant is anthelmintic, astringent, cardiotonic, diuretic, febrifuge, hemostatic, in the treatment of wounds, bleeding, piles.

4.2.40 *Rumex dentatus* L.

Rumex reticulatus (Besser) Rech.f.

English Name: Toothed dock
Local Name: Hulla

Fig. 4.40 *Rumex dentatus* L.

Family Name: Polygonaceae

Description:
An annual herb. Stem, un-branched or branched near the base. Basal leaves ovate-oblong from a truncate, flat, obtuse or slightly acute apex, petiolate; petiole shorter or as long as the lamina; stem leaves normally few, smaller and with shorter petioles as the basal one. Inflorescence whorls of many-flowered, all or at least the lower and the middle ones distant, supported by a leaf (Fig. 4.40).
Flowering period: March–May
Status and habitat: Common on waste places.
Part used: Leaves, aerial parts
Distribution:
World: Afghanistan, Iran, China, India, East Asia.
Pakistan: Hazara, Peshawar, Swat, Quetta, Ziarat, Waziristan, Parachinar, Kashmir.
Ethnobotanical uses:
Young leaves cooked in diluted milk as vegetable. Aerial parts are used as fodder.
Medicinal uses:
Young leaves are applied topically to treat stinging nettle. The root is used as an astringent application in the treatment of cutaneous disorders.

4.2.41 Rumex hastatus D. Don, Prodor.

Rumex dissectus H. Lév.

English Name: Toothed dock
Local Name: Hulla

Fig. 4.41 *Rumex hastatus* D. Don, Prodor.

Family Name: Polygonaceae

Description:
A bushy branching shrub up to 30 cm tall. Stem erect, pale green to light brown, herbaceous above, woody below. Leaves with petiolate; leaf blade hastate, lobes directed outwards, extremely variable in length and breadth. Panicles terminal with erect-divergent branches. Flowers are small, numerous, pinkish in terminal paniculate clusters. Fruit is pinkish and one-seeded nut let. Nut brown, broadest in the lower third (Fig. 4.41).
Flowering period: April–July
Status and habitat: Common on waste dry slopes and rocky places.
Part used: Leaves, roots, aerial parts
Distribution: ***World***: Afghanistan, India, Bhutan and China.
Pakistan: Hazara, Malakand, Peshawar, Chitral, Drosh, Swat, Gilgit, Murree, and Kashmir.
Ethnobotanical uses: Young leaves cooked in diluted milk as vegetable. Aerial parts are used as fodder.
Medicinal uses: Extract of fresh leaves and roots is taken orally to cure jaundice. Root are boiled in water and decoction is used as tonic to cattle.

4.2.42 *Rumex nepalensis* Spreng.

Rumex bequaertii De Wild.
Rumex peregrinus Boiss

English Name: Toothed dock
Local Name: Hulla

Fig. 4.42 *Rumex nepalensis* Spreng.

Family Name: Polygonaceae

Description:
A perennial herb up to 150 cm high, erect, branching in the upper half. Basal leaves with cordate base, hooked teeth, broad, flat, cauline. Inflorescence in whorls of many-flowers. Fruit nut, long, dark brown, thicker below the middle (Fig. 4.42).
Flowering period: July–September
Status and habitat: Common on waste places at higher altitude under shade and moist places.
Part used: Leaves
Distribution: ***World***: China, India, Afghanistan, Turkey; Java, Syria, Africa, Europe: Italy. ***Pakistan***: Abbottabad, Murree, Gilyat, Kaghan, Swat, Kalam, Shogran, Kashmir.
Ethnobotanical uses:
Young leaves cooked in diluted milk as vegetable. Plant is also used as fodder.
Medicinal uses:
Young leaves are applied topically to treat stinging nettle, relieve headache, colic. Roots are purgative, used to dislocated bones, while paste is applied to swollen gums.

4.2.43 *Portulaca quadrifida* L.

Portulaca formosana Hayata

English Name: Ten o'clock plant
Local Name: Chotta Kulfa

Fig. 4.43 *Portulaca quadrifida* L.

Family Name: Portulaceaeae

Description:
Diffuse, annual, prostate, glabrous, somewhat flashy herb, up to 5–8 cm tall. Stem yellowish green, rough, pubescent, cylindrical, filiform. Leaves opposite, entire, sessile to subsessile, fleshy, multifoliate, glabrous, pubescent or glucocus, ovate-oblong to elliptic-oblong or ovate to ovate, lanceolate, narrow base, acute-obtuse apex. Flowers: yellow, solitary terminal, bracteates (Fig. 4.43).
Flowering period: August–March
Status and habitat: Common on waste dry places.
Part used: Leaves, aerial parts
Distribution:
World: India, Tropical Africa, Asia.
Pakistan: Hazara, Swat, Murree, Kashmir, Punjab, Karachi.
Ethnobotanical uses:
Young leaves cooked in water as vegetable. Aerial parts are used as fodder.
Medicinal uses:
Fresh leaves are slightly wormed and applied topically on swelling joints.

4.2.44 *Portulaca oleracea* L.

Portulaca parviflora Haw.
Portulaca viridis Hort. ex DC.

English Name: Common purslane, Pigweed, Pursley
Local Name: Bara Kulfa

Fig. 4.44 *Portulaca oleracea* L.

Family Name: Portulaceaeae

Description: Annual or perennial, prostrate or erect, succulent, glabrous, green herb. Leaves alternate or subopposite, closely crowded below the flowers, obovate-oblong, subsessile, obtuse or truncate, thick, fleshy, glabrous, white below, green or purplish-green above; Inflorescence usually cymose, with clusters of 3–6 yellow sessile flowers rarely solitary and terminal. Fruits capsule many-seeded. Seeds shining black (Fig. 4.44).
Flowering period: March–October
Status and habitat: Common on waste dry places in cultivated fields.
Part used: Whole plant
Distribution: ***World***: India, Europe, America.
Pakistan: Hazara, Swat, Murree, Kashmir, Punjab, Karachi.
Ethnobotanical uses: Young leaves cooked in water as vegetable. Aerial parts are used as fodder.
Medicinal uses: Fresh leaves are slightly wormed and applied topically on swelling joints. The extract of stem is applied on skin against burning sensation. The plant is depurative, diuretic, febrifuge, cardiac stimulant, used in the treatment of coughs, skin infections, sores, burns, earache.

4.2.45 *Galium aparine* L.

Galium agreste P.D.Sell

English Name: False cleavers/Goose grass
Local Name: Linda

Fig. 4.45 *Galium aparine* L.

Family Name: Rubiaceae

Description:
Annual climbing herb, stem with reflexed hairs or prickles. Leaves 6–8 in a whorl, linear, narrowly obovate, oblong and narrow below the middle, upper surface mostly hispid, sessile or shortly petiolated. Inflorescence axillary, three-flowered. Flowers white. Fruit covered with dense hooked hairs (Fig. 4.45).
Flowering period: March–July
Status and habitat: Common on waste places.
Part used: Leaves, aerial parts
Distribution:
World: India, Europe, North Africa, Asia, Siberia, Iran, Afghanistan, and India.
Pakistan: Gilgit, Hazara, Swat, Murree, Kashmir, Punjab, Quetta, Kalat.
Ethnobotanical uses:
Young leaves cooked in water as vegetable. Aerial parts are used as fodder.
Medicinal uses: Fresh leaves are crushed and paste is applied topically to heal wounds, while extract of fresh leaves is taken orally to cure jaundice.
The whole plant is alterative, astringent, depurative, diaphoretic, diuretic, febrifuge, tonic used both internally and externally in the treatment of wounds, ulcers, skin problems.

4.2.46 *Veronica arvensis* L.

Cardia arvensis (L.) Dulac;

English Name: Corn speedwell
Local Name: Ghoudri

Fig. 4.46 *Veronica arvensis* L.

Family Name: Rubiaceae

Description:
Annual herb, stem hollow, erect, cylindrical. Leaves opposite, petiolate below, sessile above, broadly ovate, serrate. Inflorescence terminal bracteate raceme. Flowers deep blue-purple, united at the base (Fig. 4.46).
Flowering period: March–August
Status and habitat: Common on waste rocky places, along crop fields, roadsides.
Part used: Leaves, aerial parts
Distribution:
World: Native to Asia, Africa, and Europe.
Pakistan: Gilgit, Hazara, Swat, Kashmir, Punjab, Quetta.
Ethnobotanical uses:
Young leaves cooked in water as vegetable. Aerial parts are used as fodder.
Medicinal uses:
Fresh leaves are boiled in water and decoction is taken orally to cure skin infection, alterative, diuretic, and as blood purifier.

4.2.47 *Solanum nigrum auct.*

Solanum americanum (Mill.) O.E. Schulz

English Name: Glossy nightshade
Local Name: Kachmach

Fig. 4.47 *Solanum nigrum* auct.

Family Name: Solanaceae

Description:
Annual, erect, variously branched, 10–45 cm tall herb. Stem rough, cylindrical, pubescent. Leaves green, compound, glabrous, pubescent, lamina ovate-lanceolate to elliptical, petiolated, narrow base, acute apex, irregularly dentate margin. Flowers white, bracteates, terminal, glabrous to pubescent. Fruits Berry globose to subovoid, black or orange-red (Fig. 4.47).
Flowering period: May–August
Status and habitat: Common on waste, shady, cold places, along roadsides.
Part used: Leaves, aerial parts
Distribution:
World: Europe, India, Tropical Asia and Africa.
Pakistan: Hazara, Swat, Kashmir, Punjab.
Ethnobotanical uses:
Young leaves cooked in water as vegetable. Aerial parts are used as fodder.
Medicinal uses:
Extract of fresh leaves is applied topically to wash painful eyes. Leaves are cooked and taken orally to control body swelling. Fruits juice is used to cure sore throat and mouth gums.

4.2.48 *Solanum surattense Burm. f.*

Solanum diffusum Roxb.

English Name: Thai eggplant
Local Name: Mokhari

Fig. 4.48 *Solanum surattense* Burm. f.

Family Name: Solanaceae

Description:
A prostrate, diffuse herb; with yellow prickles. Stem and branches glabrous to stellate pubescent. Leaves elliptic-oblong, sinuate to deeply lobed, dark-green above; obtuse or acute, often toothed or lobulate. Flowers 2–4, purple. Fruits berry globose. Seeds discoid, smooth to faintly reticulate (Fig. 4.48).
Flowering period: March–October
Status and habitat: Common on waste places in dry soil, along roadsides.
Part used: Whole plant
Distribution:
World: Africa, Asia, Australia and Polynesia.
Pakistan: Hazara, Rawalpindi, salt range, Peshawar, Punjab.
Ethnobotanical uses:
Fresh stem are boiled and used as vegetable. Whole plant is used as fodder for camels.
Medicinal uses:
Stem decoction is during fever as appetizer. This plant is also effective against rheumatism, pneumonia, cough, influenza, and respiratory disorders.

4.2.49 Pimpinella diversifolia DC.

Pimpinella divisa C.B. Clarke

English Name: Sweet cumin
Local Name: Tarpakhi

Fig. 4.49 *Pimpinella diversifolia* DC.

Family Name: Umbellifereae

Description:

A perennial, pubescent, up to 1 m tall herb. Basal leaves simple, undivided, serrate, ovate to cordate; cauline leaves pinnate; leaflets 3–5, lanceolate to oval or ovate, margin serrate Flowers white, small, numerous in umbel. Fruit ovate to ovoid (Fig. 4.49).

Flowering period: August–September

Status and habitat: Common on waste dry slops in grass.

Part used: Leaves, aerial parts

Distribution:

World: Afghanistan, Himalayas, India, China, Japan.

Pakistan: Hazara, Swat, Kashmir, Chitral, Ziarat, Parachinar, Waziristan, Skardu, Gilgit, Biltistan, Murree, Margalla Hills.

Ethnobotanical uses:

Young leaves are used as spices in other vegetables. Aerial parts are grazed by cattle.

Medicinal uses:

Dried leaves are grinded along with salt and powder is taken orally with water to cure indigestion and gas trouble.

4.2.50 Torilis leptophylla (L.) Reichb.f.

Torilis erythrotricha (Rchb. f.) Zohary.

English Name: Bristle fruit hedge parsley
Local Name: Chrikanger

Fig. 4.50 *Torilis leptophylla* (L.) Reichb.f.

Family Name: Umbellifereae

Description:
Annual herb, up to 60 cm tall, hispid. Stem erect, herbaceous. Leaves bipinnate; segments linear. Influences umbels usually lateral. Flowers white-light purple. Fruit oval to ovoid (Fig. 4.50).
Flowering period: March–April
Status and habitat: Common on waste places.
Part used: Leaves, aerial parts
Distribution:
World: Europe, Africa, Central and Southern Asia, introduced in America. ***Pakistan***: Hazara, Kashmir, Chitral, Margalla Hills, Peshwar, Taxila, Punjab.
Ethnobotanical uses:
Young leaves are cooked in water as vegetables. Aerial parts are used as fodder.
Medicinal uses:
Dried leaves are grinded and powder is taken orally with water to cure gastrointestinal disorders.

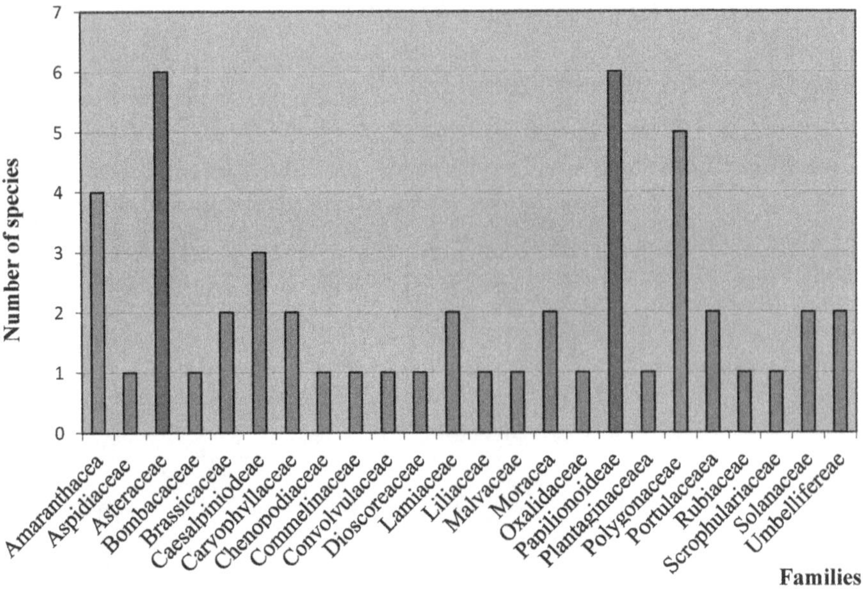

Fig. 4.51 Dominating families of wild edible vegetables

4.3 Taxonomic Diversity

The flora of Pakistan provides a great diversity of wild food resources. A total of 50 wild edible vegetable species belonging to 38 genera and 24 families are included in this book which are collected and consumed by the local inhabitants. Asteraceae and Papilionoideae were observed dominating families with six spp. each, followed by Polygonaceae (five species), Amaranthaceae (four species), whereas rest of the families were represented by either two or one species (Fig. 4.51).

4.4 Plant Parts Used and Mode of Consumption

Wild edible plant parts are consumed in different ways according to local customs. In the case of wild edible vegetables leaves are commonly utilized as plant parts both for food and medicines, followed by flowering buds, stems, fruits and tubers, etc. Wild edible vegetables are cooked as fresh in water like *Amaranthus* spp. *Bauhinia* spp. (Flower buds), *Bidens bipinnata*, *Capsella bursa-pastoris*, *Commelina benghalensis*, *Digeria muricata*, *Galium aprine*, *Lathyrus aphaca*, *Malva parviflora*, *Medicago polymorpha*, *Melilotus* spp. *Nasturtium officinale*, *Portulaca* spp. *Solanum* spp. *Stellaria media*, *Torilis leptophyll* and *Vicia* spp.; in diluted milk *Bauhinia* spp. (Flowers), *Cichorium intybus*, *Ficus* spp. *Launaea*

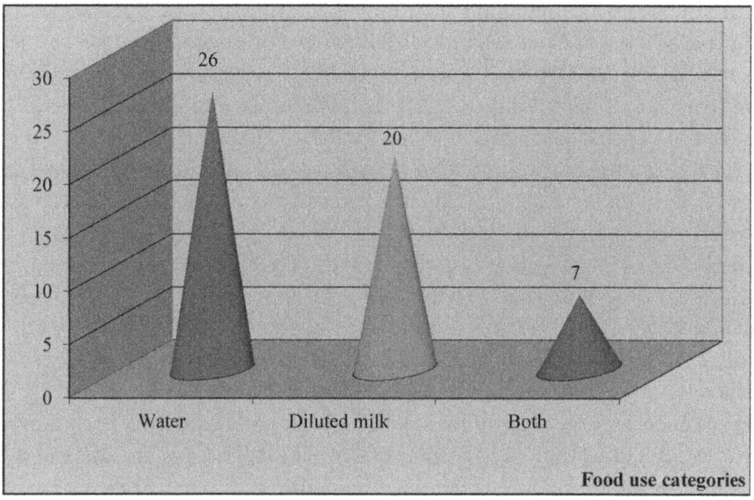

Fig. 4.52 Wild vegetables used categories

procumbens, *Persicaria amplexicaulis*, *Pimpinella diversifolia*, *Plantago lanceoplata*, *Rumex* spp. *Silene conoidea*, *Sonchus* spp. *Taraxacum officinale*, *Tulip stellata*; in both water and diluted milk (*Bauhinia* spp. *Ficus* spp. and *Sonchus* spp) as mentioned in Fig. 4.52. Flowers and rhizome of some vegetables species are also consumed fried in cooking oil or vegetable ghee (e.g., *Bombax malabaracu* and *Dioscorea deltoidea*).

4.5 Medicinal Uses of Wild Edible Vegetables

Traditional phytotherapy is a skill experienced by elder people (Herbalists) whose experiential familiarity is appreciated by everyone. They are renowned with symptoms of various common ailments and cure these maladies with the locally available plant-based remedies (Parveen et al. 2007). Right from its establishment, the documentation of conventional knowledge particularly on the medicinal uses of plants has provided many important drugs of the modern day (Fabricant and Farnsworth 2001). Even today this area holds much more concealed treasure, as almost 80 % of the human population in developing countries is dependent on plant wealth for healthcare (Farnsworth et al. 1985).

Pakistan is a rich source of medicinal plants diversity (Abbasi et al. 2012). During filed study it was emphatically noted that the trend of using traditional phytotherapies is going to decline, obviously because the younger generation usually consider the belief in plant remedies a sort of superstition and less efficient compared to contemporary medicine. Moreover, modern medical facilities are now making inroads into these traditional medicines. So, the folk belief on medicinal plants is now fading. It was, therefore, considered essential that this precious knowledge about folk medicinal uses of wild food plants be documented before these time-tested uses of herbal drugs lost forever. For instance, the well established folk herbalists (Sanasis) were dominating but now are almost being diminished.

Some species of wild edible vegetables which showed quick response during treatment and considered highly utilized by local inhabitants included: *Cichorium intybus*, *Ficus carica*, *Ficus palmata*, *Oxalis corniculata*, *Rumex dentatus*, *Rumex dentatus*, and *Solanum nigrum*, found among highly consumed species. It was noted that leaves were found to be highly utilized plant parts followed by seeds, roots, latex, bark, bulb, flowers, tubers, and rhizomes. Traditional recipes can be divided into two categories: those that prepared from (1) single plant and (2) from more than one plant species. Mostly water is used as a medium for preparation, while milk, ghee, oil, egg, and butter are used as additional dose.

Methods of preparation fall into different categories like paste, decoction, powder, extract, as vegetable, juice taken as fresh part, while mode of application falls in two categories topical as well as oral. Oral medications are taken along with water, milk, or tea. In regard to the patient condition, the preparations are applied more than two times daily until the disease is cured.

The local inhabitants identified different types of ailments including gastrointestinal disorders (abdominal pain, gas trouble, gastric ulcer, intestinal worms, constipation, vomiting, diarrhoea, dysentery), respiratory problems (asthma, flue, throat infection, cough), skin infections (measles, mouth gums, rashes, wound healing), bone fracture, rheumatism, diabetes, ear ache, tooth ache, eye infection, fever, heart problems, inflammation, jaundice, kidney problems, menstrual disorders, milk production, piles, scorpion sting, and general weakness which were treated through different plant-based remedies (Fig. 4.53).

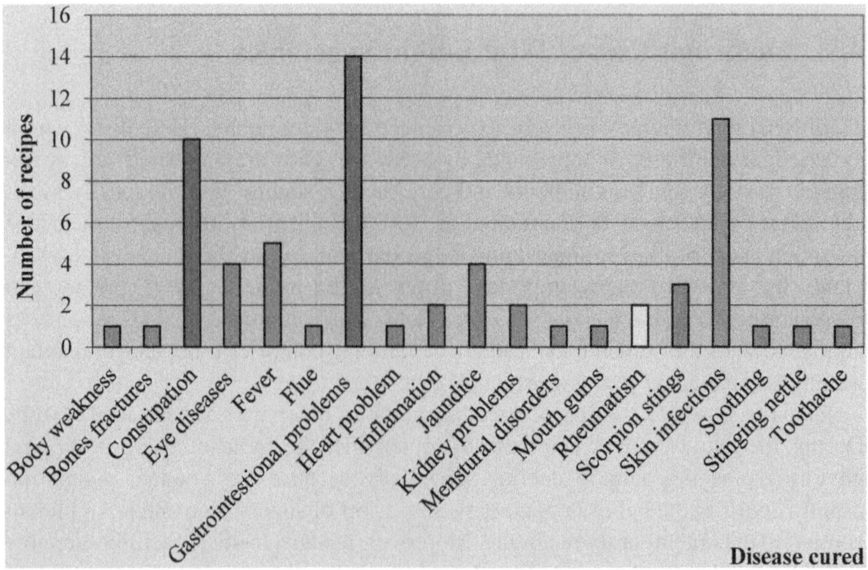

Fig. 4.53 Number of recipes and diseases cured based on wild edible vegetables

4.6 Species' Cultural Importance

The terms "cultural importance" and "relative importance" usually are used interchangeably to refer to the importance of certain plants to a given culture (Albuquerque et al. 2005). The cultural importance (CI) index explains not only the spread of the uses (number of informants) for each species, but also its worth (Tardio and Santayana 2008). It can be assumed that the CI index is a proficient tool for highlighting those species with a high-agreement for the survey culture and so to recognize the shared knowledge of the people.

Culture importance (CI) index and mean cultural index (mCI) are computed to measure the cultural values of each wild edible vegetable, which can be used to evaluate the plant consciousness between diverse cultures (Pardo-de-Santayana et al. 2007) and to study the intra-cultural variations.

The cultural importance (CI) index is calculated as the summation of the use report (UR) in every use category mentioned for a species in the locality divided by number of participants (N) in that locality. Similarly mean cultural importance (mCI) index of each species is also calculated (Pardo-de-Santayana et al. 2007). However, cultural significance of each family (CIf) is calculated by adding cultural importance (CI) index of the species from each family (Galeano 2000).

$$\mathrm{CI} = \sum_{i=1}^{i=\mathrm{NU}} \times \frac{\mathrm{UR}_i}{N}$$

On the bases of use reports (UR) the cultural importance (CI) index and mean cultural index (mCI) of wild edible vegetables within the five study localities (Margalla Hills, Haripur, Abbottabad, Murree and Mansehra) of Lesser Himalayas were calculated. It was observed that *Ficus carica* was most cited species (Fig. 4.54), followed by *Ficus palmata, Bauhinia variegata, Solanum nigrum, Amaranthus viridis, Medicago polymorpha, Chenopodium album, Cichorium intybus, Amaranthus hybridus,* and *Vicia faba* because these vegetables species are used as food, medicines, fodder, fuel wood, construction, sheltering, fencing, and making agricultural tools. All the top ten species of wild edible vegetables were cited in all five surveyed areas of Lesser Himalayas. A common cultural background may explain these similarities that can be recognized to the fact that traditional knowledge of wild edible plants and plant collection is much spread in the communities of remote areas as compared to urban localities.

A comparison between the cultural indexes of most quoted families (CIf) of wild edible vegetables mentioned in Fig. 4.55 revealed that Papilionoideae, with six species is the most quoted botanical family because members of this family are consumed locally as food, fodder, and in medicines. Other remarkable families of wild vegetables within top ten were Asteraceae, followed by Amaranthaceae, Bombacaceae, Moraceae, Polygonaceae, Caesalpiniodeae, Caryophyllaceae, Solanaceae, and Chenopodiaceae.

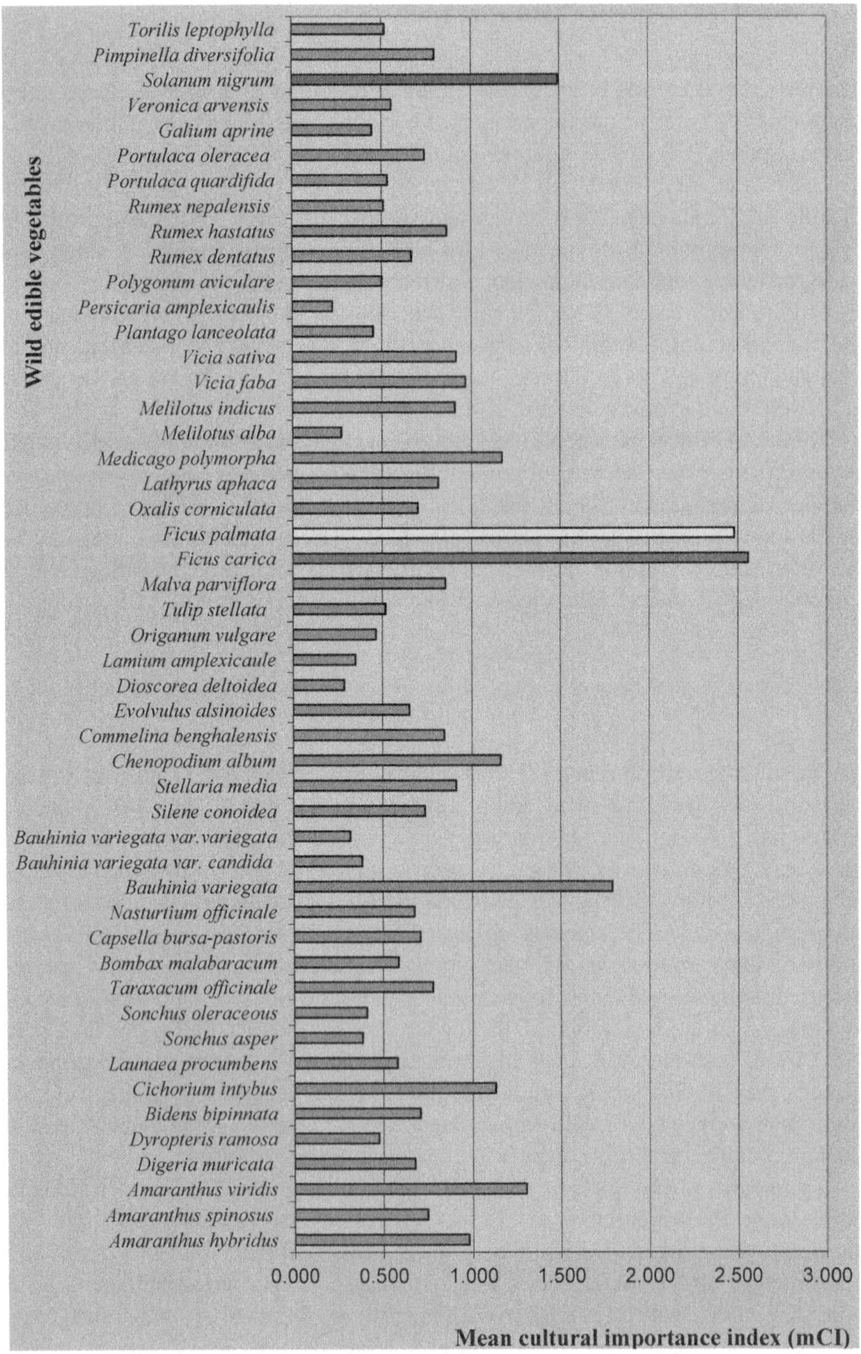

Fig. 4.54 Mean culture importance (mCI) of wild edible vegetables

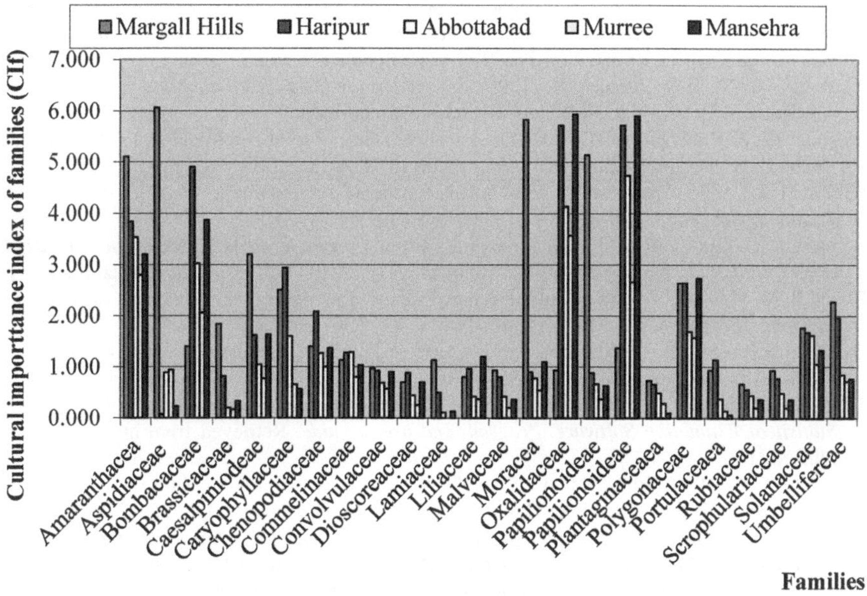

Fig. 4.55 Cultural importance index of wild edible vegetables' families (CIf)

4.7 Bio-Conservation Concerns

Various human activities may be implicated in placing some of the local flora under a state of threat within their natural habitat. Such activities include agricultural land expansion, over-harvesting, over-grazing, uncontrolled fire setting, and fuel wood collection. The perceptions that local people share regarding this phenomenon of threats to local ecological resources, i.e., wild food and medicinal plant species in particular, are examined based on interviews with key respondents during field survey. Achievement of economic payback from plant species might endorse local people's interest in protection and preservation of significant and endangered species (Balemie and Kebebew 2006). Most of the wild species in the study sites have no protection; however, few economic plant species (e.g., *Bauhiniea variegata*, *Ficus carica*, *Ficus palmata*, *Chenopodium album* and *Solanum nigrum*) are now cultivated and marketed by farmers to some extent.

References

Abbasi, A. M., Khan, M. A., Ahmad, M., & Zafar, M. (2012). *Medicinal plant biodiversity of Lesser Himalayas-Pakistan* (pp. 1–212). New York: Springer. ISBN 978-1-4614-1574-9.

Albuquerque, U. P., Andrade, L. H. C., & Silva, A. C. O. (2005). Use of plant resources in a seasonal dry forest (Northeastern Brazil). *Acta Botânica Brasílica, 19*, 27–38.

Alexiades, M. N. (1996). Collecting ethnobotanical data: An introduction to basic concepts and techniques. In M. N. Alexiades (Ed.), *Selected guidelines for ethnobotanical research: A field manual* (pp. 53–94). Bronx, NY: The New York Botanical Garden.

Alexiades, M. N., & Sheldon, J. W. (1996). *Selected guide-lines for ethnobotanical research: A field manual*. Bronx, NY: New York Botanical Garden Press.

Balemie, K., & Kebebew, F. (2006). Ethnobotanical study of wild edible plants in Derashe and Kucha Districts, South Ethiopia. *Journal of Ethnobiology and Ethnomedicine, 2*, 53.

Balick, M. J. (1996). Transforming ethnobotany for the new millennium. *Annals of the Missouri Botanical Garden, 83*, 58–66.

Bernard, R. H. (2002). Structured interviewing: cultural domain analysis. In *Research methods in anthropology: Qualitative & quantitative methods* (pp. 280–297). New York: Altamira Press.

Boom, B. M. (1990). Useful plants of the Panare Indians of the Venezuelan Guayana. *Advances in Economic Botany, 8*, 57–65.

Byg, A., & Balslev, H. (2001). Diversity and use of palms in Zahamena, eastern Madagascar. *Biodiversity and Conservation, 10*, 951–970.

Chao, A., & Shen, T. J. (2003). *Program SPADE (Species Prediction and Diversity Estimation). Statistical Estimation Software. Program and user's guide*. Retrieved from http://chao.stat.nthu.edu.tw

Chazdon, R. L. (1999). Ethnobotany of woody species in second-growth, old-growth and selectively logged forests of northern Costa Rica. *Conservation Biology, 13*, 1312–1322.

Colwell, R. K. (2005). *Estimate S 8.0: Statistical estimation of species richness and shared species from samples. User's guide and application*. Retrieved from purl.oclc.org/estimates

Cotton, C. M. (1996). *Ethnobotany: Principles and applications*. London: Wiley.

Cunningham, A. B. (2001). *Applied ethnobotany. People, wild plant use and conservation*. WWF, UNESCO, Royal Botanical Gardens, Kew. London: Earthscan.

Estabrook, G. F. (2003). *METASIG (for DOS)*. Retrieved from www.personal.umich.edu/~gfe/metasigz.exe

Fabricant, D. S., & Farnsworth, N. R. (2001). The value of plants used in traditional medicine for drug discovery. *Environment Health Perspectives, 109*, 69–75.

Farnsworth, N. R., Akerele, O., & Bingel, A. S. (1985). Medicinal plants in therapy. *Bulletin of World Health Organization, 63*, 965–981.

Friedman, J., Yaniv, Z., Dafni, A., & Palewitch, D. (1986). A preliminary classification of the healing potential of medicinal plants, based on a rational analysis of an ethnopharmacological field survey among Bedouins in the Negev Desert, Israel. *Journal of Ethnopharmacology, 16*, 275–287.

Galeano, G. (2000). Forest use at the Pacific coast of Choco, Colombia: A quantitative approach. *Economic Botany, 54*, 358–376.

Hall, P., & Bawa, K. (1993). Methods to assess the impact of extraction of non-timber tropical forest products on plant populations. *Economic Botany, 47*(3), 234–247.

Hamilton, A. C., Shengji, P., Kessy, J., Khan, A. A., Lagos-Witte, S., & Shinwari, Z. K. (2003). *The purposes and teaching of Applied Ethnobotany. People and plants working paper 11*. Godalming, UK: WWF. Retrieved from http//:www.peopleandplants.org/web-content%201/pdf/wp11.pdf.

Hammer, O., & Harper, D. A. T. (2006). *PAST: Paleontological statistics software package for education and data analysis*. Retrieved from folk.uio.no/ohammer/past/Past.exe

Hoft, M., Barik, S. K., & Lykke, A. M. (1999). Quantitative ethnobotany: Applications of multivariate and statistical analyses in ethnobotany. *People and Plants* 6. Retrieved July 2006, from http://www.fao.org/DOCREP/006/W2352E/W2352E03.htm#ch3.3

International Society of Ethnobiology (ISE): The Code of Ethics of the International Society of Ethnobiology. [http://ise.arts.ubc.ca/global_coalition/ethics.php]

Johnston, M. (1998). Tree population studies in low-diversity forests, Guyana. II. Assessments on the distribution and abundance of non-timber forest products. *Biodiversity and Conservation, 7*, 73–86.

Kremen, C., Raymond, I., & Lance, K. (1998). An interdisciplinary tool for conservation impacts in Madagascar. *Conservation Biology, 12*, 549–563.

Kvist, L. P., Andersen, M. K., Hesselsoe, M., & Vanclay, J. K. (1996). Estimating use values and relative importance of Amazonian flood plain trees and forests to local inhabitants. *Commonwealth Forestry Review, 74*, 293–300.

Lawrence, A., Phillips, O. L., Ismodes, A. R., Lopez, M., Rose, S., Wood, D., & Farfan, A. J. (2005). Local values for harvested forest plants in Madre de Dios, Peru: Towards a more contextualized interpretation of quantitative ethnobotanical data. *Biodiversity and Conservation, 14*, 45–79.

Lozada, M., Ladio, A., & Weigandt, M. (2006). Cultural transmission of ethnobotanical knowledge in a rural community of northwestern Patagonia, Argentina. *Economic Botany, 60*, 374–385.

Lykke, A. M., Kristensen, M. K., & Ganaba, S. (2004). Valuation of local use and dynamics of 56 woody species in the Sahel. *Biodiversity and Conservation, 13*, 1961–1990.

Martin, G. J. (1995). *Ethnobotany. A "people and plants" conservation manual. World wide fund for nature*. London: Chapman & Hall.

Martin, G. J. (2004). *Ethnobotany: A methods manual*. London: Earthscan.

Martin, E., & Polivka, A. E. (1995). Diagnostics for redesigning survey questionnaires. *Public Opinion Quarterly, 59*(4), 547–567.

McClatchey, W., Sirikolo, M. Q. B., Kaleveke, L., & Pitanapi, C. (2006). Differential conservation of two species of *Canarium* (Burseraceae) among the Babatana and Ririo of Lauru (Choiseul), Solomon Islands. *Economic Botany, 60*, 212–226.

Pardo-de-Santayana, M., Tardo, J., Blanco, E., Carvalho, A. M., Lastra, J. J., San Miguel, E., & Morales, R. (2007). Traditional knowledge of wild edible plants used in the northwest of the Iberian Peninsula (Spain and Portugal): a comparative study. *Journal of Ethnobiology and Ethnomedicine, 3*, 27.

Parveen, B., Upadhyay, B., Roy, S., & Kumar, A. (2007). Traditional uses of medicinal plants among the rural communities of Churu district in the Thar Desert, India. *Journal of Ethnopharmacology, 113*, 387–399.

Paz, Y., Mino, C. G., Blaslev, H., & Valencia, R. (1995). Useful lianas of the Siona-Secoya Indians from Amazonian Ecuador. *Economic Botany, 49*, 269–275.

Peters, C. M. (1996). Beyond nomenclature and use: A review of ecological methods for ethnobotanists. In M. N. Alexiades (Ed.), *Selected guidelines for ethnobotanical research: A field manual* (pp. 241–276). Bronx, NY: The New York Botanical Garden.

Phillips, O. L. (1996). Some quantitative methods for analysing ethnobotanical knowledge. In M. N. Alexiades (Ed.), *Selected guidelines for ethnobotanical research: A field manual* (pp. 171–197). Bronx, NY: The New York Botanical Garden.

Phillips, O., & Gentry, A. H. (1993a). The useful plants of Tambopata, Peru: I. Statistical hypotheses tests with a new quantitative technique. *Economic Botany, 47*, 15–32.

Phillips, O., & Gentry, A. H. (1993b). The useful plants of Tambopata, Peru: II. Additional hypothesis testing in quantitative ethnobotany. *Economic Botany, 47*, 33–43.

Phillips, O. L., Gentry, A. H., Reynel, C., Wilkin, P., & Galvez-Durand, C. (1994). Quantitative ethnobotany and Amazonian conservation. *Conservation Biology, 8*, 225–248.

Piñedo-Vasquez, M., Zarin, D., Jipp, P., & Chota-Inuma, J. (1990). Use values of tree species in a communal forest reserve in northeastern Peru. *Conservation Biology, 4*, 405–416.

Prance, G. T., Balee, W., Boom, B. M., & Carneiro, R. L. (1987). Quantitative ethnobotany and the case for conservation in Amazonia. *Conservation Biology, 1*, 296–310.

Reyes-García, V., Huanca, T., Vadez, V., Leonard, W., & Wilkie, W. (2006). Cultural, practical, and economic value of wild plants: A quantitative study in the Bolivian Amazon. *Economic Botany, 60*, 62–74.

Stoffle, R. W., Halmo, D. B., Evans, M. J., & Olmsted, J. E. (1990). Calculating the cultural significance of American Indian plants: Paiute and Shoshone ethnobotany at Yucca Mountain, Nevada. *American Anthropologist, 92*, 416–432.

Tardio, J., & Santayana, P. D. (2008). Cultural importance indices: A comparative analysis based on the useful wild plants of sout.

Townsley, P. (1996). *Rapid rural appraisal, participatory rural appraisal and aquaculture*. FAO Fisheries Technical Paper 358, Rome

Turner, N. J. (1988). "The importance of a rose": Evaluating the cultural significance of plants in Thompson and Lillooet Interior Salish. *American Anthropologist, 90*, 272–290.

Wong, J. L. G. (2000). *The biometrics of non-timber forest product resource assessment: A review of current methodology*. Paper presented at European Tropical Forest Research Network (ETFRN) workshop, Rome, Italy, 4–5 May 2000. Retrieved from www.etfrn.org/etfrn/workshop/ntfp/text.pdf

Chapter 5 Nutritional Contents of Wild Edible Vegetables

5.1 Nutritional Investigation

An ever-growing gap between human population and food supply is a great challenge to human survival, mainly in the developing world. However, this gap could be reduced through research and development focused on less important edible plants. Wild edible plants are those species which are neither cultivated nor domesticated, but are accessible from their natural habitat and used as sources of food (Beluhan and Ranogajec 2010). In addition to their role in reducing food gaps during periods of drought or scarcity, these species are playing significant role in maintaining livelihood security for many people in developing countries (Afolayan and Jimoh 2009), where natural disasters, underdeveloped economies, political instability, population explosion, high prices of food commodities, poor implementation of agricultural policies, and inadequate food intake among people are common (Adebooye and Phillips 2006). In these countries, starch-based foods are major sources of nutrition (Ladeji et al. 1995). In order to improve the situation, efforts should be focused on sustainable use of underutilized and lesser-known wild plants as sources of nutrient supplements in enhancing food and nutrition security. Results of nutritional analysis provide clues to aid the promotion of those wild edible species which have the best nutritional values helping to ensure dietetic diversity and combat food insecurity (Tardio et al. 2006).

Although various methods are used by different workers with certain modifications to estimate the nutritional potential of both cultivated and wild edible plants, here we analyze the plant samples for their moisture, dry matter, total protein contents, crude fats, crude fibers, carbohydrates, energy value, and ash contents using standard methods as outlined by Association of Official Analytical Chemists (AOAC 1990, 1995, 2000; AOCS 2005).

A.M. Abbasi et al., *Wild Edible Vegetables of Lesser Himalayas*,
DOI 10.1007/978-3-319-09543-1_5

5.1.1 Sample Processing

Plant material collection and processing is an important step prior to nutritional assessment. Although chemical composition in plants vary with age, season, time, soil conditions, climate, and geography, it is necessary to collect vegetable samples in proper time and season particularly when they are ready to be used by local communities. Amount of collected plant material depends on techniques used during analysis and quantity of sample required for the analysis. During present study about 2–10 kg fresh plant material was selected from several plants of each species collected from different localities, then fresh edible parts were cleaned with tap water followed by distilled water and dried at room temperature to remove residual moisture. In order to save time cleaned samples were also packed in paper envelope/news papers and dried at 55–65 °C for 24 h in an oven (Abuye et al. 2003; Wahab et al. 2008). After complete drying, the samples were grinded into fine powder using a pestle and mortar and sieved through 20-mesh sieve. The powder samples were packed in clean and labelled plastic bottles or envelop and stored at room temperature in desiccators (Meena et al. 2010; Street et al. 2008) until further analysis.

5.1.2 Moisture and Dry Matter Content

About 50–100 g of freshly collected vegetables material is measured then packed in paper envelops and placed in an electric oven at 100–105 ºC for 12–18 h (AOAC 1995). The weight of dried samples is measured and the percentage loss in weight is expressed as percentage moisture contents.

Calculation

$$\begin{aligned} \%\ \text{Dry matter (DM)} &= \frac{W_3 - W_1}{W_2} \times 100 \\ \%\ \text{Moisture content (MC)} &= 100 - \text{DM} \end{aligned}$$

where $W_1 =$ weight of empty paper envelop, $W_2 =$ weight of sample, and $W_3 =$ weight of dry matter and paper envelop.

5.1.3 Ash Content

Ash content is determined by following (AOAC 1995) methods. Clean and dried crucibles are first heated on hot burner for one minute, cooled in desiccators and weigh. About 5–10 g of dried powder of vegetable samples is weighted in these crucibles and placed in a muffle furnace at 550 ± 5 °C for 8–10 h until ash is obtained. After ash formation crucibles are removed from furnace and placed in desiccators and weighted again. The ash content is intended by measuring the weight loss percentage.

Calculation

$$\% \text{ Ash } = \frac{W_3 - W_1}{W_2} \times 100$$

where W_1 = weight of oven-dried empty crucible, W_2 = weight of sample, and W_3 = weight of ash and crucible.

5.1.4 Estimation of Crude Protein

The crude protein contents of wild edible vegetables are estimated by following AOAC (2000) using auto Macro Kjeldhal apparatus. Estimation of crude protein is based on three steps, digestion, distillation, and titration.

- *Digestion*: Mix 1 g of dried powder sample and 3 g of digesting mixture in a 100 mL Kjeldahl digestion flask. Then add 30 mL of concentrated sulphuric acid (H_2SO_4) into it and digest the mixture in Kjeldahl digester at 400 °C for 80 min until the mixture becomes clear.

  Digestion mixture: contains 96.5 g sodium sulphate (Na_2SO_4), 2.5 g copper sulphate ($CuSO_4$), and 1 g of selenium (Se) in 100 g mixture.
- *Distillation and titration*: After complete digestion, solution is filtered into a 100 mL volumetric flask and diluted up to the mark with distilled water. Ammonia in the digest is steam-distilled from 10 mL of the digest to which 20 mL of 45 % sodium hydroxide (NaOH) solution is added. The liberated ammonia is collected in 50 mL of 20 % boric acid (H_3BO_3) solution containing a mixed indicator (0.01 g of methyl red and 0.03 g of bromocresol green in 100 mL of alcohol). Ammonia is estimated by titrating with standard 0.01 M hydrochloric acid (HCl) solution. Blank determination is also carried out by same way. Crude protein is estimated by multiplying the value obtained for percentage nitrogen content by a factor of 6.25.

5.1.5 Estimation of Crude Fats

Crude fats extraction is carried out following the official procedure (AOCS 2005) employing crude fats extractor containing petroleum ether as solvent. About 2–3 g of powder sample is added into the filtered bag, and then the filter bags are sealed with heat sealer within 4 mm of the top to encapsulate the samples. Sealed bags are then placed in oven at 80 °C for 3 h. After drying samples are cooled in desiccant pouch and weighed. These sample bags are then placed into bag holder or carousel of fats extractor at 100 °C for 60 min. After complete extraction, the samples are placed in oven again for 15–30 min, then cooled in desiccant pouch and weighed third time. Crude fat contents (%CF) are calculated as

Calculation

$$\% \text{ Crude fats } = \frac{W_2 - W_3}{W_1} \times 100$$

where W_1 = original weight of the sample, W_2 = weight of pre-dried sample with the filter bag, and W_3 = weight of dried sample and filter bag after extraction.

5.1.6 Estimation of Crude Fiber

Crude fibers contents of wild edible vegetables are estimated by acid–base digestion (AOAC 1990; Aberoumand 2009) with 1.25 % sulphuric acid "H_2SO_4" (7.2 mL of 94 % H_2SO_4/L distilled water) and 1.25 % sodium hydroxide "NaOH" (12.5 g/L distilled water) solutions. Approximately 1–2 g dried powder sample is added into a 600 mL beaker and 200 mL of boiling 1.25 % H_2SO_4 is mixed in it. This mixture is boiled for 30 min, cooled, filtered and the residue is washed thrice with 50 mL aliquots of boiling water. The washed residue is then brought back to the original beaker and further digested by boiling in 200 mL of 1.25 % NaOH for 30 min. The digest is filtered to obtain the residue and washed three times with 50 mL aliquots of boiling water and finally with 25 mL ethanol. The washed residue is dried in an oven at 130 °C to constant weight and cooled in desiccators. The residue is scraped into a pre-weighed crucible and ashed at 550 °C for 2 h, cooled in desiccators and reweighed. Crude fiber contents (% CFb) are expressed as percentage loss in weight on ignition.

Calculation

$$\% \text{ Crude fibers } = \frac{W_2 - W_3}{W_1} \times 100$$

where W_1 = weight of sample, W_2 = weight of dried residue, and W_3 = weight of ash.

5.1.7 Estimation of Available Carbohydrates

The carbohydrate contents in the samples of wild edible vegetables presented here are obtained by indirect or difference method (AOAC 1990; Bamigboye et al. 2010), by subtracting the values obtained for moisture, crude protein, crude fats, crude fiber, and ash from 100.

$$\text{Carbohydrates} = 100 - (\%\text{Moisture} + \%\text{CP} + \%\text{CF} + \%\text{CFb} + \%\text{Ash})$$

5.1.8 Nutritive Value/Energy Value

Energy value of vegetable samples in kilocalorie per gram (kcal/g) can be calculated by multiplying the percentage content of crude proteins, crude fats, and carbohydrates by the recommended factors of 4, 9, and 4, respectively, first and then by taking the sum of the these values. Resultant value can be converted to kilojoules by multiplying with factor 4.2 (AOAC 1990, 1995, 2000; AOCS 2005; Abuye et al. 2003; Wahab et al. 2008; Meena et al. 2010; Street et al. 2008; Aberoumand 2009; Bamigboye et al. 2010; Effiong and Udo 2010).

$$\text{Energy value (kcal/g)} = (\text{CP} \times 4) + (\text{CF} \times 9) + (\text{Carb.} \times 4)$$

5.2 Nutritional Contents of Wild Edible Vegetables

5.2.1 *Amaranthus hybridus* L.

Fresh weight basis: Moisture content 81.20 %, dry matter contents 18.80 %.

Dry weight basis: Carbohydrates 46.79 %, crude proteins 25.14 %, crude fats 1.160 %, crude fibers 9.250 %, ash content 17.66 %, Energy value 298.1 Kcal/100 g.

5.2.2 *Amaranthus spinosus* L.

Fresh weight basis: Moisture contents 80.34 %, dry matter contents 19.67 %.

Dry weight basis: Carbohydrates 43.34 %, crude proteins 24.79 %, crude fats 2.180 %, crude fibers 11.72 %, ash content 17.97 %, Energy value 292.1 Kcal/100 g.

5.2.3 *Amaranthus viridis* L.

Fresh weight basis: Moisture contents 76.77 %, dry matter contents 23.24 %.

Dry weight basis: Carbohydrate 45.24 %, crude proteins 26.22 %, crude fats 2.640 %, crude fibers 7.270 %, ash content 18.49 %, Energy value 310.1 Kcal/100 g.

5.2.4 *Digera muricata (L.) Mart.*

Fresh weight basis: Moisture contents 79.48 %, dry matter contents 20.52 %.
Dry weight basis: Carbohydrates 45.24 %, crude proteins 22.48 %, crude fats 1.240 %, crude fibers 10.92 %, ash content 20.12 %, Energy value 282.0 Kcal/ 100 g.

5.2.5 *Dryopteris ramosa (Hope) C. Chr.*

Fresh weight basis: Moisture contents 80.75 %, dry matter contents 19.25 %.
Dry weight basis: Carbohydrates 34.21 %, crude proteins 22.69 %, crude fats 5.640 %, crude fibers 26.68 %, ash content 10.78 %, Energy value 278.3 Kcal/ 100 g.

5.2.6 *Bidens bipinnata L.*

Fresh weight basis: Moisture content 82.78 %, dry matter contents 17.22 %.
Dry weight basis: Carbohydrates 52.40 %, crude proteins 19.97 %, crude fats 1.610 %, crude fibers 12.33 %, ash content 13.70 %, Energy value 303.9 Kcal/ 100 g.

5.2.7 *Centaurea calcitrapa L.*

Fresh weight basis: Moisture content 83.12 %, dry matter contents 16.89 %.
Dry weight basis: Carbohydrates 39.68 %, crude proteins 32.34 %, crude fats 3.435 %, crude fibers 10.00 %, ash content 14.54 %, Energy value 319.0 Kcal/ 100 g.

5.2.8 *Cichorium intybus L.*

Fresh weight basis: Moisture content 86.52 %, dry matter contents 13.48 %.
Dry weight basis: Carbohydrates 7.680 %, crude proteins 54.90 %, crude fats 3.73 %, crude fibers 13.28 %, ash content 20.42 %, Energy value 283.8 Kcal/ 100 g.

5.2.9 *Launaea procumbens (Roxb).*

Fresh weight basis: Moisture contents 83.06 %, dry matter contents 16.94 %.

Dry weight basis: Carbohydrates 47.41 %, crude proteins 15.22 %, crude fats 6.860 %, crude fibers 15.59 %, ash content 14.92 %, Energy value 312.2 Kcal/100 g.

5.2.10 *Sonchus asper L.*

Fresh weight basis: Moisture contents 86.70 %, dry matter contents 13.30 %.

Dry weight basis: Carbohydrates 51.26 %, crude proteins 15.19 %, crude fats 4.290 %, crude fibers 15.85 %, ash content 13.41 %, Energy value 304.4 Kcal/100 g.

5.2.11 *Sonchus oleraceous L.*

Fresh weight basis: moisture content 87.84 %, dry matter contents 12.16 %.

Dry weight basis: carbohydrates 49.12 %, crude proteins 18.62 %, crude fats 4.640 %, crude fibers 14.25 %, ash content 13.37 %, Energy value 312.7 Kcal/100 g.

5.2.12 *Taraxacum officinale L.*

Fresh weight basis: Moisture contents 91.13 %, dry matter contents 8.870 %.

Dry weight basis: Carbohydrates 43.42 %, crude proteins 23.07 %, crude fats 4.150 %, crude fibers 14.80 %, ash content 14.55 %, Energy value 303.6 Kcal/100 g.

5.2.13 *Bombax malabaracum DC.*

Fresh weight basis: Moisture content 84.68 %, dry matter contents 15.32 %.

Dry weight basis: Carbohydrates 55.65 %, crude proteins 15.97 %, crude fats 2.530 %, crude fibers 17.15 %, ash content 8.710 %, Energy 309.2 Kcal/100 g.

5.2.14 *Capsella bursa-pastoris (L.) Medic.*

Fresh weight basis: Moisture contents 94.25 %, dry matter contents 5.750 %.
Dry weight basis: Carbohydrates 24.86 %, crude proteins 40.36 %, crude fats 3.160 %, crude fibers 10.30 %, ash content 21.32 %, Energy 289.3 Kcal/100 g.

5.2.15 *Nasturtium officinale R. Br.*

Fresh weight basis: Moisture content 92.79 %, dry matter contents 7.210 %.
Dry weight basis: Carbohydrates 28.64 %, crude proteins 37.08 %, crude fats 4.300 %, crude fibers 13.05 %, ash content 16.93 %, Energy value was estimated at 301.5 Kcal/100 g.

5.2.16 *Bauhinia variegata L.*

Fresh weight basis: Moisture content 84.07 %, dry matter contents 15.93 %.
Dry weight basis: Carbohydrates 57.42 %, crude proteins 19.89 %, crude fats 4.800 %, crude fibers 11.61 %, ash contents 6.290 %, Energy value was at 301.5 Kcal/100 g.

5.2.17 *Bauhinia variegata var. candida Voigt*

Fresh weight basis: Moisture content 85.29 %, dry matter contents 14.71 %.
Dry weight basis: Carbohydrates 55.91 %, crude proteins 21.49 %, crude fats 4.147 %, crude fibers 13.34 %, ash contents 4.837 %, Energy value was at 349.4 Kcal/100 g.

5.2.18 *Bauhinia variegata L. var. variegata*

Fresh weight basis: Moisture content 84.39 %, dry matter contents 15.62 %.
Dry weight basis: Carbohydrates 53.23 %, crude proteins 21.38 %, crude fats 4.026 %, crude fibers 14.65 %, ash contents 6.714 %, Energy value was at 334.7 Kcal/100 g.

5.2.19 *Silene conoidea* L.

Fresh weight basis: Moisture content 80.67 %, dry matter contents 19.34 %.
Dry weight basis: Carbohydrates 41.87 %, crude proteins 21.69 %, crude fats 3.130 %, crude fibers 15.90 %, ash contents 17.41 %, Energy value 282.4 Kcal/100 g.

5.2.20 *Stellaria media* (L.) Cyr.

Fresh weight basis: Moisture content 88.30 %, dry matter contents 11.71 %.
Dry weight basis: Carbohydrates 36.05 %, crude proteins 21.26 %, crude fats 3.020 %, crude fibers 15.04 %, ash contents 24.63 %, Energy value 256.4 Kcal/100 g.

5.2.21 *Chenopodium album* L.

Fresh weight basis: Moisture contents 87.70 %, dry matter contents 12.30 %.
Dry weight basis: Carbohydrates 7.680 %, crude proteins 54.90 %, crude fats 3.730 %, crude fibers 13.28 %, ash contents 20.42 %, Energy value 283.8 Kcal/100 g.

5.2.22 *Commelina benghalensis* L.

Fresh weight basis: Moisture content 88.50 %, dry matter contents 11.50 %.
Dry weight basis: Carbohydrates 43.14 %, crude proteins 25.26 %, crude fats 2.370 %, crude fibers 13.01 %, ash contents 16.23 %, Energy value 294.87 Kcal/100 g.

5.2.23 *Evolvulus alsinoides* L.

Fresh weight basis: Moisture content 70.63 %, dry matter contents 29.37 %.
Dry weight basis: Carbohydrates 38.61 %, crude proteins 27.29 %, crude fats 2.560 %, crude fibers 11.42 %, ash contents 20.12 %, Energy value 286.6 Kcal/100 g.

5.2.24 *Dioscorea deltoidea Wall. ex, Griseb*

Fresh weight basis: Moisture contents 76.52 %, dry matter contents 23.48 %.
Dry weight basis: Carbohydrates 79.76 %, crude proteins 12.87 %, crude fats 0.690 %, crude fibers 2.780 %, ash content 3.900 %, Energy value 376.7 Kcal/100 g.

5.2.25 *Lamium amplexicaule L.*

Fresh weight basis: Moisture content 75.84 %, dry matter contents 24.16 %.
Dry weight basis: Carbohydrates 35.86 %, crude proteins 25.05 %, crude fats 6.020 %, crude fibers 15.70 %, ash content 17.38 %, Energy value 297.9 Kcal/100 g.

5.2.26 *Origanum vulgare subsp. Hirtum (Link) Letsw.*

Fresh weight basis: Moisture content 71.76 %, dry matter contents 28.24 %.
Dry weight basis: Carbohydrates 62.61 %, crude proteins 14.83 %, crude fats 3.230 %, crude fibers 9.14 %, ash content 10.19 %, Energy value 338.8 Kcal/100 g.

5.2.27 *Tulipa stellata var. clusiana Hk. f.*

Fresh weight basis: Moisture content 88.60 %, dry matter contents 11.41 %.
Dry weight basis: Carbohydrates 59.30 %, crude proteins 16.63 %, crude fats 8.870 %, crude fibers 10.25 %, ash content 4.960 %, Energy value 383.5 Kcal/100 g.

5.2.28 *Malva parviflora L.*

Fresh weight basis: Moisture contents 90.31 %, dry matter contents 9.700 %.
Dry weight basis: Carbohydrates 27.37 %, proteins contents 40.43 %, crude fats 2.960 %, crude fibers 10.06 %, ash content 19.17 %, Energy value was found at 297.8 Kcal/100 g.

5.2.29 *Ficus carica* L.

Fresh weight basis: Moisture content 74.61 %, dry matter contents 24.50 %.
Dry weight basis: Carbohydrates 47.96 %, crude proteins 26.19 %, crude fats 2.780 %, crude fibers 13.82 %, ash content 9.250 %, Energy value 321.5 Kcal/100 g.

5.2.30 *Ficus palmata* Forssk.

Fresh weight basis: Moisture content 76.45 %, dry matter contents 23.55 %.
Dry weight basis: Carbohydrates 51.66 %, crude proteins 20.76 %, crude fats 3.200 %, crude fibers 14.48 %, ash content 9.900 %, Energy value 318.4 Kcal/100 g.

5.2.31 *Oxalis corniculata* L.

Fresh weight basis: Moisture content 89.48 %, dry matter contents 10.52 %.
Dry weight basis: Carbohydrates 45.01 %, crude proteins 21.71 %, crude fats 6.600 %, crude fibers 15.02 %, ash content 11.66 %, Energy value 326.2 Kcal/100 g.

5.2.32 *Lathyrus aphaca* L.

Fresh weight basis: Moisture content 77.70 %, dry matter contents 22.30 %.
Dry weight basis: Carbohydrates 44.17 %, crude proteins 20.72 %, crude fats 2.079 %, crude fibers 23.36 %, ash content 9.471 %, Energy value 273.89 Kcal/100 g.

5.2.33 *Medicago polymorpha* L.

Fresh weight basis: Moisture content 89.99 %, dry matter contents 10.01 %.
Dry weight basis: Carbohydrates 40.35 %, crude proteins 29.51 %, crude fats 3.290 %, crude fibers 14.79 %, ash content 12.07 %, Energy value 309.0 Kcal/100 g.

5.2.34 *Melilotus alba Ledeb.*

Fresh weight basis: Moisture content 81.47 %, dry matter contents 18.53 %.
Dry weight basis: Carbohydrates 41.34 %, crude proteins 30.22 %, crude fats 5.070 %, crude fibers 13.56 %, ash contents 9.810 %, Energy value 301.5 Kcal/100 g.

5.2.35 *Melilotus indicus (L.) All.*

Fresh weight basis: Moisture content 70.95 %, dry matter contents 29.06 %.
Dry weight basis: Carbohydrates 48.22 %, crude proteins 26.90 %, crude fats 3.470 %, crude fibers 12.78 %, ash contents 8.620 %, Energy value 331.7 Kcal/100 g.

5.2.36 *Vicia faba L.*

Fresh weight basis: Moisture content 81.78 %, dry matter contents 18.23 %.
Dry weight basis: Carbohydrates 41.19 %, crude proteins 27.61 %, crude fats 2.310 %, crude fibers 20.07 %, ash contents 8.820 %, Energy value 295.9 Kcal/100 g.

5.2.37 *Vicia sativa L.*

Fresh weight basis: Moisture content 82.79 %, dry matter contents 17.22 %.
Dry weight basis: Carbohydrates 49.01 %, crude proteins 19.74 %, crude fats 1.230 %, crude fibers 15.56 %, ash contents 14.47 % Energy value 286.05 Kcal/100 g.

5.2.38 *Plantago lanceolata L.*

Fresh weight basis: Moisture content 83.99 %, dry matter contents 16.01 %.
Dry weight basis: Carbohydrates 54.88 %, crude proteins 16.54 %, crude fats 2.720 %, crude fibers 12.62 %, ash contents 13.24 %, Energy value 310.1 Kcal/100 g.

5.2.39 *Persicaria amplexicaulis (D. Don) Ronse Decr.*

Fresh weight basis: Moisture content 81.47 %, dry matter contents 18.53 %.
Dry weight basis: Carbohydrates 47.65 %, crude proteins 15.08 %, crude fats 4.770 %, crude fibers 18.41 %, ash contents 14.09 %, Energy value 293.8 Kcal/100 g.

5.2.40 *Polygonum aviculare L.*

Fresh weight basis: Moisture content 68.33 %, dry matter contents 31.67 %.
Dry weight basis: Carbohydrates 40.60 %, crude proteins 29.91 %, crude fats 3.519 %, crude fibers 14.35 %, ash content 11.62 %, Energy value 313.4 Kcal/100 g.

5.2.41 *Rumex dentatus L.*

Fresh weight basis: Moisture content 88.43 %, dry matter contents 11.57 %.
Dry weight basis: Carbohydrates 34.11 %, crude proteins 36.68 %, crude fats 3.190 %, crude fibers 10.77 mg/kg, ash content 15.26 %, Energy value 311.8 Kcal/100 g.

5.2.42 *Rumex hastatus D. Don, Prodor.*

Fresh weight basis: Moisture content 89.18 %, dry matter contents 10.82 %.
Dry weight basis: Carbohydrates 55.24 %, crude proteins 12.92 %, crude fats 3.320 %, crude fibers 13.18 %, ash content 15.34 %, Energy value 302.5 Kcal/100 g.

5.2.43 *Rumex nepalensis Spreng.*

Fresh weight basis: Moisture content 87.84 %, dry matter contents 12.16 %.
Dry weight basis: Carbohydrates 39.40 %, crude proteins 32.89 %, crude fats 3.279 %, crude fibers 11.06 %, ash contents 13.37 %, Energy value was at 318.7 Kcal/100 g.

5.2.44 *Portulaca quadrifida* L.

Fresh weight basis: Moisture content 88.31 %, dry matter contents 11.69 %.

Dry weight basis: Carbohydrates 28.38 %, crude proteins 28.94 %, crude fats 4.080 %, crude fibers 13.44 %, ash content 25.16 %, Energy 266.0 Kcal/100 g.

5.2.45 *Portulaca oleracea* L.

Fresh weight basis: Moisture content 92.77 %, dry matter contents 7.235 %.

Dry weight basis: Carbohydrates 29.84 %, crude proteins 33.27 %, crude fats 4.171 %, crude fibers 7.602 %, ash contents 5.215 %, Energy value was at 290.0 Kcal/100 g.

5.2.46 *Galium aparine* L.

Fresh weight basis: Moisture content 92.30 %, dry matter contents 7.700 %.

Dry weight basis: Carbohydrates 33.59 %, crude proteins (25.63 %), crude fats 1.320 %, crude fibers 21.43 %, ash content 18.03 %, Energy value 248.5 Kcal/ 100 g.

5.2.47 *Veronica arvensis* L.

Fresh weight basis: Moisture content 93.05 %, dry matter contents 6.960 %.

Dry weight basis: Carbohydrates 47.96 %, crude proteins 18.81 %, crude fats 2.210 %, crude fibers 15.86 %, ash content 15.16 %, Energy value 286.9 Kcal/ 100 g.

5.2.48 *Solanum nigrum auct.*

Fresh weight basis: Moisture content 87.88 %, dry matter contents 12.12 %.

Dry weight basis: Carbohydrates 52.56 %, crude proteins 22.72 %, crude fats 3.830 %, crude fibers 8.250 %, ash content 12.65 %, Energy value 335.5 Kcal/ 100 g.

5.2.49 *Pimpinella diversifolia (Wall.) DC. Prodor.*

Fresh weight basis: Moisture content 80.11 %, dry matter contents 19.89 %.
Dry weight basis: Carbohydrates 61.06 %, crude proteins 13.01 %, crude fats 2.940 %, crude fibers 12.31 %, ash content 10.69 %, Energy value 322.6 Kcal/ 100 g.

5.2.50 *Torilis leptophylla (L.) Reichb. f.*

Fresh weight basis: Moisture content 89.48 %, dry matter contents 10.52 %.
Dry weight basis: Carbohydrates 58.11 %, crude proteins 15.35 %, crude fats 1.380 %, crude fibers 10.56 %, ash content 14.60 %, Energy value 306.2 Kcal/ 100 g.

5.3 Comparative Assessment of Nutrients in Wild Edible Vegetables

The results of proximate composition of wild edible vegetables show variations in concentration of moisture and dry matter contents, carbohydrates, proteins, fats, fibers, and energy value on dry weight basis except moisture contents and dry matter which are estimated on wet basis.

According to Mitchel et al. (1976) increase in moisture content reduces fats, proteins, and carbohydrates level and energy values in vegetables. Comparative analysis of moisture and dry matter contents of wild edible vegetables mentioned in Figs. 5.1 and 5.2 indicated that significantly high moisture content is observed in *Capsella bursa-pastoris* (94.25 %), followed by *Veronica arvensis* (93.05 %), *Portulaca olerecea* (92.77 %), and *Nasturtium officinale* (92.69 %) while, lowest moisture content is estimated in *Evolvulus alsinoides* (70.63 %), followed by *Melilotus indicus* (70.95 %), *Origanum vulgare* (71.76 %), and *Ficus carica* (74.61 %).

Maximum dry matter content is recorded in *Evolvulus alsinoides* (29.37 %), followed by *Melilotus indicus* (29.06 %), *Origanum vulgare* (28.24 %), and *Ficus carica* (25.40 %). However, lowest levels of dry matter contents are observed in *Capsella bursa-pastoris* (5.750 %), *Veronica arvensis* (6.955 %), *Portulaca olerecea* (7.235 %), and *Nasturtium officinale* (7.210 %).

Among investigated wild edible vegetables carbohydrates content is found maximum in *Dioscorea deltoidea* (79.76 %) as mentioned in Fig. 5.3, followed by *Origanum vulgare* (62.61 %), *Pimpinella diversifolia* (61.16 %), and *Tulipa stellata* (59.30 %), whereas lowest levels are estimated in *Cichorium intybus* and

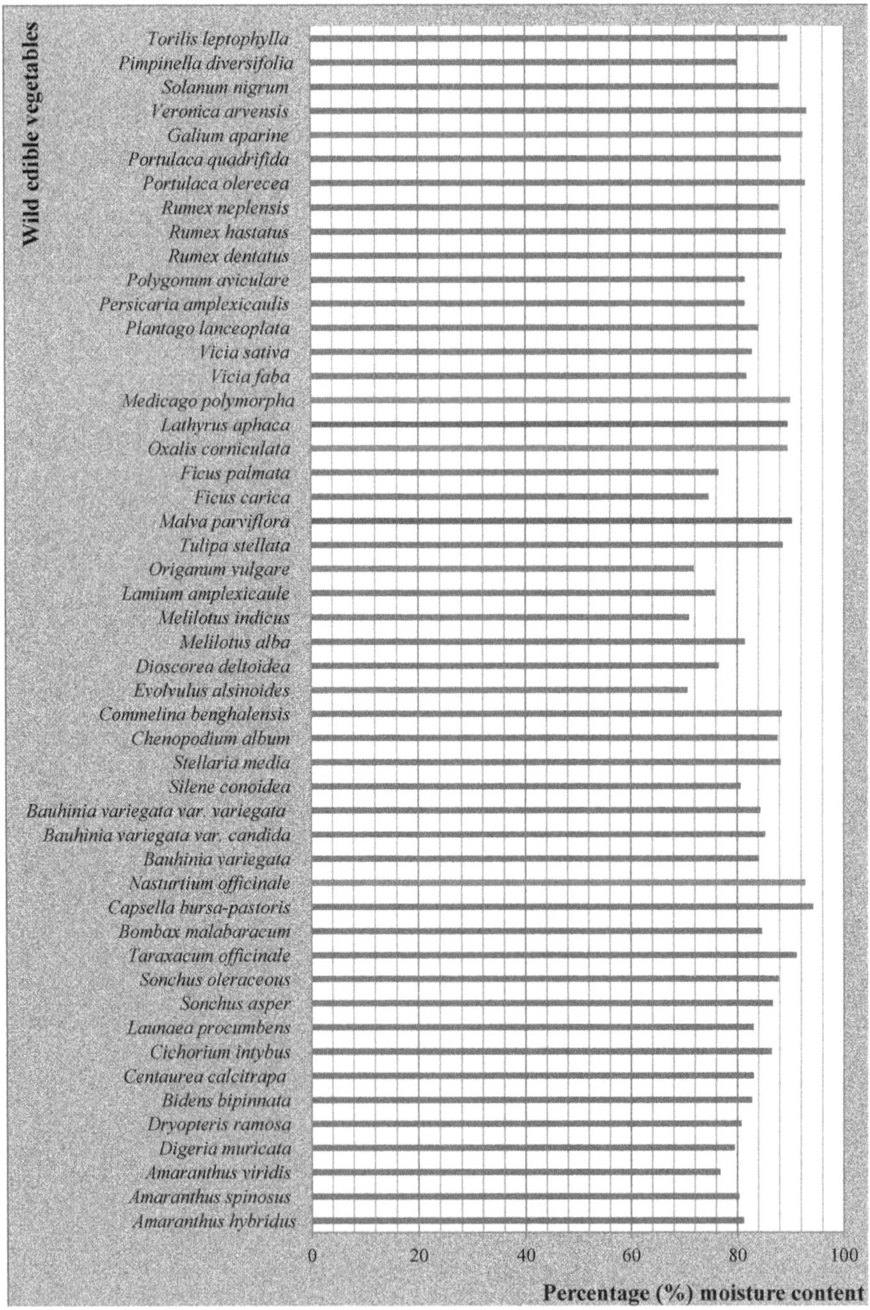

Fig. 5.1 Moisture content of wild edible vegetables

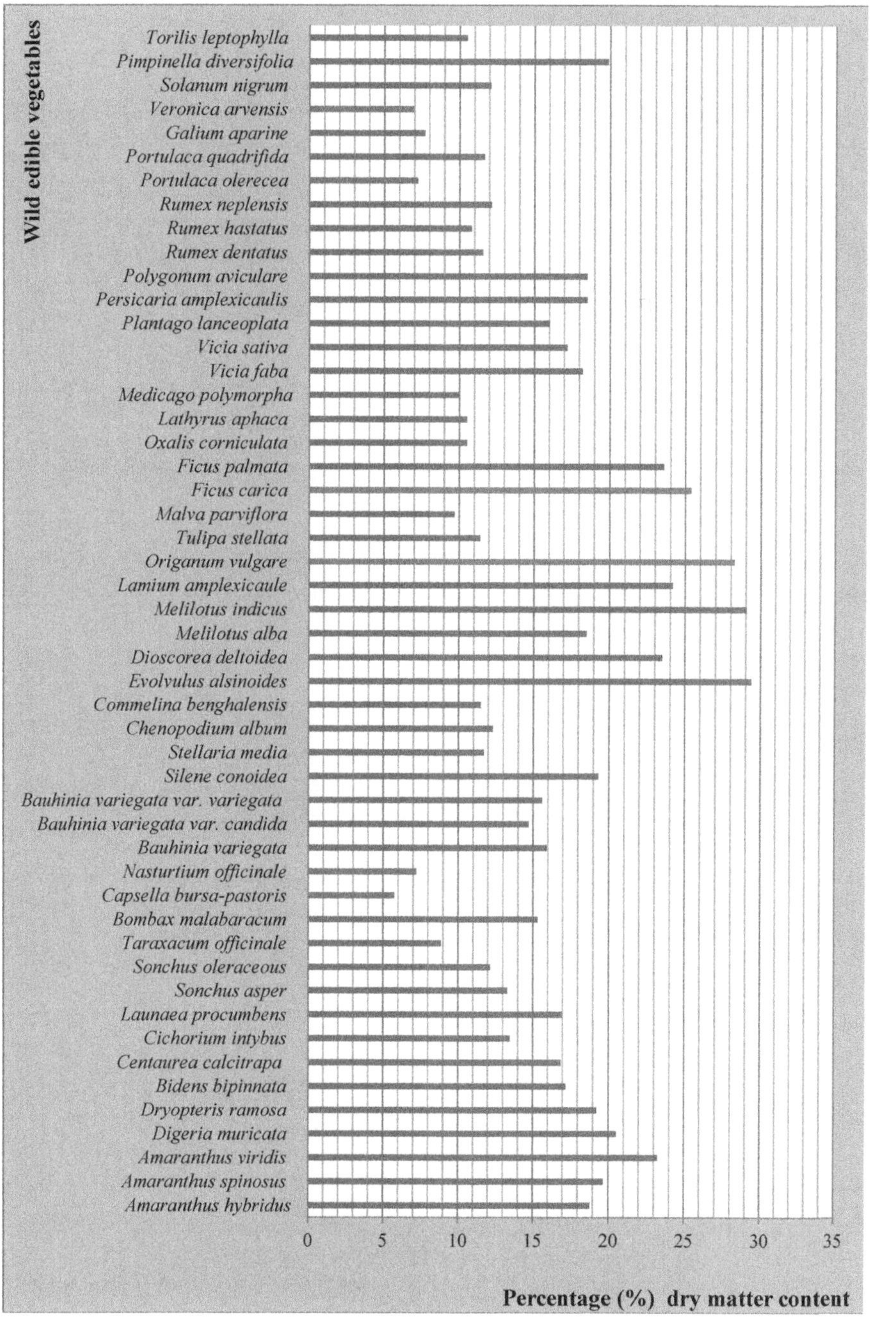

Fig. 5.2 Dry matter content of wild edible vegetables

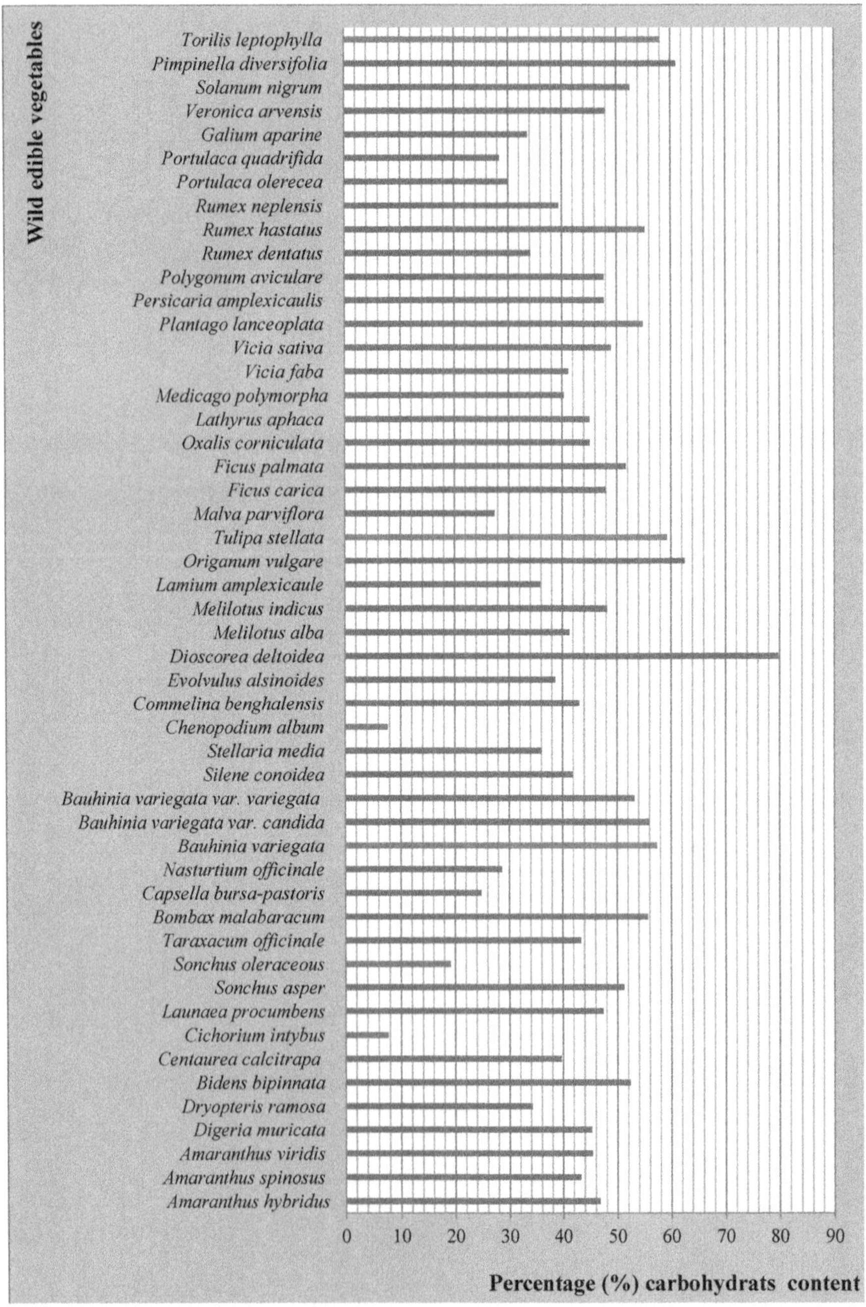

Fig. 5.3 Carbohydrates content of wild edible vegetables

Chenopodium album (7.676 %), followed by *Sonchus oleraceous* (19.13 %) and *Capsella bursa-pastoris* (24.86 %).

A protein's malnutrition deficiency is a major factor responsible in nutritional pathology. Plant-based food which provide more than 12.00 % of its calorific value from proteins are considered a good source of proteins (Roger et al. 2005; Pearson 1976). The highest value of crude proteins in wild food resources persuade their use as high protein sources in some food formulations. Present investigation revealed that crude proteins contents of wild edible vegetables are ranged from 12.87 % in *Dioscorea deltoidea* to 54.90 % in *Cichorium intybus* and *Chenopodium album* (54.90 %) (Fig. 5.4). Significant concentrations of crude proteins are also noted in *Malva parviflora* (40.43 %), *Capsella bursa-pastoris* (40.36 %), *Nasturtium officinale* (37.08 %), and *Rumex dentatus* (36.68 %).

Wild edible vegetables under investigation revealed that in all samples, crude fat contents are ranged from 0.690 % in *Dioscorea deltoidea* to 8.870 % in *Tulipa stellata* (Fig. 5.5). Significant values of crude fats are also observed in *Launaea procumbens* (6.857 %), *Oxalis corniculata* (6.601 %), and *Dryopteris ramosa* (5.637 %).

Intake of dietary fibers can lower the blood cholesterol level, risk of cardiovascular diseases, hypertension, constipation, diabetes, colon, and breast cancer (Ishida et al. 2000). The American Dietetic Association (ADA) recommended an intake of 20–35 g of fiber per day (Duyff 2002). Crude fibers contents of wild edible vegetables ranged from 2.780 to 26.68 % being lowest in *Dioscorea deltoidea* and highest in *Dryopteris ramosa* (Fig. 5.6). *Galium aparine* (21.43 %), *Vicia faba* (20.07 %), *Polyganum aviculare* (18.41 %), *Persicaria amplexicaulis* (18.42 %), and *Bombax malabaracum* (17.15 %) showed significant fiber composition as compared to other samples of wild edible vegetables.

Ash content is an index of mineral contents in biota (Hussain et al. 2010). In the case of wild edible vegetables ash contents are ranged from 3.900 % in *Dioscorea deltoidea* to 25.16 % in *Portulaca quadrifida* (Fig. 5.7), whereas significant value of ash contents is observed in *Stellaria media* (24.63 %), *Capsella bursa-pastoris* (21.32 %), *Cichorium intybus*, and *Chenopodium album* (20.42 %).

Looking at the results obtained from energy value, among vegetables energy value is estimated highest in *Tulipa stellata* (383.5 Kcal/100 g) and lowest in *Galium aparine* (248.7 Kcal/100 g). Significant energy value is observed in *Dioscorea deltoidea* (376.7 Kcal/100 g), *Bauhinia variegata* (352.4 Kcal/100 g), and *Solanum nigrum* (335.5 Kcal/100 g) as compared to other vegetables species (Fig. 5.8).

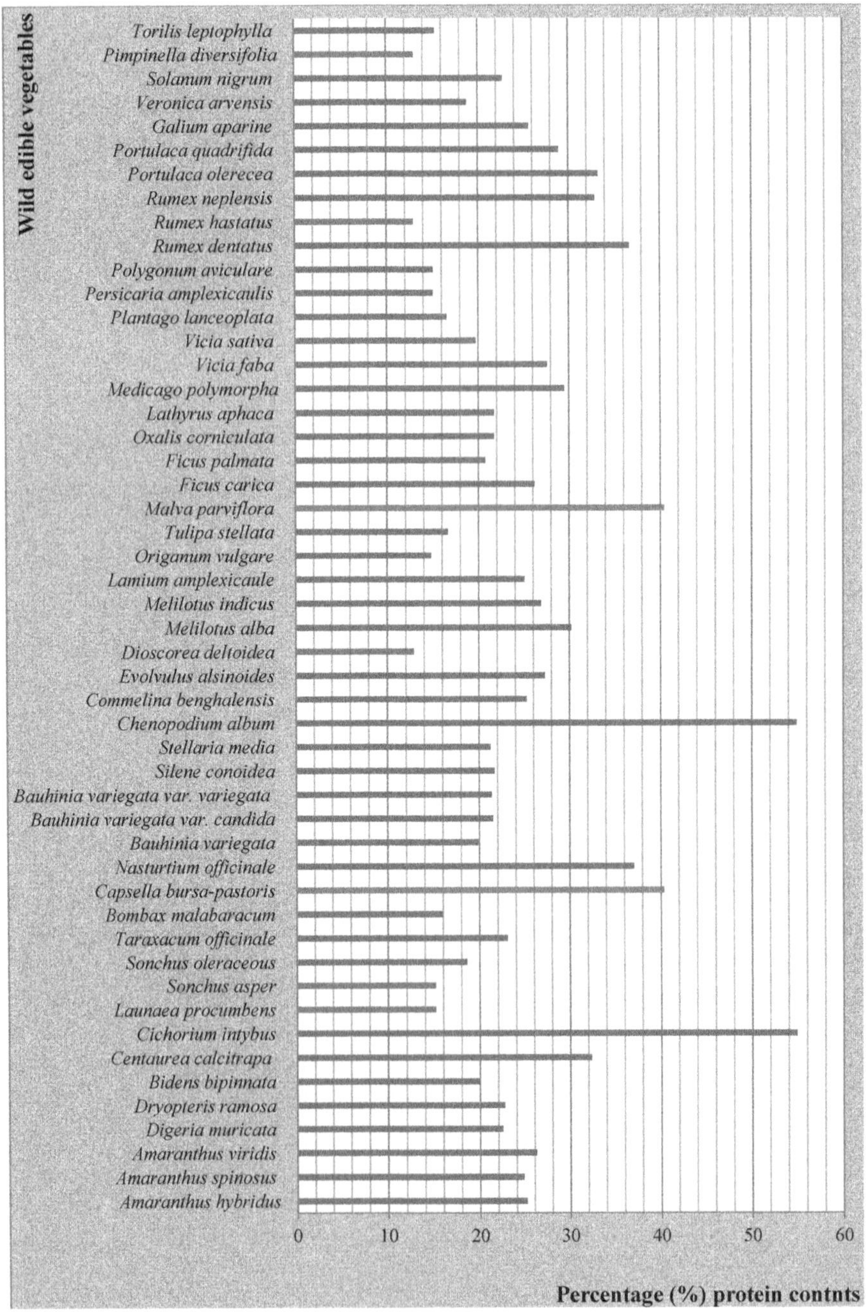

Fig. 5.4 Proteins content of wild edible vegetables

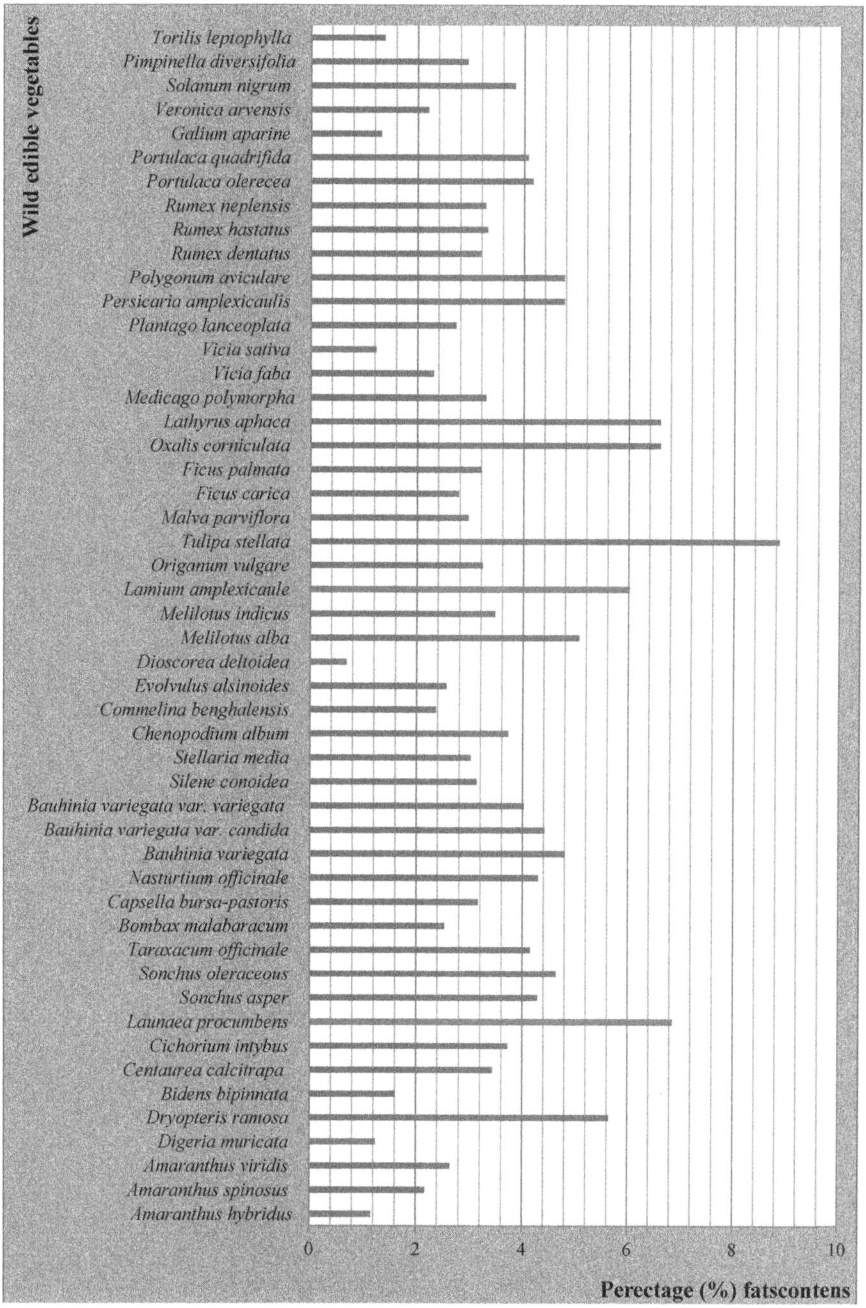

Fig. 5.5 Crude fats content of wild edible vegetables

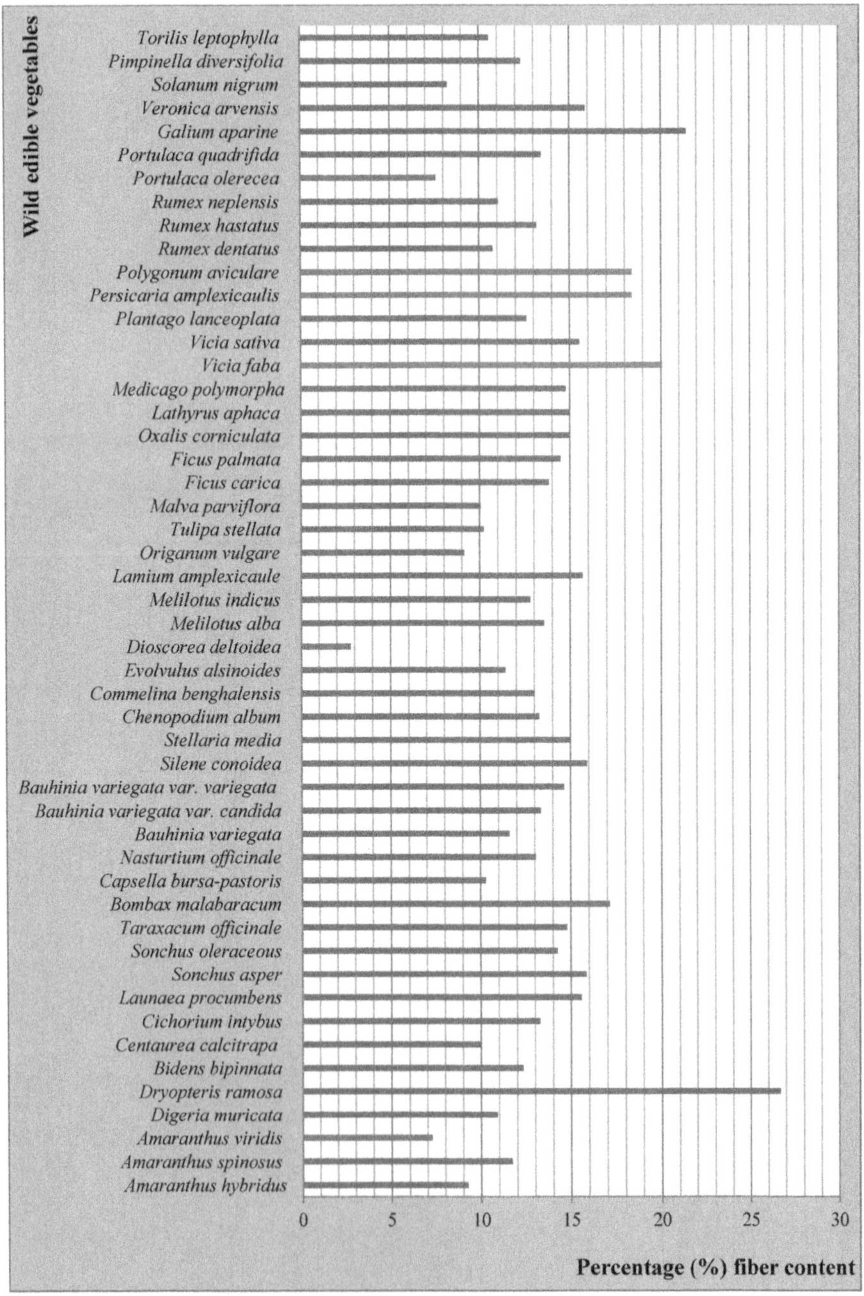

Fig. 5.6 Crude fibers content of wild edible vegetables

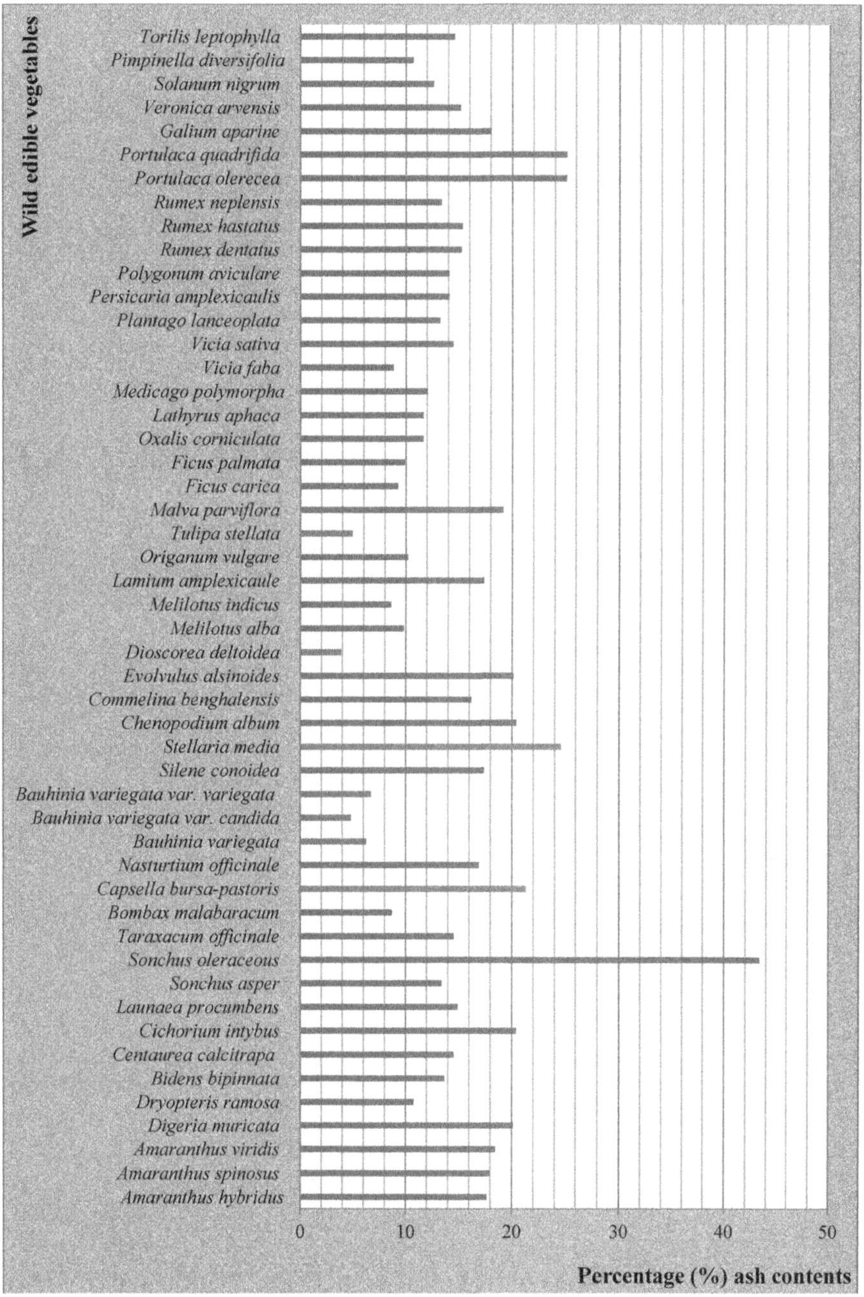

Fig. 5.7 Ash content of wild edible vegetables

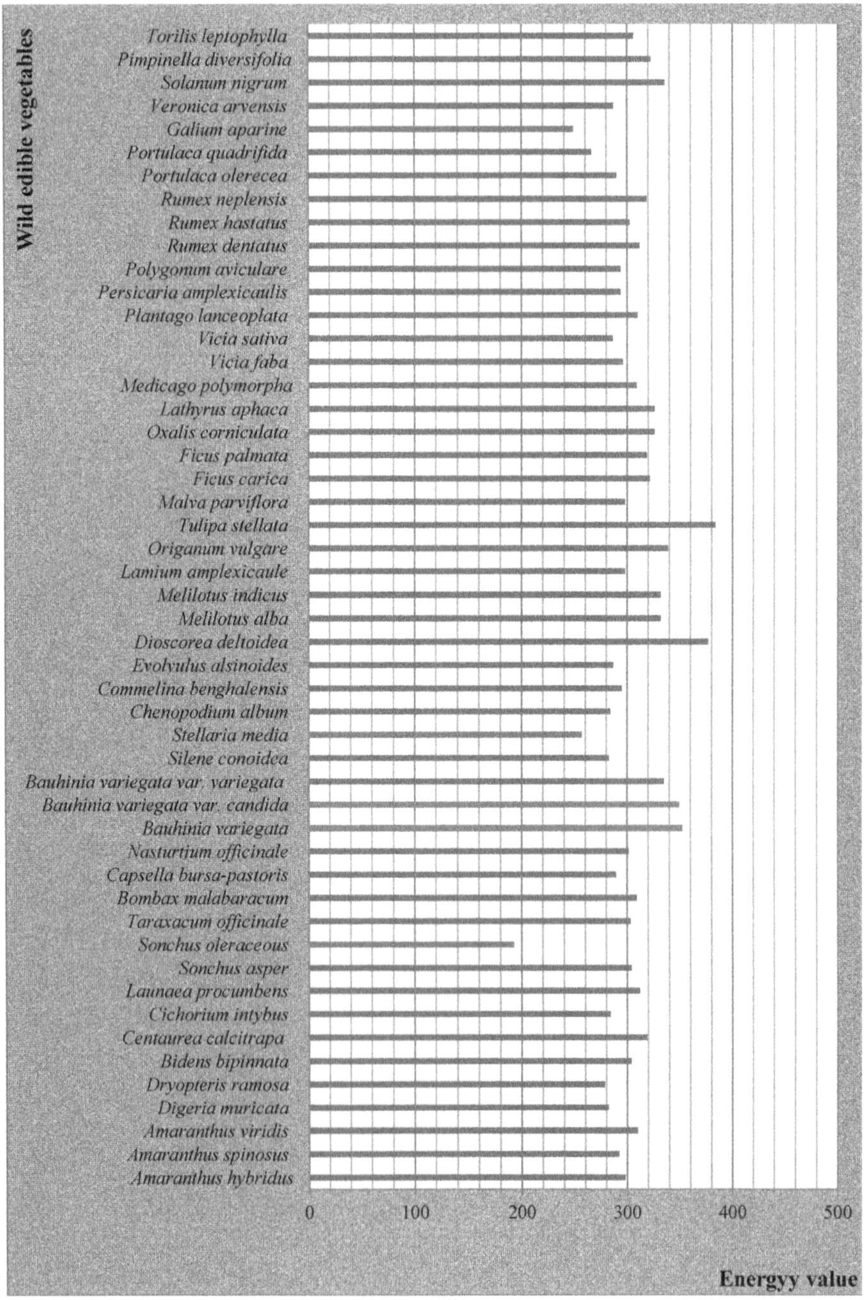

Fig. 5.8 Energy value (Kcal/100 g) of wild edible vegetables

Table 5.1 Statistical distribution parameters for nutritional contents (dry weight) of wild edible vegetables in Lesser Himalayas

	Min	Max	Mean	Median	SD	SE	Kurtosis	Skewness
MC (%)	70.63	94.25	84.34	84.53	5.970	0.84	−0.318	−0.531
DM (%)	5.750	29.37	15.66	15.47	5.970	0.844	−0.318	0.531
CP (%)	12.87	54.90	24.57	22.58	9.385	1.327	2.761	1.473
CF (%)	0.688	8.866	3.558	3.284	1.631	0.231	1.344	0.848
Carb (%)	7.676	79.76	43.68	45.13	13.27	1.877	1.463	−0.492
Fib (%)	2.785	26.68	13.35	13.28	3.850	0.545	2.734	0.570
Ash (%)	3.897	43.37	14.84	14.50	6.561	0.928	6.158	1.625
EV (Kcal/100 g)	192.7	383.5	305.0	303.6	30.88	4.367	3.322	−0.466

MC moisture content, *DM* Dry matter, *CP* Crude proteins, *CF* Crude fats, *Carb* Carbohydrates, *CFb* Crude fibers, *EV* Energy value, *Min* Minimum, *Max* Maximum

5.3.1 Distribution Parameters of the Nutrients in Wild Edible Vegetables

Basic statistical parameters related to the distribution of nutrients levels (% and kcal/100 g, dry weight) in wild edible vegetables from different areas of Lesser Himalayas are presented in Table 5.1. Among wild vegetables dominant mean levels are shown by energy value (305.0 kcal/100 g DW (Dry weight)), followed by moisture content (84.34 % FW (Fresh weight)), carbohydrates (43.68 %), crude proteins (24.57 %), dry matter (15.66 % FW), ash content (14.48 %), crude fibers (13.35 % DW), and crude fats (3.558 % DW).

5.3.2 Correlation Study of the Nutrients in Wild Edible Vegetables

Correlation coefficients between nutritional contents are calculated to examine their inter-relationships in the wild edible vegetables of Lesser Himalayas (Table 5.2); such study indicated mutual relationship among the various nutrients. In the case of wild vegetables strong correlations are noted for following pairs; carbohydrates-energy value ($r = 0.666$), dry matter-carbohydrates ($r = 0.360$), and crude proteins-ash value ($r = 0.306$). Negative correlations are also observed in different pairs; ash-energy value ($r = -0.863$), carbohydrates-proteins ($r = -0.820$), carbohydrates-ash ($r = -0.691$), fibers-energy value ($r = -0.413$). All other correlations show either weak positive or weak negative relationships.

Table 5.2 Correlation coefficient[a] matrix for nutritional contents in wild edible vegetables

	MC	DM	CP	CF	Carb.	Fib	Ash	EV
MC	1.000							
DM	−1.000	1.000						
CP	0.257	−0.257	1.000					
CF	0.152	−0.152	−0.033	1.000				
Carb	−0.360	**0.360**	**−0.820**	−0.120	1.000			
Fib	0.076	−0.076	−0.121	0.240	−0.221	1.000		
Ash	0.279	−0.279	**0.306**	−0.101	**−0.691**	−0.027	1.000	
EV	−0.235	0.235	−0.209	0.230	**0.666**	**−0.413**	**−0.863**	1.000

[a]Bold r-values are significant at $p < 0.01$

References

Aberoumand, A. (2009). Nutritional evaluation of edible *Portulaca oleracia* as plant food. *Food Analytical Methods, 2*, 204–207.

Abuye, C., Urga, K., Knapp, H., Selmar, D., Omwega, A. M., & Imungi, J. K. (2003). A compositional study of *Moringa stenopetala* leaves. *East African Medical Journal, 80*, 247–252.

Adebooye, O. C., & Phillips, O. T. (2006). Studies of seed characteristics and chemical composition of three morphotypes of *Mucuna urens* (L.) Medikus-*Fabaceae. Food Chemistry, 95*, 658–663.

Afolayan, A., & Jimoh, F. (2009). Nutritional quality of some wild leafy vegetables in South Africa. *International Journal of Food Science and Nutrition, 60*(5), 424–431.

AOAC. (1990). *Official methods of analysis* (14th ed.). Washington, DC: Association of Official Analytical Chemists.

AOAC. (1995). *Official methods of analysis* (15th ed.). Washington, DC: Association of Official Analytical Chemists.

AOAC. (2000). *Official methods of analysis*. Washington, DC: Association of Official Analytical Chemists.

AOCS. (2005). *Official procedure, Approved Procedure Am 5-04. Rapid determination of oil/fat utilizing high temperature solvent extraction*. Urbana, IL: American Oil Chemists Society.

Bamigboye, A. Y., Okafor, A. C., & Adepoju, O. T. (2010). Proximate and mineral composition of whole and dehulled Nigerian sesame seed. *African Journal of Food Science and Technology, 1*, 76–81.

Beluhan, S., & Ranogajec, A. (2010). Chemical composition and non-volatile components of Croatian wild edible mushrooms. *Food Chemistry, 124*, 1076–1082.

Duyff, R. (2002). American dietetic association's. *Complete Food and Nutrition Guide, 2*, 130–142.

Effiong, G. S., & Udo, I. F. (2010). Nutritive values of four indigenous wild fruits in southeastern Nigeria. *Electronic Journal of Environmental, Agricultural and Food Chemistry, 9*(7), 1168–1176.

Hussain, J., Rehman, N. U., Khan, A. L., Hamayun, M., Hussain, S. M., & Shinwari, Z. K. (2010). Proximate and nutrients evaluation of selected vegetables species from Kohat Region Pakistan. *Pakistan Journal of Botany, 42*(4), 2847–2855.

Ishida, H., Suzuno, H., Sugiyama, N., Innami, S., Todokoro, T., & Maekawa, A. (2000). Nutritional evaluation of chemical component of leaves, stalks and stems of sweet potatoes (*Ipomea batatas* poir). *Food Chemistry, 68*, 359–367.

Ladeji, O. Z., Okoye, S., & Ojobe, T. (1995). Chemical evaluation of the nutritive value of leaf of fluted pumpkin (*Telferia occidentalis*). *Food Chemistry, 53*, 353–355.

Meena, A. K., Bansal, P., Kumar, S., Rao, M. M., & Garg, V. K. (2010). Estimation of heavy metals in commonly used medicinal plants: A market basket survey. *Environmental Monitoring and Assessment, 170*, 657–660.

Mitchel, H. S., Rynbergen, S., Henderika, J., Andreson, L., & Dibble, M. V. (1976). *Nutrition in health and disease*. New York: JB Lipincott.

Pearson, D. (1976). *The chemical analysis of food* (7th ed., p. 142). New York: Churchill Livingston.

Roger, P., Elie, F., Rose, L., Martin, F., Jacop, S., Mercy, A. B., & Felicite, M. T. (2005). Methods of preparation and nutritional evaluation of dishes consumed in a malaria endemic zone in Cameroon (Ngali II). *African Journal of Biotechnology, 4*(3), 273–278.

Street, R. A., Kulkarni, M. G., Stirk, W. A., Southway, C., & Staden, V. J. (2008). Variation in heavy metals and microelements in South African medicinal plants obtained from street markets. *Food Additives and Contaminants, 25*(8), 953–960.

Tardio, J., Pardo-de-Santayana, M., & Morales, R. (2006). Ethnobotanical review of wild edible plants in Spain. *Botanical Journal of the Linnean Society, 152*, 27–71.

Wahab, O. A., Rjoob, E., Massadeh, A. M., & Omari, M. N. (2008). Evaluation of Pb, Cu, Zn, Cd, Ni and Fe levels in *Rosmarinus officinalis* labaiatae (Rosemary) medicinal plant and soils in selected zones in Jordan. *Environmental Monitoring and Assessment, 140*, 61–68.

Chapter 6
Metal Levels in Wild Edible Vegetables

6.1 Trace Metals

Trace metals are the elements that normally occur at very low levels in the environment and are required in very small amount to living organisms as their higher levels can be toxic. These metals are mostly released from rocks through natural process or human activities that lead to rise in their levels. Natural processes include breakdown of rocks, volcanic activities, and spreading of mid-ocean ridges, while human activities that release trace metals into the environment include mining, smelting, burning of coal/fuel, and wastewater disposal. The metals mining can release large amounts of trace metals into the surrounding environment. This often occurs due to improper disposal of contaminated waste or when a lot of dust from the mine sites blow around. Generally, human contributions of trace metals to the environment have been approaching, or even exceeding the natural inputs (Morel and Price 2003). Toxicity of trace metals depends on:

- Their levels in the environment.
- Where they are found in the environment (water, soil or air).
- What are their sources (anthropogenic or natural weathering).
- Acidic nature of the environment as trace metals are more problematic in acidic areas.
- Existence of the metals, either they exist by themselves or as part of larger chemical compounds.

The role of trace metals in human metabolism is of prime importance. Their deficiency causes diseases, whereas their presence in excess may result in toxicity to human life. Some trace metals are essential for chemical, biological, biochemical, metabolic, catabolic, and enzymatic activities taking place in the living cells of plants, animals, and human beings. Trace metals have an ability to accumulate in the vital human organs over prolonged period of time. Injury to vegetation caused by trace metals has been well recognized because of many botanical and chemical

A.M. Abbasi et al., *Wild Edible Vegetables of Lesser Himalayas*,
DOI 10.1007/978-3-319-09543-1_6

investigations during past 100 years. More than 60 elements in various parts of human body have been detected, among these at least 25 elements are essential to human health out of which 14 are termed as trace metals (Schwarz 1977).

6.2 Quantification of Selected Metals in the Vegetables

Determination of metal levels in the vegetables comprises two major steps:

- Collection and digestion of the samples
- Quantification of the metal levels in the samples

6.2.1 Collection and Digestion of the Samples

First step of metal quantification is the collection and digestion of vegetable samples. Fresh vegetables samples were collected in pre-cleaned polythene bags and transported to the laboratory. Prior to digestion each vegetable sample was carefully washed with distilled water and the edible parts were cut into small pieces and then oven dried at 70–80 ºC to attain constant weight (Arora et al. 2008; Tiwari et al. 2011; Yang et al. 2011). The dried samples were crumbled and crushed with a porcelain pestle and mortar and the resulting fine powder was sieved through muslin cloth. For each vegetable, three samples were accurately weighed (1.000 g each) in clean and dried conical flasks. Then powdered sample was digested in nitric acid (HNO_3), sulphuric acid (H_2SO_4), and perchloric acid ($HClO_4$) solution in a ration of (20:4:2) for 35–40 min on hot plate until the reddish brown fumes disappeared (Zafar et al. 2010; Jan et al. 2011) and a clear solution was obtained. The samples were cooled to room temperature and filtered through a fine (0.45 μm) filter paper. A blank digest containing all reagents except the samples was also carried out in the same way with every batch of eight samples. Each sample solution was diluted up to the mark (50 mL) with doubly distilled water and stored in a refrigerator before the analysis.

6.2.2 Quantification of the Metal Levels

The quantification of selected essential and toxic metals (Ca, Cd, Co, Cr, Cu, Fe, K, Li, Mg, Mn, Na, Pb, Sr, and Zn) in the vegetables samples was carried out using anatomic absorption spectrophotometer under optimum analytical conditions as shown in Table 6.1. Calibration line method was adopted for the measurement of metal levels in the samples. Samples were appropriately diluted whenever required. The reagents and standard solutions used during the investigation were of AAS

Table 6.1 Optimum analytical conditions for the metal analysis using air-acetylene flame on atomic absorption spectrophotometer

	Wavelength (nm)	HC lamp current (mA)	Slit width (nm)	Fuel-gas flow rate (L/min)	Limit of detection (μg/L)
Ca	422.7	6.0	0.5	2.0	4
Cd	228.8	4.0	0.3	1.8	4
Co	240.7	6.0	0.2	2.2	5
Cr	357.9	5.0	0.5	2.6	6
Cu	324.8	3.0	0.5	1.8	4
Fe	248.3	8.0	0.2	2.0	6
K	766.5	5.0	0.5	1.9	4
Li	670.7	4.0	0.5	1.6	3
Mg	285.2	4.0	0.5	1.6	1
Mn	279.5	5.0	0.4	1.9	3
Na	589.0	6.0	0.5	1.6	1
Pb	217.0	7.0	0.3	1.8	10
Sr	460.7	4.0	0.5	1.6	5
Zn	213.9	4.0	0.5	2.0	2

grade (>99.99 % purity). Standard stock solutions of 1,000 mg/L of the metals were used to prepare the working standards afresh on the day of analysis throughout this work. The accuracy of the method was evaluated using certified reference materials (NIST-SRM 1515 & 2709). During this study, inter-laboratory comparison of the data was also exercised at an independent laboratory, and normally, a maximum of ±2 % deviation was observed in the results of the two laboratories. All measurements were made in triplicate.

6.3 Concentration of Selected Metals in the Vegetables

The quantified metal levels (mg/kg, dry weight) in wild edible vegetables are shown in various tables and figures. Comparison of the average metal levels in *Amaranthus hubridus* (Fig. 6.1) revealed that Ca showed highest concentration, followed by K, Fe, Mg, Na, and Sr, while other metals exhibited relatively lower levels. In the case of *Amaranthus spinosus* again Ca showed highest contribution, followed by K, Na, Mg, Fe, and Sr (Fig. 6.2). Potassium showed highest concentration in the leaves of *Amaranthus viridis*, followed by Ca, Mg, Na, Fe, and Zn, while comparatively lower levels were observed for Sr, Cu, Li, Cr, Cd, and Pb (Fig. 6.3). Metal levels measured in the leaves of *Digera muricata* are shown in Fig. 6.4 which demonstrated that Ca showed highest concentration, followed by K, Mg, Fe, Na, Sr, Mn, and Zn.

Average metal contents in *Dryopteris ramosa* are mentioned in Fig. 6.5, which revealed highest concentration of K, followed by Ca, Na, Fe, Mg, Sr, Zn, and Mn,

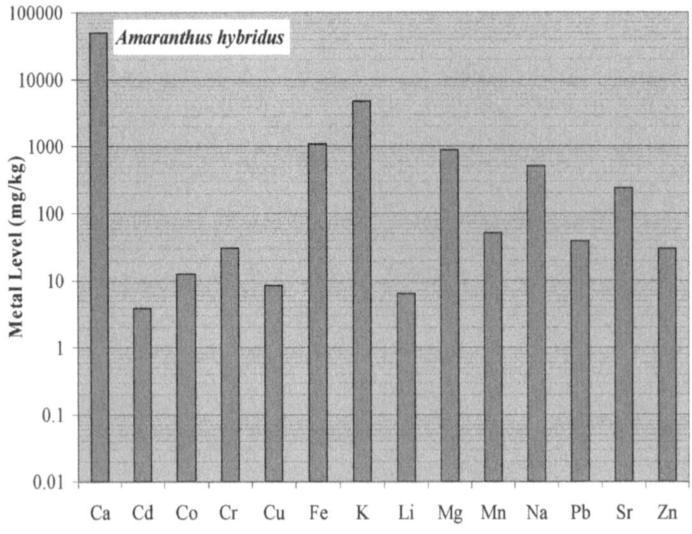

Fig. 6.1 *Amaranthus hybridus*

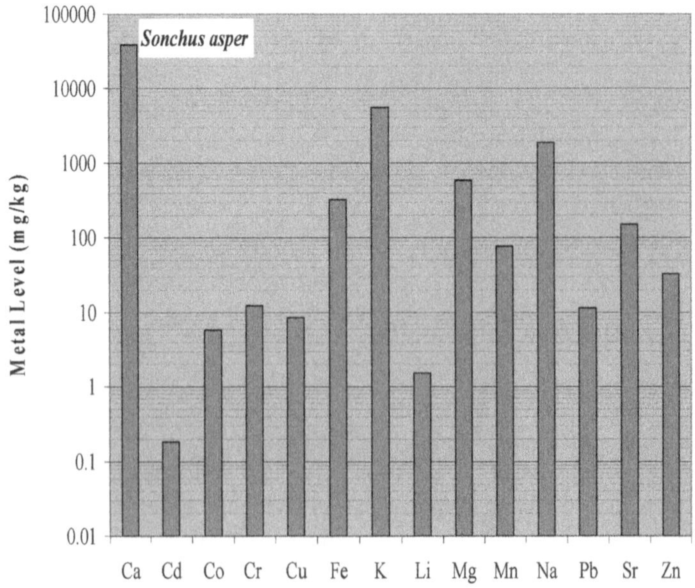

Fig. 6.2 *Amaranthus spinosus*

while relatively lower concentrations were recorded for Li, Cu, Co, Cr, and Cd. Measurement of metal levels in *Bidens bipinnata* indicated that maximum concentration was shown by Ca, followed by K, Mg, Fe, Na, Sr, Mn, and Zn, while other metals exhibited moderately lower levels (Fig. 6.6). Selected metal

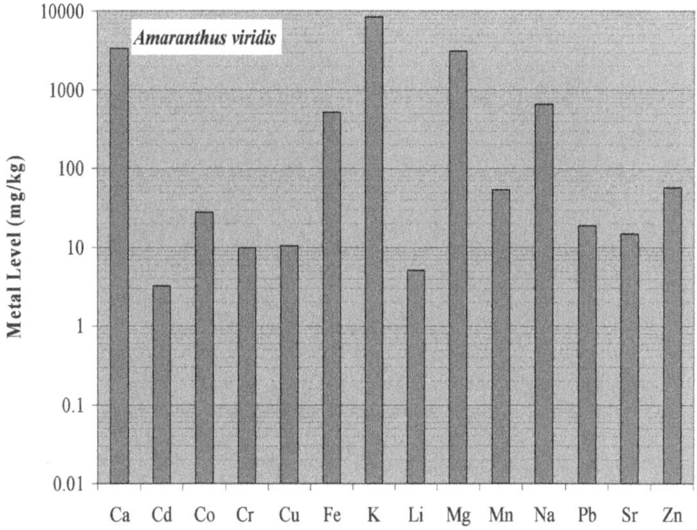

Fig. 6.3 *Amaranthus viridis*

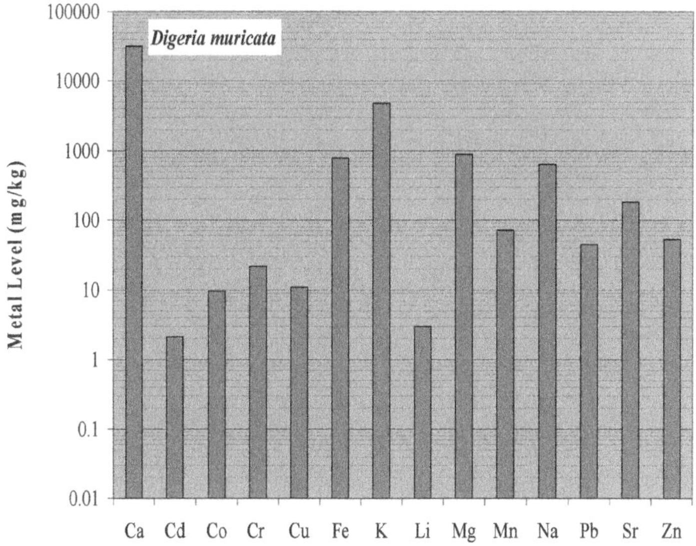

Fig. 6.4 *Digeria muricata*

levels in the leaves of *Centaurea calcitrapa* as shown in Fig. 6.7 indicated maximum concentration of Ca, followed by K, Na, Mg, Fe, Sr, Mn, and Zn, whereas somewhat lower levels were noted for Cd, Co, Cr, Cu, Li, and Pb. Measured levels of the metals in *Cichorium intybus* are given in Fig. 6.8, which exhibited highest

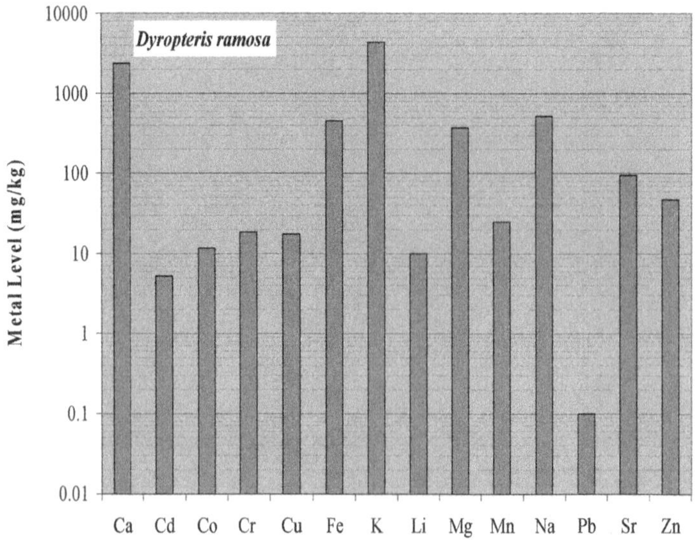

Fig. 6.5 *Dryopteris ramosa*

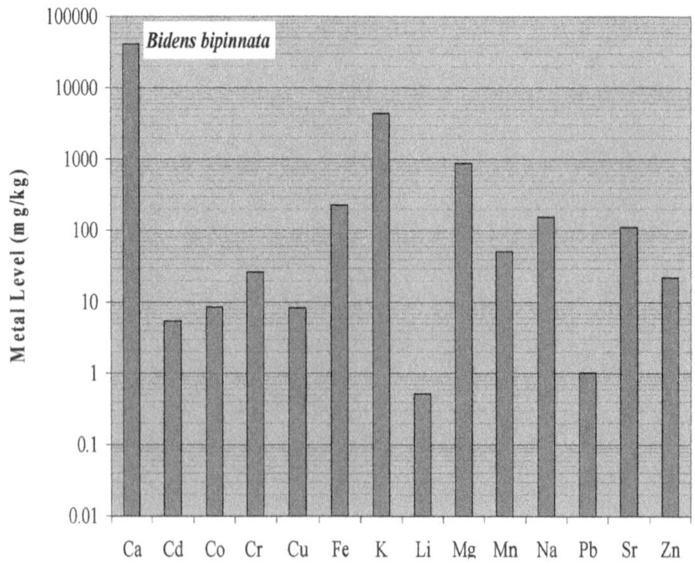

Fig. 6.6 *Bidens bipinnata*

concentration of K, followed by Ca, Fe, Mg, Na, Co, Zn, and Mn. Among rest of metals relatively lower levels were observed for Sr, Cu, Li Cr, and Pb and Cd. Elevated levels of Co and Mn in the leaves of *C. intybus* may exert adverse health effects to the consumers.

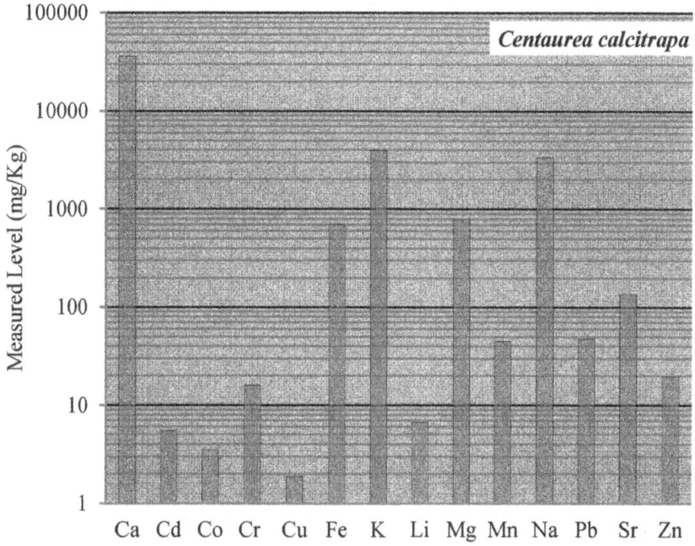

Fig. 6.7 *Centaurea calcitrapa*

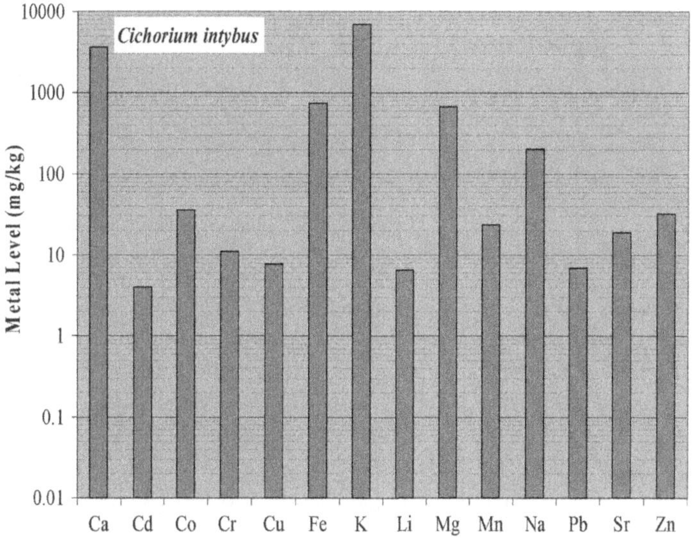

Fig. 6.8 *Cichorium intybus*

Average concentrations of selected metals measured in the leaves of *Launaea procumbens* are shown in Fig. 6.9, which indicated that Ca once again showed highest contribution, followed by K, Mg, Na, Fe, Sr, Zn, and Mn. However, comparatively lower concentrations were noted for Cu, Li, and Cr which were in the range of 2.970–13.91 mg/kg. On the other hand, concentration of Cd was found to be below the detection limit of instrument. Measurement of the selected metal

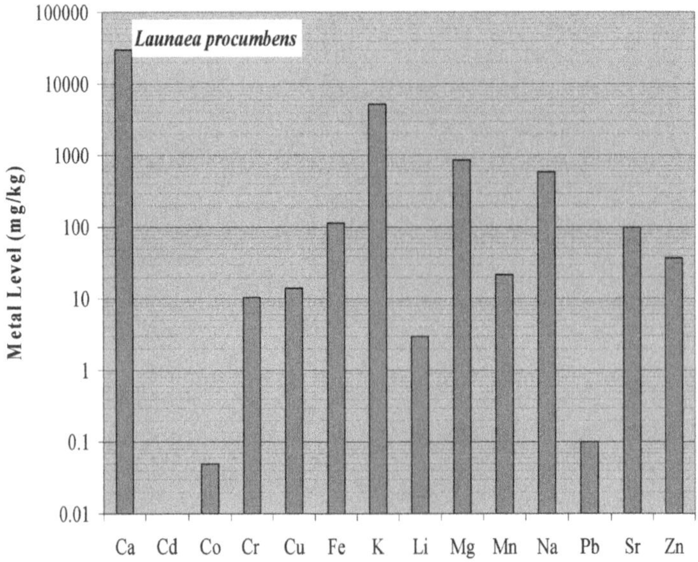

Fig. 6.9 *Launaea procumbens*

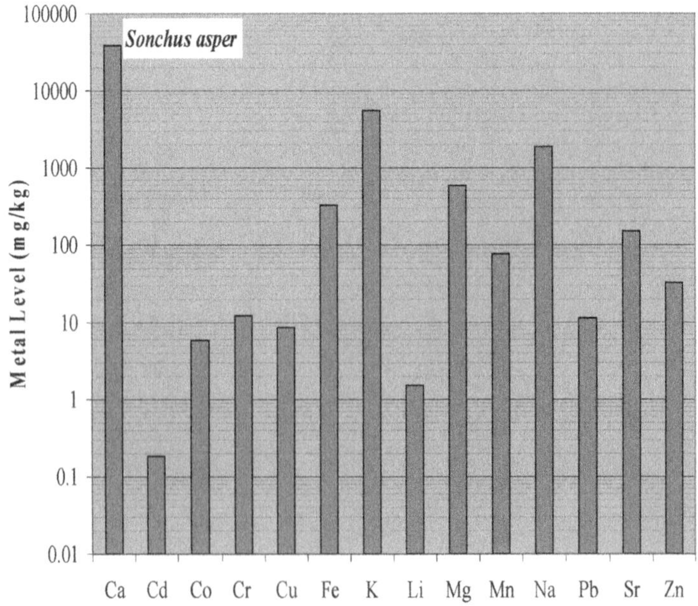

Fig. 6.10 *Sonchus asper*

levels in *Sonchus asper* revealed that among essential metals, Ca showed highest contents, followed by K, Na, Mg, Fe, Sr, Mn, and Zn. Among rest of the metals, Cu, Li, Co, Cr, and Cd concentrations were noted in the range of 1.527 mg/kg to 12.35 mg/kg (Fig. 6.10).

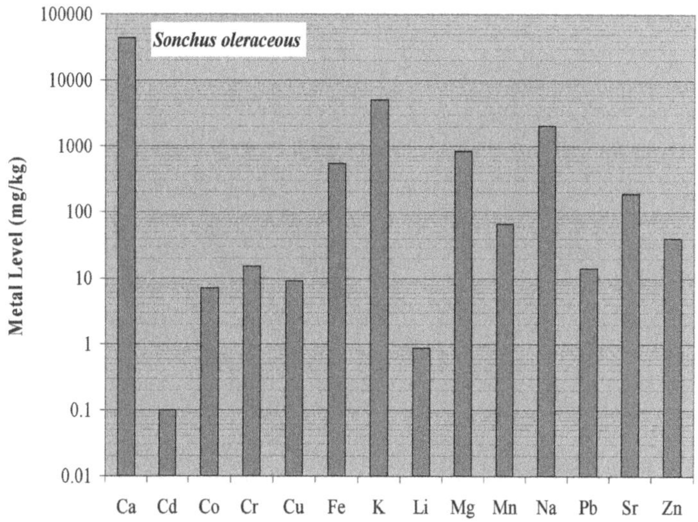

Fig. 6.11 *Sonchus oleraceous*

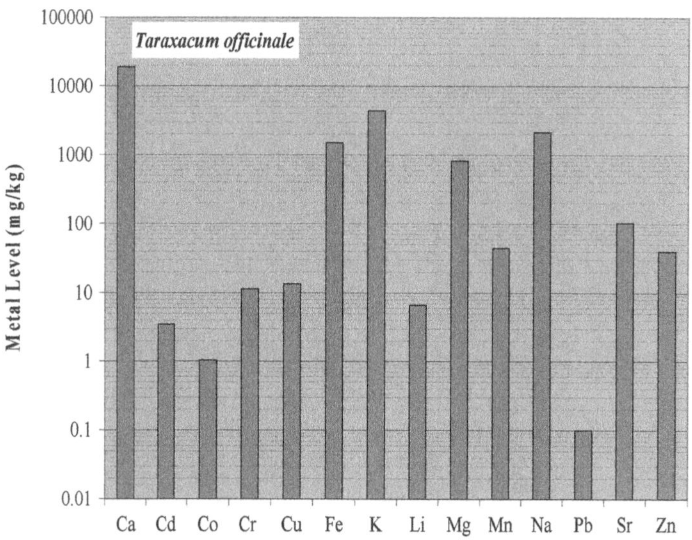

Fig. 6.12 *Taraxacum officinale*

Average concentrations of the metals in leaves of *Sonchus oleraceous* are shown in Fig. 6.11, which indicated maximum concentration of Ca, followed by K, Na, Mg, Fe, Sr, Mn, and Zn. Rest of the metals (Cu, Co, Cr, Pb, and Li) showed rather lower levels. Comparison of the metals levels in *Taraxacum officinale* leaves revealed that Ca exhibited highest level, followed by K, Na, Fe, Mg, Sr, Mn, and Zn. Relatively lower concentrations were recorded for Cu, Li, Cr, Co, Cd, and Pb (Fig. 6.12). Quantified metal levels in case of *Bombax malabaracum* are mentioned

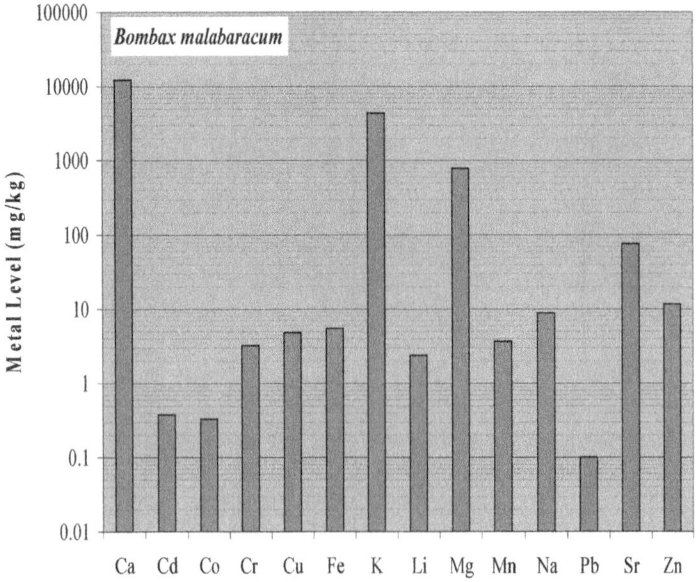

Fig. 6.13 *Bombax malabaracum*

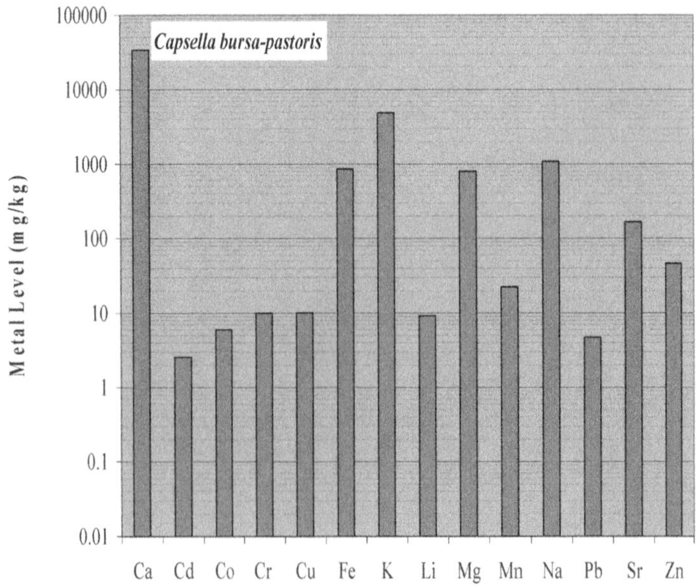

Fig. 6.14 *Capsella bursa-pastoris*

in Fig. 6.13, which indicated highest concentration of Ca, followed by K, Mg, Sr, Zn, Na, and Fe. Comparatively lower contributions were observed for Cu, Li, Cr Co, Cd, and Pb. Average concentrations of selected metals measured in the leaves of *Capsella bursa-pastoris* are given in Fig. 6.14, which showed the highest

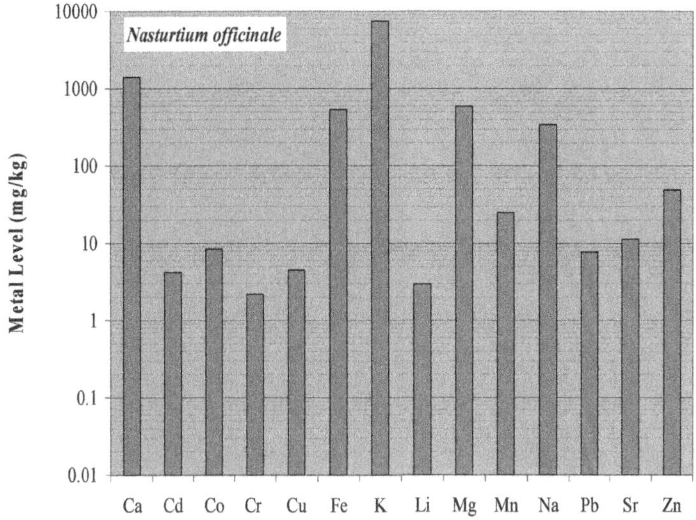

Fig. 6.15 *Nasturtium officinale*

contribution for Ca at 33,680 mg/kg, followed by K, Na, Fe, Mg, and Zn. Among rest of the metals, measured concentrations of Sr, Mn, Cu, Co, Li, Cr, Cd, and Pb were in the range of 2.549–165.1 mg/kg. Mean metal levels evaluated in the leaves of *Nasturtium officinale* are shown in Fig. 6.15. A comparative evaluation of the metals levels revealed highest contribution of K (7,490 mg/kg), followed by Ca, Mg, Fe, Na, Zn, and Mn. Fairly lower levels were noted for Sr, Cu, Co, Li, Cd, and Pb; however, lowest concentration was recorded for Cr (2.176 mg/kg).

Comparative average levels of the metals in flower buds of *Bauhinia variegata* are shown in Fig. 6.16, which indicated highest contents of K, followed by Ca, Mg, Na, Fe, Zn, and Mn. However, relatively lower levels were noted for Sr, Cu, Li, Cr, and Pb. Lowest concentration was noted for Cd, while it was observed that somewhat higher concentration of Co (29.10 mg/kg) may be associated with some health risk to the consumers. Mean metal levels measured in the flowering buds of *Bauhinia variegata* var. *candida* as mentioned in Fig. 6.17 showed significantly higher levels of K, followed by Ca, Mg, Na, Fe, and Sr. However, slightly lower concentrations were noted for Zn, Pb, Li, Cr, Cu, and Cd. Average concentrations of selected metals measured in the flowering buds of *Bauhinia variegata* var. *variegata* are given in Fig. 6.18, which demonstrated that again K showed highest concentration, followed by Ca, Mg, Na, Fe, Sr, and Zn. Moderately lower levels were noted for Cd, Co, Cr, and Pb. Comparison of the metals levels in leaves of *Silene conoidea* indicated highest concentration of K(5,838 mg/kg), followed by Ca, Mg, Fe, Na, Co, Mn, and Zn. Among other metals, Li showed the lowest levels (0.242 mg/kg), whereas Sr, Cu, Cd, Cr, and Pb concentrations were found to be in the range of 3.007–13.43 mg/kg as shown in Fig. 6.19.

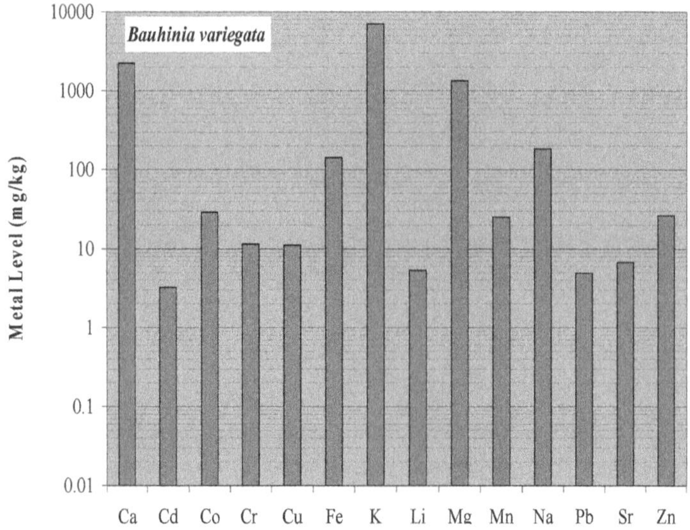

Fig. 6.16 *Bauhinia variegate*

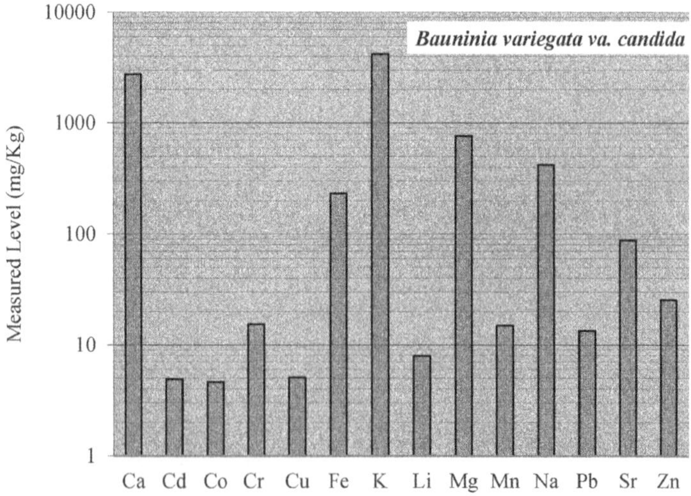

Fig. 6.17 *Bauhinia variegata* var. *candida*

Mean metal levels measured in *Stellaria media* are shown in Fig. 6.20 for comparative evaluation. An examination of the data given in figure showed highest concentration of K (8,600 mg/kg), followed by Ca, Mg, Na, Fe, Zn, and Mn. Among other metals, Co exhibited relatively higher levels while Pb showed lowest levels. However, the average concentrations of Sr, Cu, Li Cd, and Cr ranged from 4.459 to 8.640 mg/kg. Measured levels of the metals in leaves of *Chenopodium album* revealed highest contributions for K, followed by Ca, Mg, Fe, Na, Zn, and

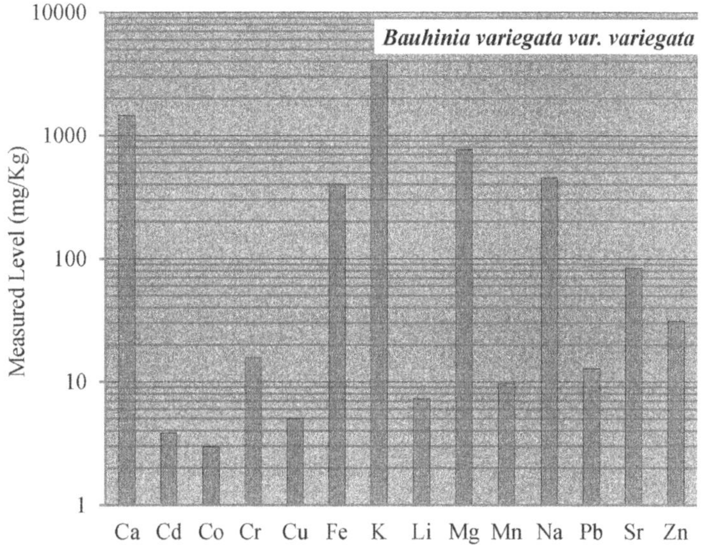

Fig. 6.18 *Bauhinia variegata* var. *variegate*

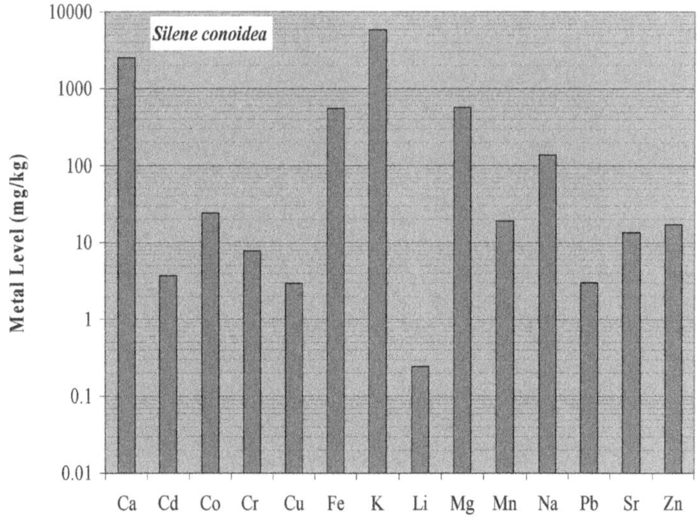

Fig. 6.19 *Silene conoidea*

Mn. Among rest of the metals, notable contributions were recorded for Pb while lowest levels were found for Cr. Mean levels of Sr, Cu, Co, Li, and Cd were found in the range of 3.030–5.132 mg/kg as shown in Fig. 6.21. Comparison of mean metal levels for *Commelina benghalensis* is mentioned in Fig. 6.22, which indicated highest concentration of Ca, followed by K, Mg, Fe, Na, Sr, Mn, and Zn.

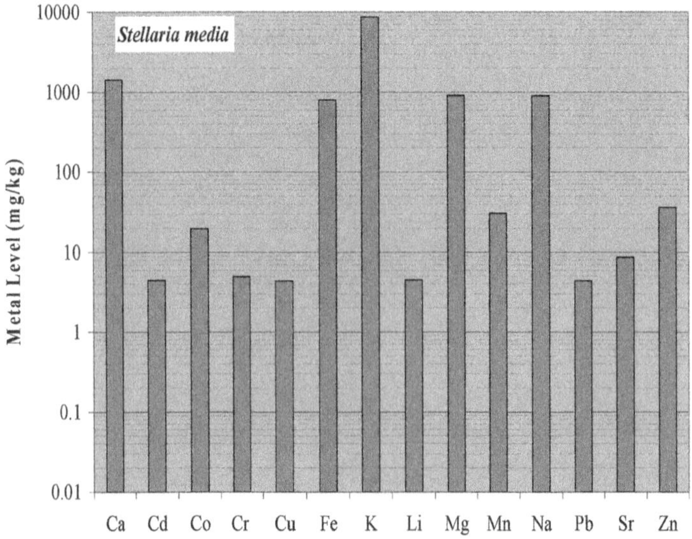

Fig. 6.20 *Stellaria media*

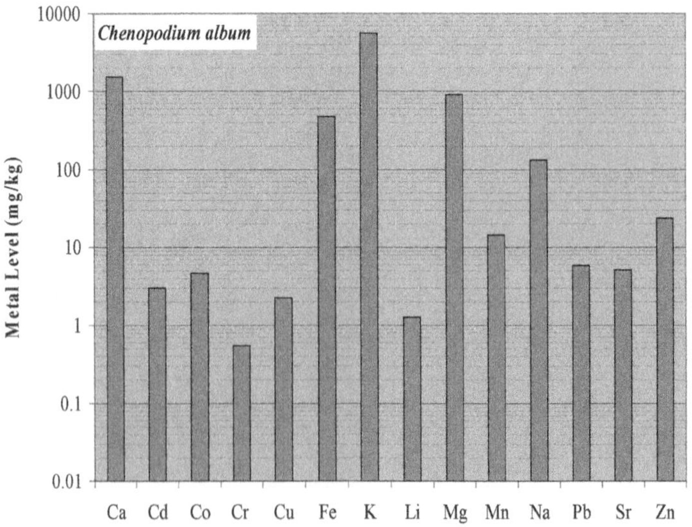

Fig. 6.21 *Chenopodium album*

Among other metals, Li was found at lowest levels, whereas moderately lower concentrations were observed for Cu, Co Cr, Cd, and Pb. Average metal levels in the leaves of *Evolvulus alsinoides* revealed highest concentration for Ca (27,066 mg/kg), followed by K, Fe, Mg, Na, Sr, Zn, and Mn, whereas Cu, Li, Co, Cr, and Cd levels were found to be comparatively lower (Fig. 6.23).

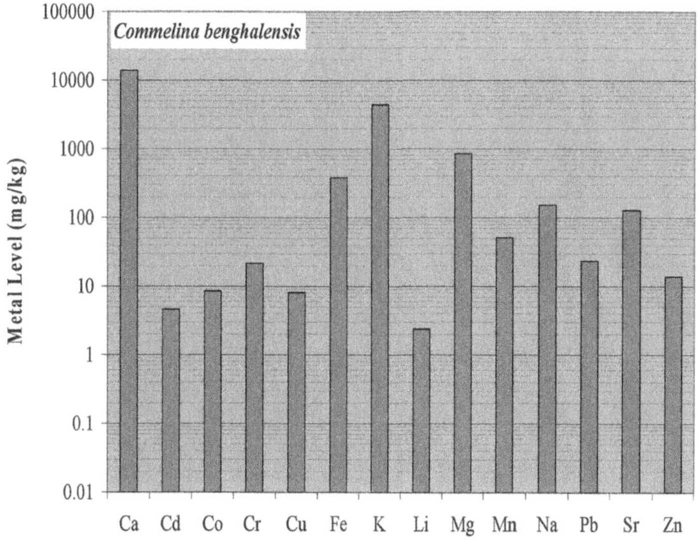

Fig. 6.22 *Commelina benghalensis*

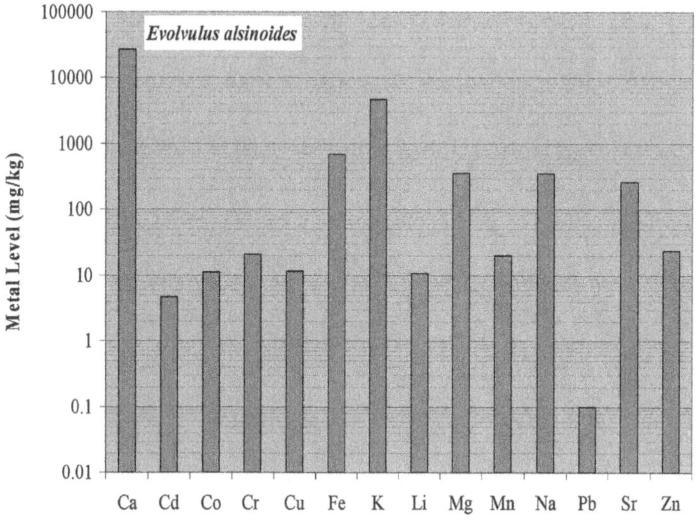

Fig. 6.23 *Evolvulus alsinoides*

Concentrations of selected metals measured in the tubers of *Dioscorea deltoidea* are shown in Fig. 6.24, which demonstrated highest concentration in favor of Ca, followed by Ka, Mg, Sr, Fe, Na, Mn, and Zn. Lowest level was recorded for Li (0.769 mg/kg), whereas Cu, Co, Cr, Cd, and Pb levels were found in the range of 5.288–31.97 mg/kg. Measurement of the metal levels, in leaves of *Lamium*

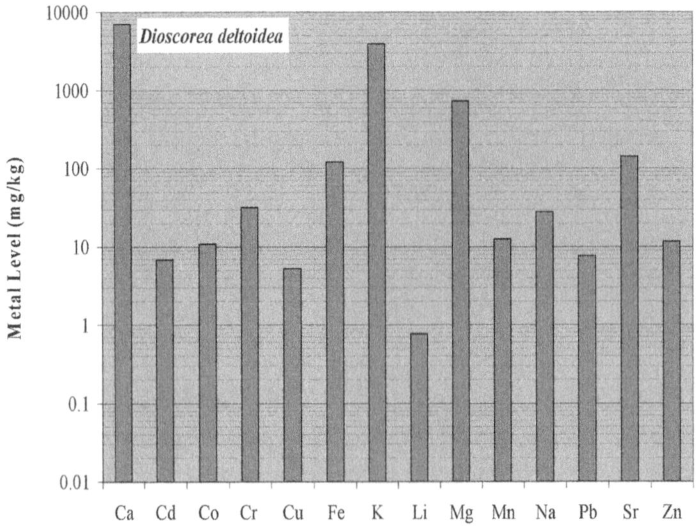

Fig. 6.24 *Dioscorea deltoidea*

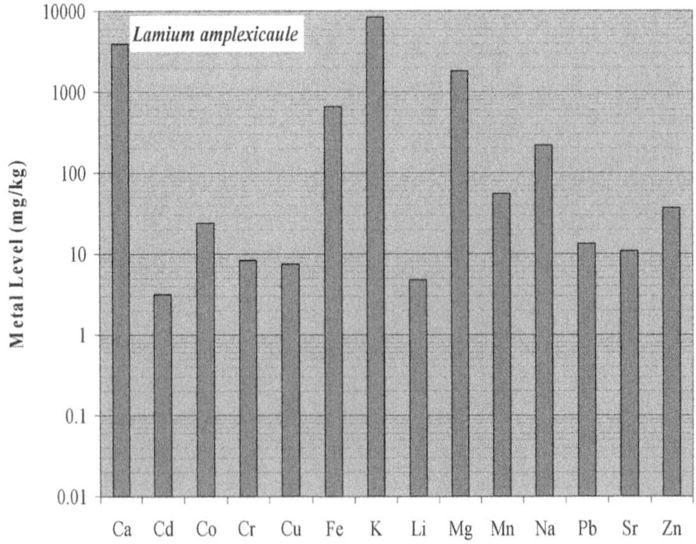

Fig. 6.25 *Lamium amplexicaule*

amplexicaule, revealed highest contribution for K (8,380 mg/kg) on dry weight basis, followed by Ca, Mg, Fe, Na, Mn, and Zn. Among rest of the metals notable contribution was recorded for Co (23.79 mg/kg), while lowest concentration was found for Cd (3.149 mg/kg). Mean concentrations of Sr, Cu, Li, Cr, and Pb levels were recorded in the range of 4.748 mg/kg to 13.28 mg/kg as shown in Fig. 6.25.

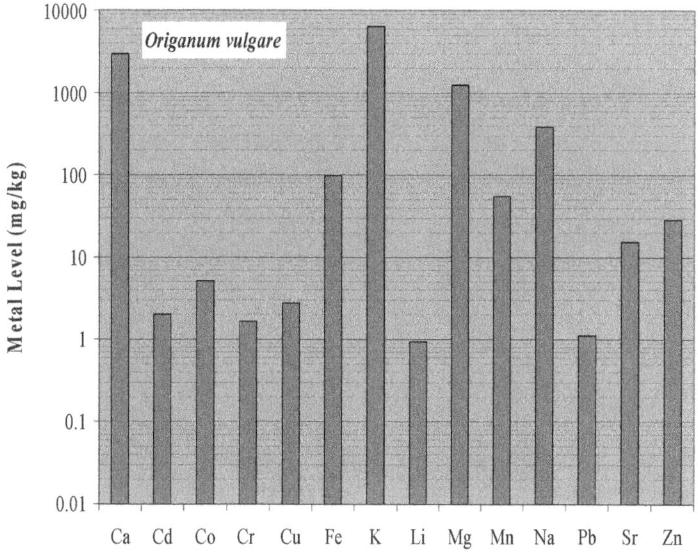

Fig. 6.26 *Origanum vulgare*

Comparison of the metal levels in leaves of *Origanum vulgare* on the average basis is shown in Fig. 6.26. As noted in most of the previous cases, highest contribution among the metals was shown by K, followed by significant contributions from Ca, Mg, Na, Fe, Mn, Sr, and Zn. Among rest of the metals, Li exhibited the lowest concentration (0.938 mg/kg), whereas moderately lower levels were recorded for Cu, Co, Cr, Cd, and Pb which were in the range of 1.116–5.134 mg/kg as shown in the figure. Overall, most of the essential metals revealed significantly higher contributions in the wild edible vegetables; nonetheless, elevated levels of some toxic metals may be of concern to the consumers.

Average metal levels in the leaves of *Tulipa stellata var. clusiana* Hk. f. are shown in Fig. 6.27 which indicated that K exhibited maximum concentration (3,888 mg/kg), followed by Ca, Mg, Fe, Na, Sr, Zn, and Mn. Among rest of the metals Cd and Pb showed rather lowest levels, while Cu, Co, Li, and Cr concentrations were found in the range of 2.525–8.267 mg/kg. Mean metal contents in leaves of *Malva parviflora* revealed highest contribution of K (7,705 mg/kg), followed by Ca, Mg, Na, Fe, Mn, and Zn. Among rest of the metals, highest concentration was noted for Sr (30.65 mg/kg) and lowest for Cd (3.838 mg/kg), whereas Cu, Co, Li, Cr, and Pb levels ranged from 3.930 to 17.36 mg/kg as shown in Fig. 6.28. Comparison of mean concentrations of selected metals in the leaves of *Ficus carica* is shown in Fig. 6.29, which indicated highest contribution for Ca (23,734 mg/kg), followed by K, Mg, Na, Fe, Sr, Mn, and Zn. Among rest of the metals Cu exhibited relatively lower concentration at 2.039 mg/kg, whereas moderately lower levels were observed for Co, Cd, Li, Cr, and Pb. Some of the metals (Cd, Cu, Co, and Li) exhibited almost comparable levels in the vegetable samples.

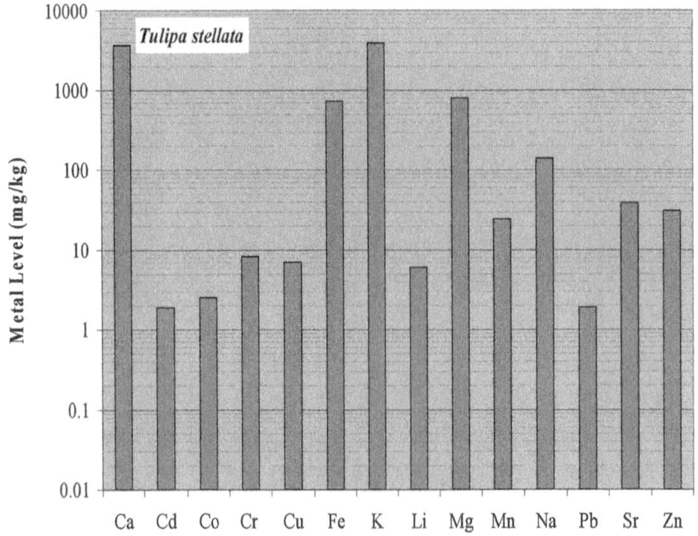

Fig. 6.27 *Tulipa stellata*

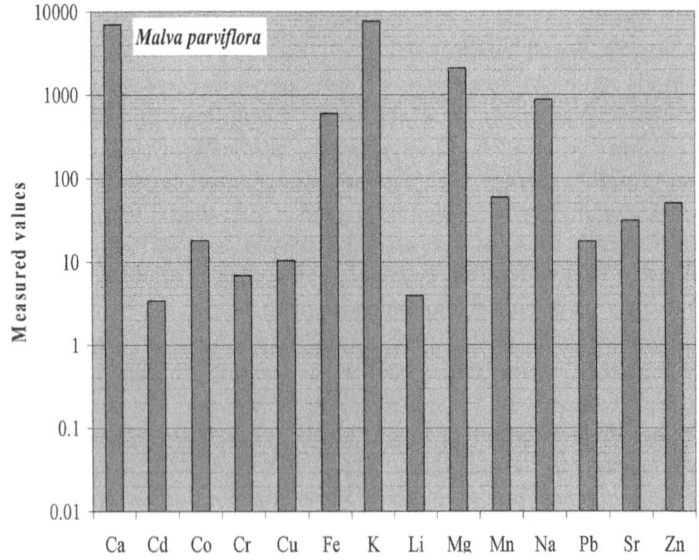

Fig. 6.28 *Malva parviflora*

Selected metal levels measured in the leaves of *Ficus palmata* are shown in Fig. 6.30, which again showed highest concentration for Ca, followed by K, Mg, Na, Fe, Mn, and Zn. Relatively higher level was noted for Sr (32.62 mg/kg), while moderately lower levels were observed for Li, Co Cr, Cd, and Pb. However, lowest concentration was recorded for Cu at 0.534 mg/kg.

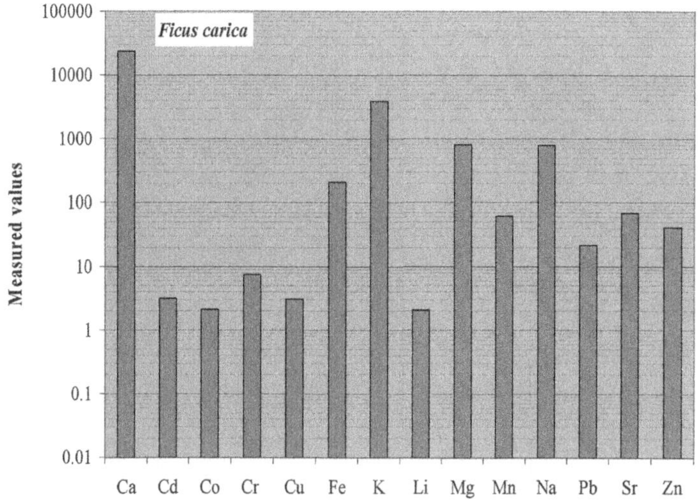

Fig. 6.29 *Ficus carica*

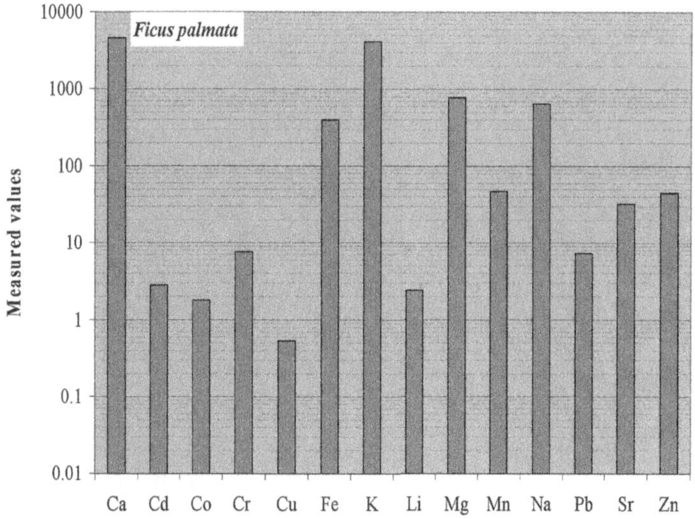

Fig. 6.30 *Ficus palmata*

Evaluation of average metal levels in the leaves of *Oxalis corniculata* revealed significantly elevated concentration of K (8,373 mg/kg), followed by Ca, Mg, Na, Fe, Mn, and Zn. Among rest of the metals, Sr showed slightly higher level (19.71 mg/kg) while Li exhibited lowest concentration (1.393 mg/kg) among all metals. The mean levels of Cu, Cr, Co, Cd, and Pb were found in the range of 2.838–9.236 mg/kg (Fig. 6.31). Mean metal levels in the leaves of *Lathyrus aphaca* indicated highest concentration for Ca (25,954 mg/kg), followed by K, Mg, Fe, Sr,

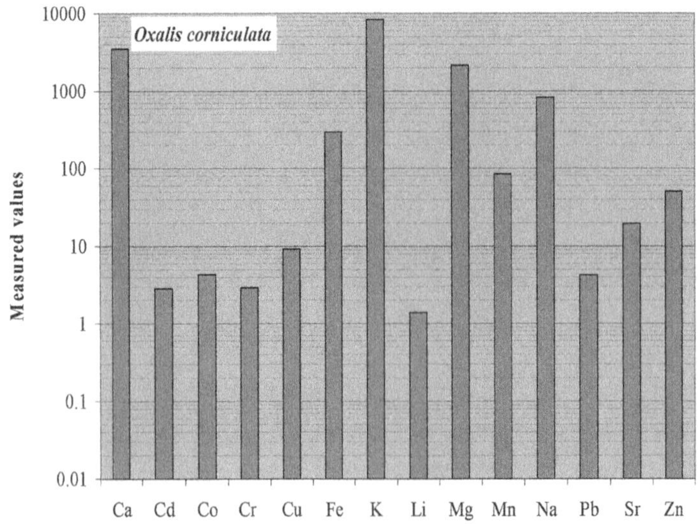

Fig. 6.31 *Oxalis corniculata*

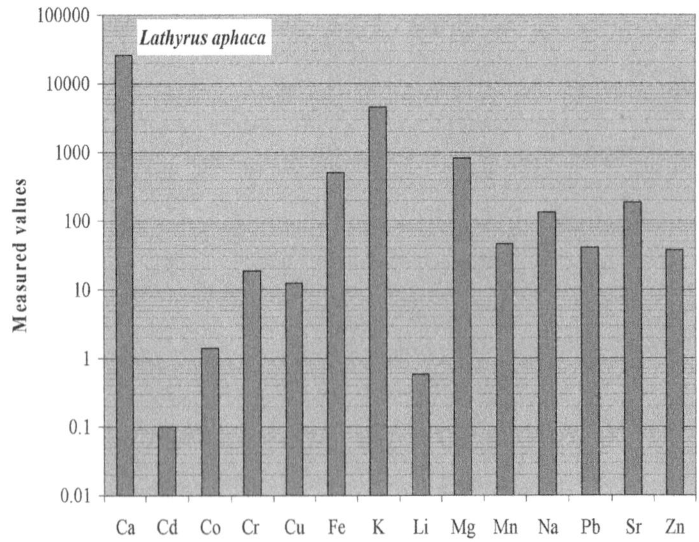

Fig. 6.32 *Lathyrus aphaca*

Na, Mn, and Zn as portrayed in Fig. 6.32. Among other metals, Cd was found at lowest level (0.101 mg/kg), while somewhat lower levels were observed for Cu, Co, Cr, Pb, and Cd. Average metal contents of *Medicago polymorpha* are displayed in Fig. 6.33. On the average basis, highest contribution was noted for K, followed by Ca, Mg, Na, Fe, Mn, and Zn. Moderately higher level was noted for Sr (21.36 mg/kg), while minimum concentration was recorded for Cd (2.745 mg/kg) which was

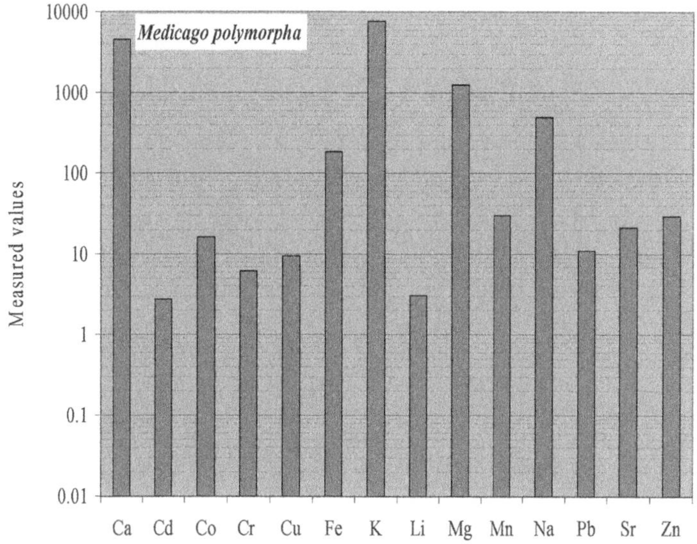

Fig. 6.33 *Medicago polymorpha*

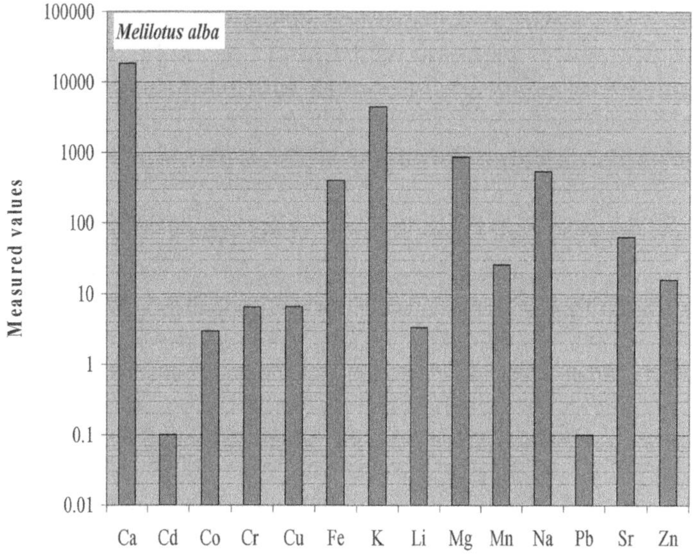

Fig. 6.34 *Melilotus alba*

somewhat comparable to that of Li. Mean levels of Cu, Co, Li, Cr, and Pb were found in the range of 3.044–11.03 mg/kg.

Concentrations of selected metals measured in the leaves of *Melilotus alba* are shown in Fig. 6.34, on the average basis. Comparison of the data revealed dominant

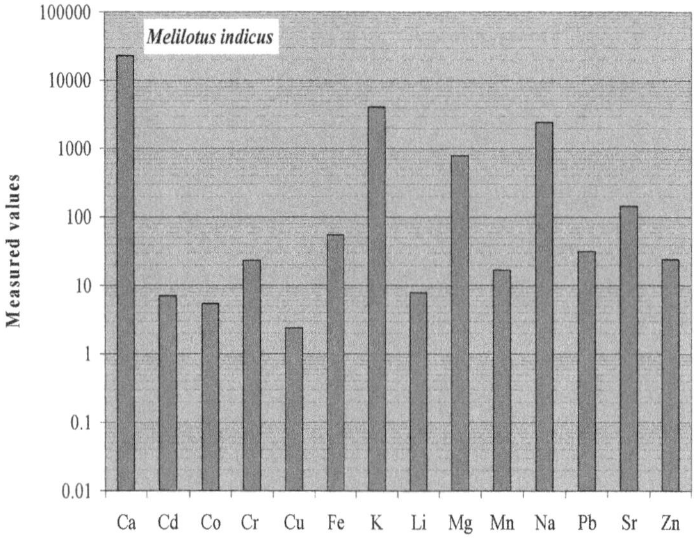

Fig. 6.35 *Melilotus indicus*

contribution of Ca (18,328 mg/kg), followed by K, Mg, Na, Fe, Sr, Mn, and Zn. Among other metals lowest contributions were shown by Cd and Pb, while slightly higher contents were observed for Cu, Cr, Li, and Co. Most of the toxic metals exhibited rather lower accumulation in this case. Comparison of average metal level in the leaves of *Melilotus indicus* (Fig. 6.35) demonstrated highest concentration of Ca (22,892 mg/kg), followed by K, Na, Mg, Sr, Fe, Zn, and Mn. Among rest of the metals Cu showed lowest contribution (2.379 mg/kg), whereas mean concentrations of Li, Cr, Co, Cd, and Pb were found in the range of 5.388–31.75 mg/kg.

Average concentrations of the metals measured in *Vicia faba* are shown in Fig. 6.36 which once again revealed highest contribution of Ca (5,388 mg/kg), followed by K, Mg, Na, Fe, Mn, and Zn. Among rest of the metals, Sr exhibited comparatively higher level (37.33 mg/kg), while lowest contents were observed for Co (2.079 mg/kg). Moderately lower levels were recorded for Cu, Li, Cr, Cd, and Pb. Comparison of selected metals in *Vicia sativa* revealed overwhelmingly higher concentration of Ca, on the average basis, followed by K, Na, Mg, Sr, Fe, Mn, and Zn as depicted in Fig. 6.37. Lowest level among the metals was noted for Cd at 0.962 mg/kg, whereas mean levels of Cu, Li, Co, Cr, and Pb were found in the range of 5.577–11.88 mg/kg.

Mean levels of selected metals measured in *Plantago lanceolata* are shown in Fig. 6.38. Potassium exhibited highest concentration among the selected metals (7,700 mg/kg, on dry weight basis), followed by Ca, Mg, Na, Fe, Zn, Co, and Mn, whereas comparatively lower concentrations were observed for Cu, Cr, Li, Cd, Pb, and Sr. Mean metal levels in the leaves of *Persicaria amplexicaulis* are mentioned in Fig. 6.39, which demonstrated highest contribution for Ca (14,233 mg/kg),

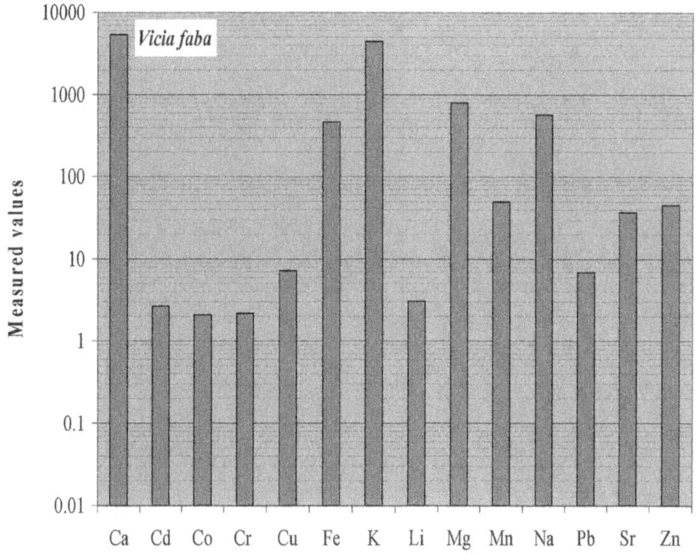

Fig. 6.36 *Vicia faba*

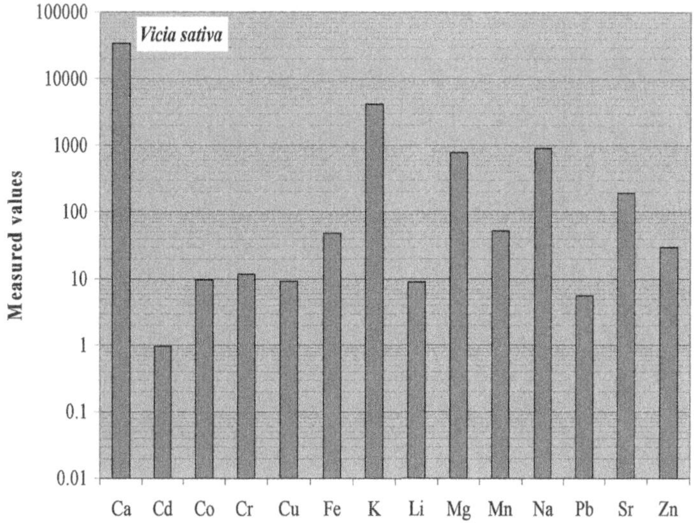

Fig. 6.37 *Vicia sativa*

followed by K, Fe, Mg, Mn, Na, and Zn. Among rest of the metals, notable contribution was observed for Sr (44.50 mg/kg), while relatively lower levels were recorded for Cu, Cr, Co, Cd, and Pb. The lowest contribution was noted for Li (0.589 mg/kg).

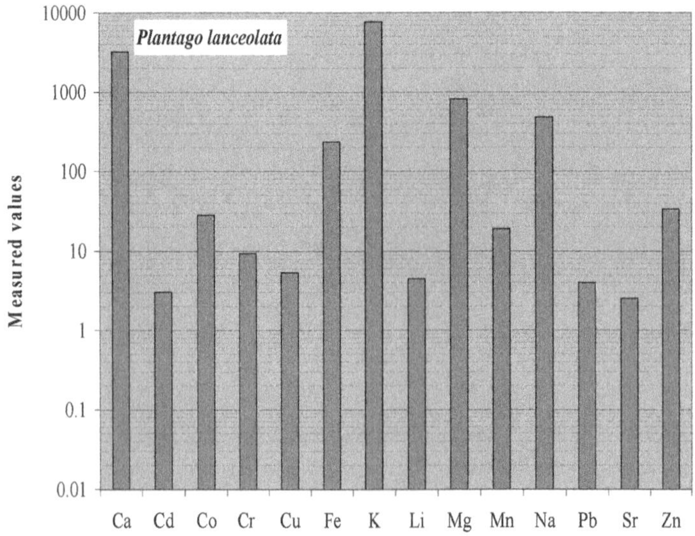

Fig. 6.38 *Plantago lanceolata*

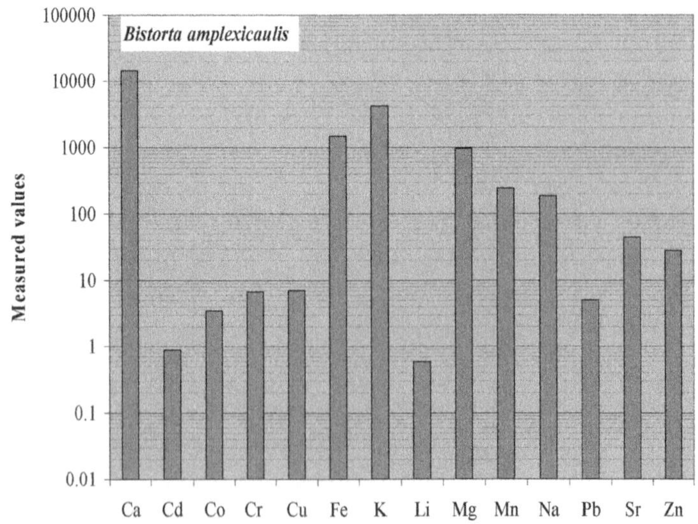

Fig. 6.39 *Persicaria amplexicaulis*

Comparison of selected metal levels in the leaves of *Polygonum aviculare* indicated highest concentration for Ca (20,473 mg/kg), followed by K, Mg, Fe, Na, Zn, and Mn as mentioned in Fig. 6.40. Among other metals, Sr showed comparatively higher concentration, while Cd exhibited lowest contents (0.882 mg/kg). The average concentrations of Cu, Li, Cr, Co, and Pb in the

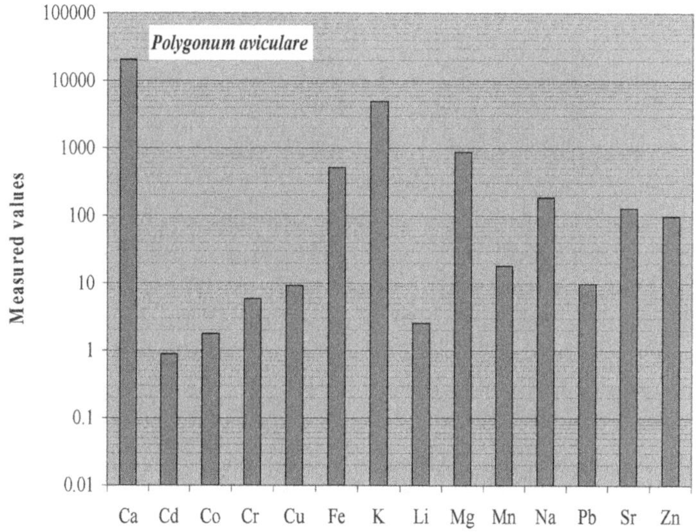

Fig. 6.40 *Polygonum aviculare*

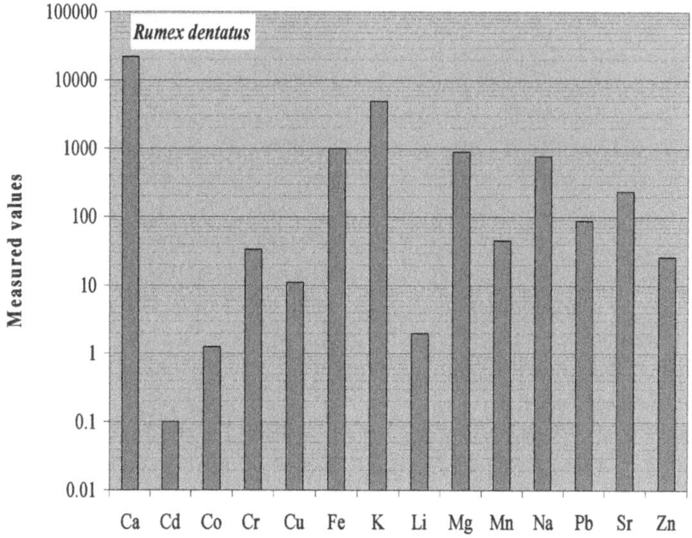

Fig. 6.41 *Rumex dentatus*

vegetable samples varied from 1.765 to 9.118 mg/kg as shown in the figure. More or less comparable levels (about 9 mg/kg) were noted for Pb and Cu among the metals.

Comparison of average concentrations of the metals measured in the leaves of *Rumex dentatus* is shown in Fig. 6.41 which revealed highest contribution for Ca at

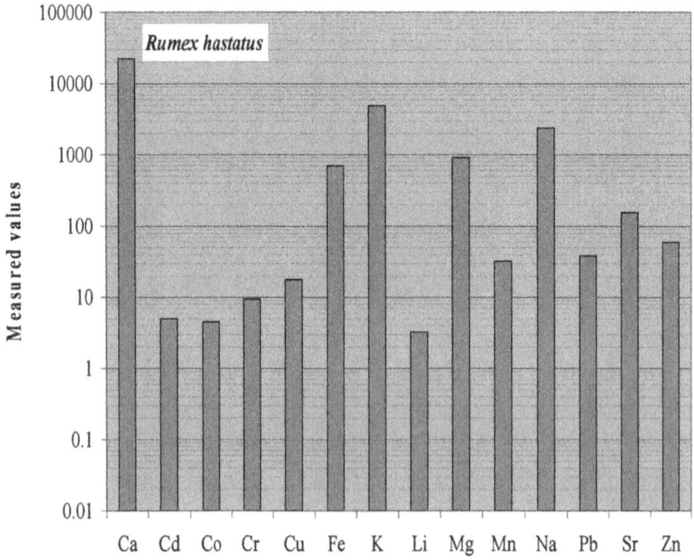

Fig. 6.42 *Rumex hastatus*

22,122 mg/kg, followed by K, Fe, Mg, Na, Sr, and Pb. Among rest of the metals, Mn exhibited relatively higher level, almost comparable to those of Zn and Cr, while least contribution was noted for Cd. Moderately lower contributions were recorded for Cu, Li, and Co. Average metal levels measured in *Rumex hastatus* are given in Fig. 6.42, which demonstrated highest concentration of Ca (22,328 mg/kg), followed by K, Na, Mg, Fe, and Sr. Moderately higher levels were noted for Zn, Pb, and Mn, while lowest concentration was observed for Li (3.200 mg/kg). Mean levels of Cu, Co, Cr and Cd were found in relatively lower range. Selected metal levels measured in *Rumex nepalensis* are depicted in Fig. 6.43, which showed overwhelmingly elevated levels for Ca than other selected metals. Somewhat higher concentrations were noted for K, Na, Mg, Fe, Sr, Mn, Pb, and Zn in the vegetable samples; however, rather lower levels were observed for Cd, Co, Cr, Cu, and Li, on average basis.

Average metal concentrations measured in the leaves of *Portulaca oleracea* are shown in Fig. 6.44. Once again dominant contribution was shown by Ca, followed by K, Mg, Fe, Na, Mn, Sr, Pb, and Zn, whereas comparatively lower concentrations were observed for Cd, Co, Cr, Cu, and Li. Average contributions of the metals in the edible parts of *Portulaca quadrifida* indicated highest concentration of Ca at 35,601 mg/kg, followed by K, Mg, Fe, Na, Mn, Sr, Pb, and Zn as shown in Fig. 6.45. Among other metals, Li concentration was found lowest, whereas Cu, Cr, Co, and Cd concentrations were found to be in the range of 2.324–23.21 mg/kg. Measured levels of selected metals in the edible parts of *Gallium aparine* exhibited maximum concentration of Ca (27,070 mg/kg), followed by K, Fe, Mg, Na, and Sr. Moderately higher levels were observed for Pb, Mn, and Zn, followed by Cr;

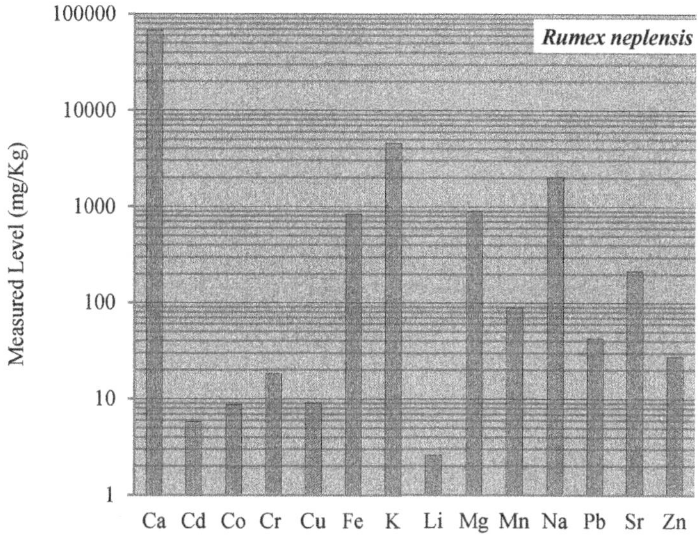

Fig. 6.43 *Rumex nepalensis*

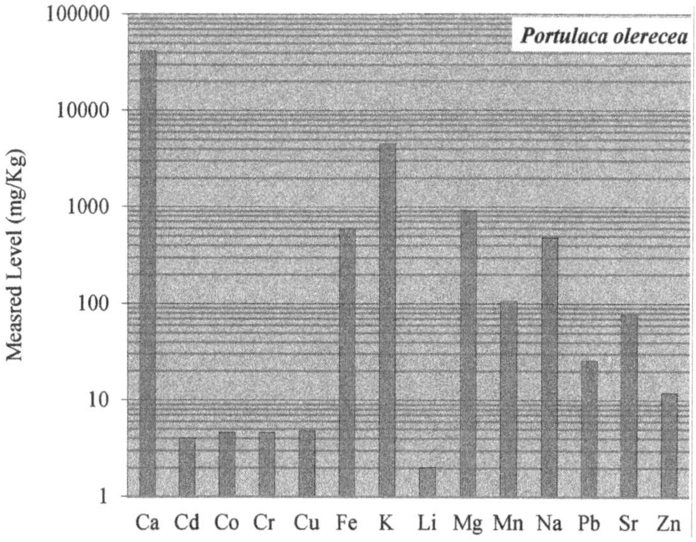

Fig. 6.44 *Portulaca oleracia*

however, relatively lower contributions were noted for Cr, Cu, Li, Co, and Cd as portrayed in Fig. 6.46.

Comparison of mean metal levels in the leaves of *Veronica arvensis* revealed maximum concentration of K (6,798 mg/kg), followed by Ca, Fe, Mg, Na, Zn, Co, Sr, and Mn. Other metals (Pb, Cd, Cr, Cu, and Li) exhibited somewhat comparable

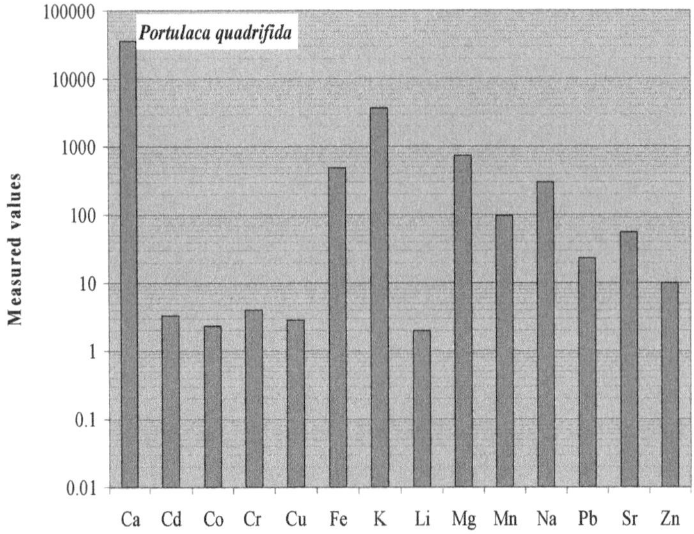

Fig. 6.45 *Portulaca quadrifida*

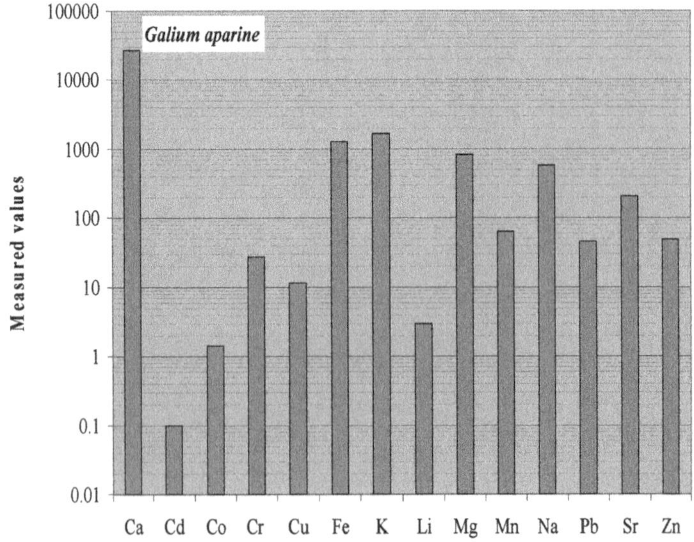

Fig. 6.46 *Gallium aparine*

concentrations as shown in Fig. 6.47. Average concentrations of the metals in *Solanum nigrum* indicated significantly higher levels (>1,000 mg/kg) for K, Na, Ca, and Mg (Fig. 6.48). Moderately higher levels were observed for Fe and Zn, followed by Mn, Sr, Co, Cr, Cu, and Pb. Lowest concentrations were noted for Cd and Li in the vegetable samples.

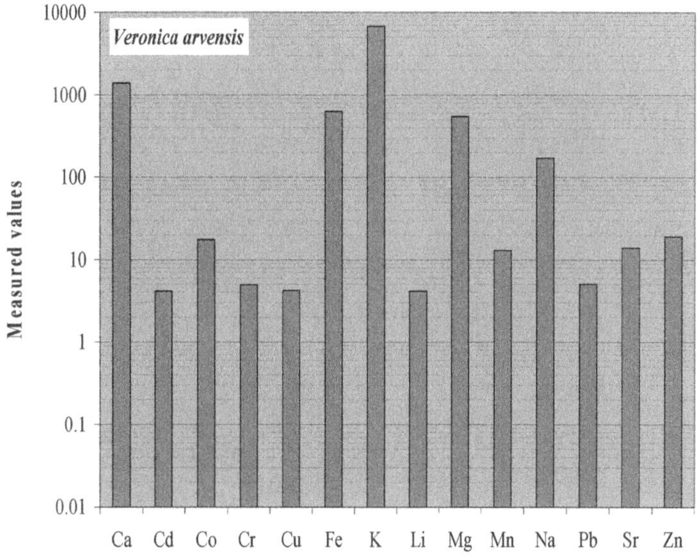

Fig. 6.47 *Veronica arvensis*

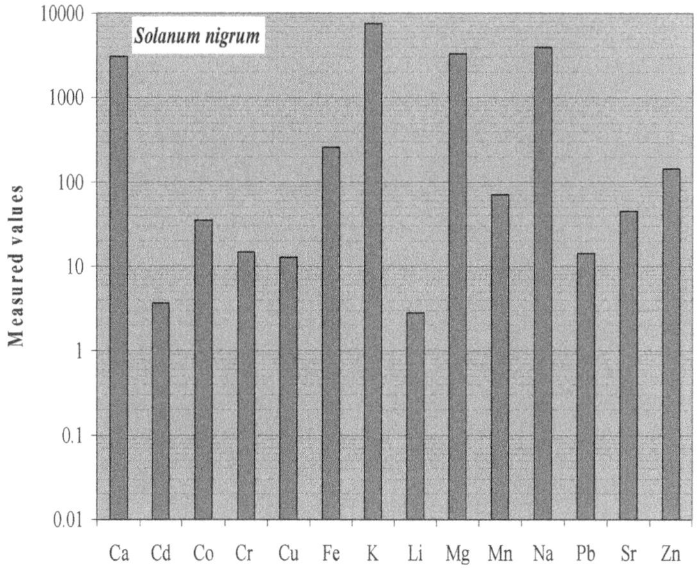

Fig. 6.48 *Solanum nigrum*

Comparison of average metal levels measured in *Pimpinella diversifolia* is shown in Fig. 6.49. Once more, highest contribution was observed for K at 6,329 mg/kg, followed by Ca and Fe. Somewhat higher levels were noted for Mg and Na (>100 mg/kg), while relatively lower concentrations were recorded for Co,

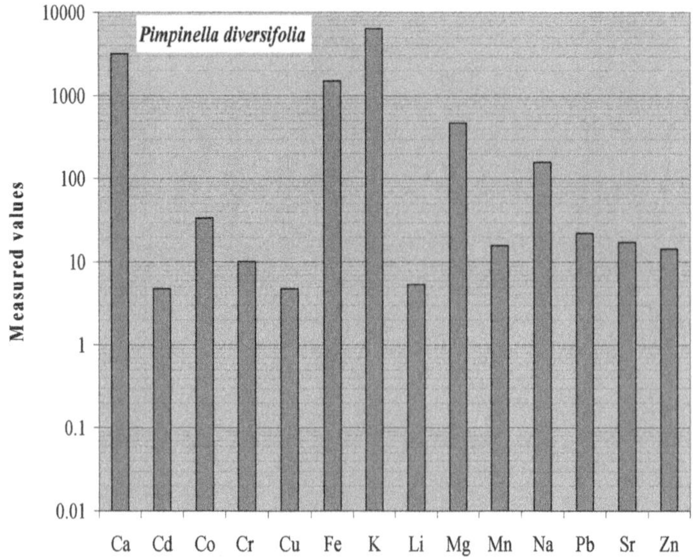

Fig. 6.49 *Pimpinella diversifolia*

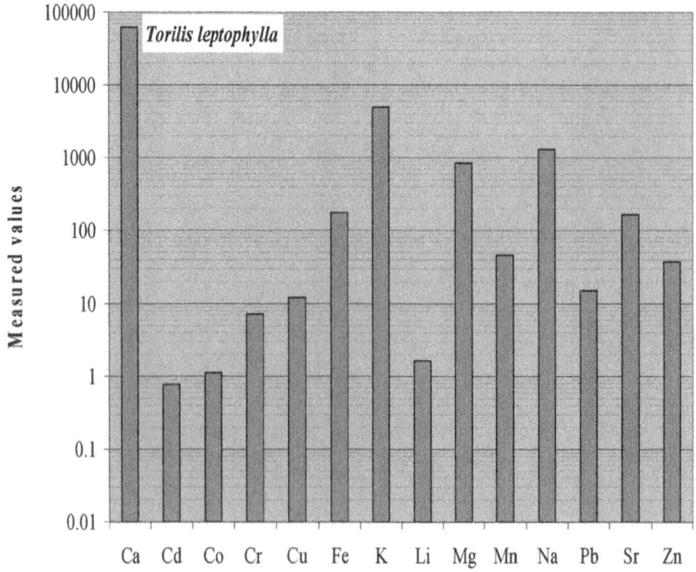

Fig. 6.50 *Torilis leptophylla*

Pb, Sr, Zn, Mn, and Cr. Approximately comparable average concentrations were shown by Cu, Cd, and Li, which were lowest among the metals. Average concentrations of the metals in *Torilis leptophylla* revealed highest concentration for Ca (62,827 mg/kg), followed by K, Na, Mg, Fe, Sr, Mn, and Zn as illustrated in Fig. 6.50. Among rest of the metals, notable contributions were recorded for Pb,

Cu, and Cr, whereas lowest concentration was observed for Cd in the vegetable samples. Overall, most of the essential metals (Ca, Mg, K, Na, Fe, Zn, etc.) revealed significantly higher concentrations in various wild edible vegetables, indicating their nutritional importance in the study area. Nevertheless, some carcinogenic metals such as Pb and Cd also exhibited considerably higher levels in some vegetables, which might be associated with adverse health effects to the humans. This aspect of the study would be explored in detail in subsequent sections by evaluating health risk assessment.

6.4 Comparative Evaluation of Metal Contents in the Vegetables

Foregoing discussion revealed that the vegetables are good sources of essential minerals. Average concentrations of each metal would be compared in the wild edible vegetables in this section. Calcium is considered as an essential structural component and in majority of industrialized countries dairy products provide 50–80 % of dietary Ca, while plant-based foods contribute about 25 % of the requirement (Hussain et al. 2011). In the present study, highest concentration of Ca was found in the leaves of *Rumex neplensis* (67,486 mg/kg), followed by *Torilis leptophylla, Amaranthus hybridus, Sonchus oleraceous,* and *Partulica oleracea*, while relatively lower concentrations were measured in the edible parts of *Veronica arvensis Nasturtium officinale, Bauhinia variegata var. variegata, Silene conoidea, Stellaria media,* and *Chenopodium* as shown in Fig. 6.51. For rest of the vegetables, Ca levels were found in the range of 1,417–50,148 mg/kg. Present findings were in good agreement with the reported levels of Ca in the edible parts of *Amaranthus hybridus, Amaranthus viridis, Amaranthus spinosus, Chenopodium album, Cichorium intybus, Nasturtium officinale, Pimpinella* spp., *Portulaca oleracea, Rumex hastatus, Rumex dentatus, Solanum nigrum, Sonchus asper, Sonchus oleraceous* and *Stellaria media* from various parts of the world (Jan et al. 2011; Hussain et al. 2009, 2011; Bahadur et al. 2011; Jimoh et al. 2011; Seal 2011; Aberoumand 2009; Hameed et al. 2008; Akubugwo et al. 2007; Imran et al. 2007; Odhav et al. 2007; Ozcan and Akbulut 2007; Trichopoulou et al. 2000; Wilman and Derrick 1994).

Potassium plays significant role in the activation of some enzymes, particularly co-enzyme for normal growth and muscle function (Birch and Padgham 1994). Highest concentration of K in this study was recorded in *Stellaria media* (8,600 mg/kg), followed by *Amaranthus viridis, Oxalis corniculata, Lamium amplexicaule, Malva parviflora, Medicago polymorpha,* and *Plantago lanceolata*. However, comparatively lower levels of K were noted in the edible parts of *Galium aparine, Portulaca quadrifida, Tulipa stellata, Ficus carica,* and *Dioscorea deltoidea* (Fig. 6.52). Previous studies indicated that in different wild edible vegetables (*Amaranthus hybridus, Amaranthus viridis, Nasturtium officinale, Cichorium*

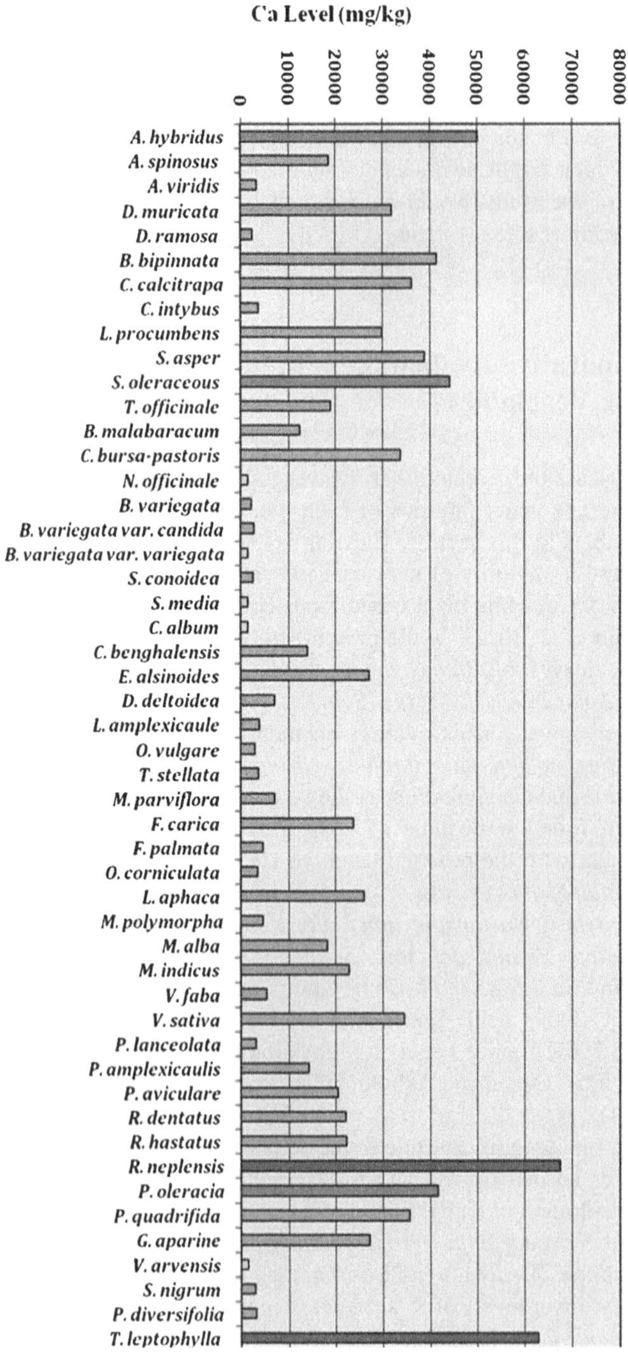

Fig. 6.51 Calcium level (mg/kg) in wild edible vegetables

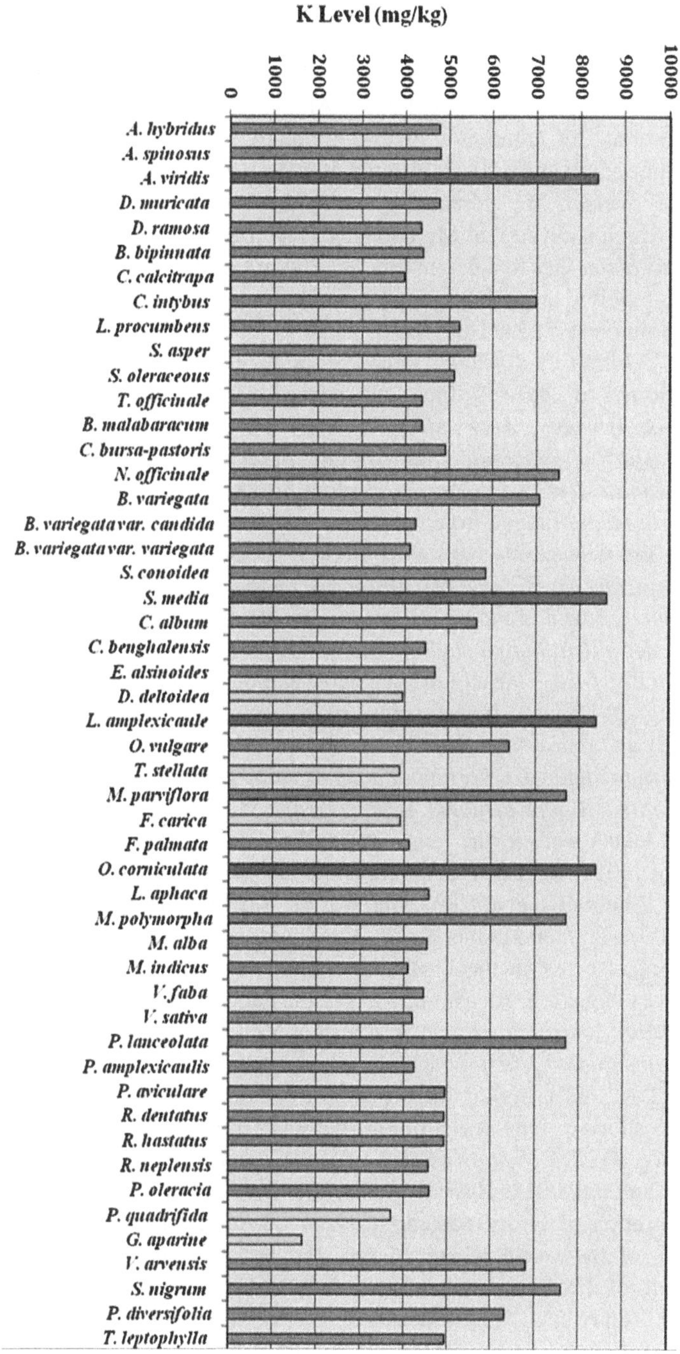

Fig. 6.52 Potassium level (mg/kg) in wild edible vegetables

intybus, Pimpinella spp., *Portulaca oleracea, Rumex hastatus, R. dentatus, Sonchus asper, Sonchus oleraceous* and *Stelleria media*) K levels were found ranging from 23.66 to 34,600 mg/kg (Jan et al. 2011; Hussain et al. 2009, 2011; Bahadur et al. 2011; Jimoh et al. 2011; Seal 2011; Aberoumand 2009; Hameed et al. 2008; Akubugwo et al. 2007; Imran et al. 2007; Odhav et al. 2007; Ozcan and Akbulut 2007; Trichopoulou et al. 2000; Wilman and Derrick 1994).

Average levels of Mg estimated in the wild edible vegetables are mentioned in Fig. 6.53, which showed that Mg contents ranged from a minimum of 360 mg/kg in *Evolvulus alsinoides* to a maximum of 3,294 mg/kg in *Solanum nigrum*. Numerous studies reported around the world (Bahadur et al. 2011; Jimoh et al. 2011; Seal 2011; Aberoumand 2009; Hussain et al. 2009; Hameed et al. 2008; Akubugwo et al. 2007; Imran et al. 2007; Odhav et al. 2007; Ozcan and Akbulut 2007; Trichopoulou et al. 2000; Wilman and Derrick 1994) showed that Mg levels in *Amaranthus hybridus, Amaranthus viridis, Amarantrhus spinosus, Cichorium intybus, Nasturtium officinale, Medicago denticulata, Portulaca oleracea, Rumex hastatus, Rumex dentatus, Solanum nigrum, Sonchus asper, Sonchus oleraceous,* and *Stellaria media* ranged from 6.100 to 18,420 mg/kg.

Among the wild edible vegetables highest mean level of Na was observed in *Solanum nigrum* (3,982 mg/kg), followed by *Centaurea calcitrapa, Melilotus indicus, Rumex hastatus* and lowest levels were observed in *Bombax malabaracum, Dioscorea deltoidea, Tulipa stellata, Lathyrus aphaca,* and *Pimpinella diversifolia* as shown in Fig. 6.54. In rest of the vegetable samples, Na levels ranged from 28.07 to 2,426 mg/kg. Previous investigations revealed that in various wild edible vegetables, such as, *Amaranthus hybridus, Amaranthus viridis, Amaranthus spinosus, Chenopodioum album, Cichorium intubus, Nasturtium officinale, Soalnum nigrum, Rumex hastatus, Rumex dentatus, Sonchus asper, Sonchus oleraceous,* and *Stellari media*, Na levels were in the range of 0.500–1,210 mg/kg (Bahadur et al. 2011; Jimoh et al. 2011; Seal 2011; Aberoumand 2009; Hussain et al. 2009; Hameed et al. 2008; Akubugwo et al. 2007; Imran et al. 2007; Odhav et al. 2007; Ozcan and Akbulut 2007; Trichopoulou et al. 2000; Wilman and Derrick 1994).

Mean levels of Sr in the wild edible vegetables are shown in Fig. 6.55 for comparative evaluation. An examination of the data revealed highest concentration of the metal in *Amaranthus spinosus* (277.8 mg/kg) and lowest level was found in *Plantago lanceolata* (2.520 mg/kg). In other vegetables samples, Sr levels ranged from 5.132 to 263.1 mg/kg. Few previous studies reporting Sr levels in wild vegetables showed its concentration ranging from 55.0 to 1,021 mg/kg in *Amaranthus viridis, Nasturtium officinale, Cichorium intybus,* and *Portulaca oleracea* (Jan et al. 2011; Hussain et al. 2011; Imran et al. 2007).

Concentration of Fe in food might be of nutritional importance especially in those parts of the world where anemia and Fe deficiency are common. Daily requirement of Fe for an adult is 20 mg and for a child is 10 mg (Imran et al. 2010; Reddy et al. 1987). Measured levels of Fe in the wild edible vegetables during present study are given in Fig. 6.56, which showed that highest Fe concentration was recorded in the leaves of *Taraxacum officinale* (1,502 mg/kg), followed by *Pimpinella diversifolia, Persicaria amplixicaulis, Gallium aparine,* and

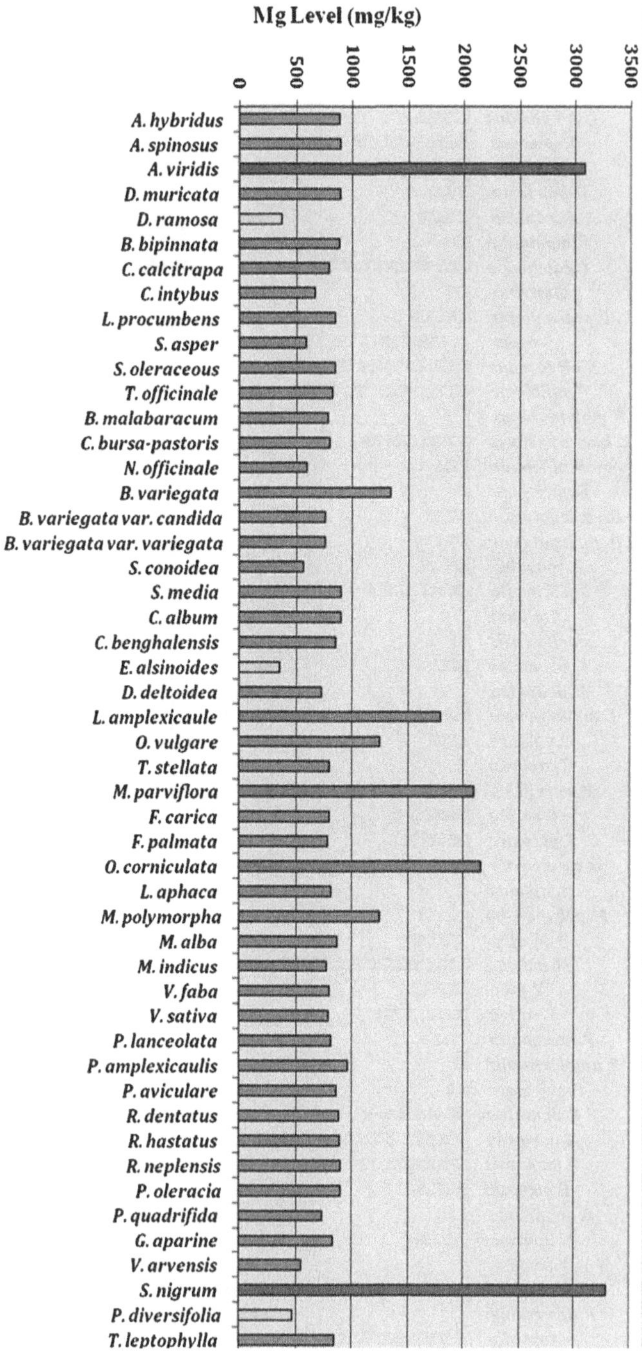

Fig. 6.53 Magnesium level (mg/kg) in wild edible vegetables

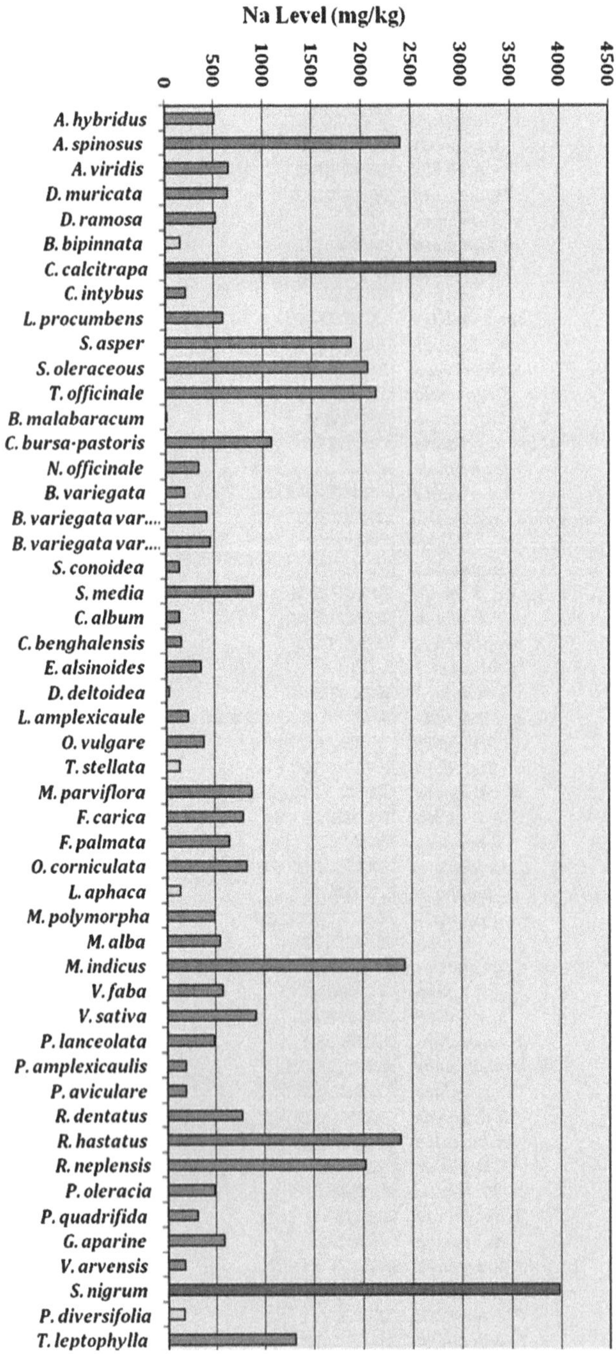

Fig. 6.54 Sodium level (mg/kg) in wild edible vegetables

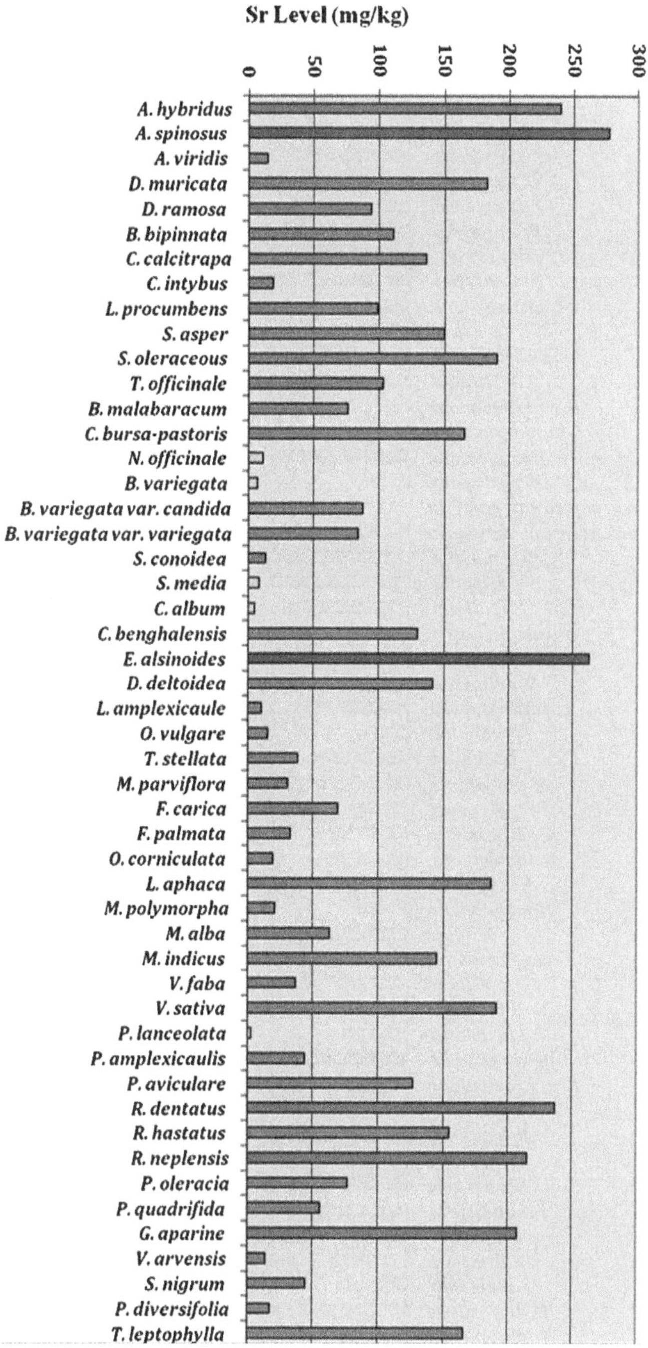

Fig. 6.55 Strontium level (mg/kg) in wild edible vegetables

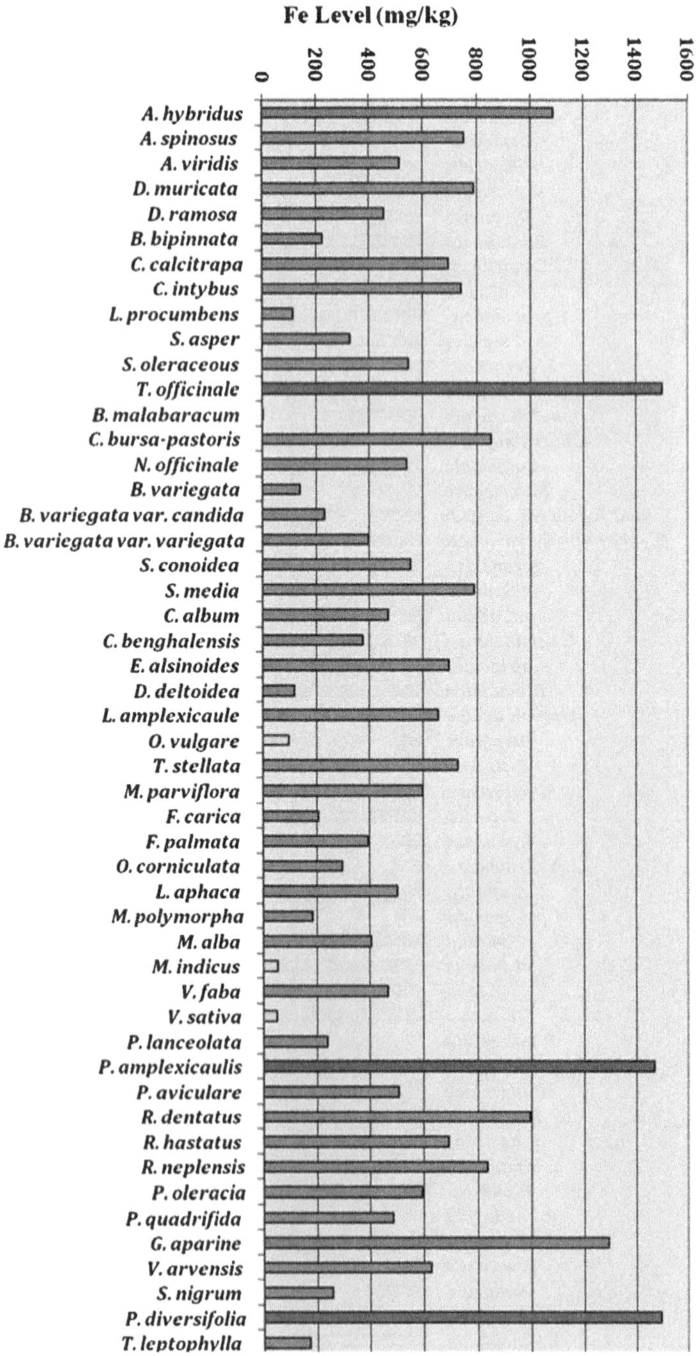

Fig. 6.56 Iron level (mg/kg) in wild edible vegetables

Amaranthus hybridus. However, relatively lower concentrations were observed in the edible parts of *Bombax malabaracum, Vicia sativa, Melilotus indicus,* and *Origanum vulgare*. In previously reported studies, Fe contents were found in the range of 9.240–1,810 mg/kg in the edible parts of *Amaranthus hybridus, Amaranthus spinosus, Amaranthus viridis, Medicago denticulata, Nasturtium officinale, Chenopodium album, Solanum nigrum, Portulaca oleracea, Rumex hasttatus, Rumex dentatus Prunus domestica, Prunus aremenica, Pyrus communis, Sonchus arvensis Sonchus asper,* and *Sonchus oleraceous* (Bahadur et al. 2011; Jimoh et al. 2011; Seal 2011; Aberoumand 2009; Hussain et al. 2009; Hameed et al. 2008; Akubugwo et al. 2007; Imran et al. 2007; Odhav et al. 2007; Ozcan and Akbulut 2007; Trichopoulou et al. 2000; Zahoor et al. 2003).

Average levels of Zn found in various wild vegetables during the present study are shown in Fig. 6.57. Highest concentration of Zn (145 mg/kg) was found in the leaves of *Solanum nigrum*, while lowest level was observed in the edible parts of *Portulaca quadrifida* (9.982 mg/kg). However, in rest of the wild edible vegetables Zn contents were noted in the range of 11.61–98.82 mg/kg. Some earlier studies reporting the concentration of Zn in *Amaranthus hybridus, Amaranthus spinosus, Amaranthus viridis, Bidens pilosa, Malva sylvestris, Chenopodium album, Cichorium intybus, Medicago denticulata, Nasturtium officinale, Pimpinella anisum, Partulaca oleracea, Solanum nigrum, Sonchus arvensis, Sonchus asper,* and *Sonchus oleraceous* revealed the metal levels ranging from 35 to 168 mg/kg from different parts of the world (Bahadur et al. 2011; Jimoh et al. 2011; Seal 2011; Aberoumand 2009; Hussain et al. 2009; Hameed et al. 2008; Akubugwo et al. 2007; Imran et al. 2007; Odhav et al. 2007; Ozcan and Akbulut 2007; Trichopoulou et al. 2000; Wilman and Derrick 1994; Luo et al. 2011; Khan et al. 2010).

Copper contents in the vegetables measured during the present study revealed highest concentration in *Rumex hastatus* (17.75 mg/kg), followed by *Dyropteris ramose* and *Amaranthus spinosus*, whereas comparatively lower concentrations were observed in the edible parts of *Ficus palmata, Centurea calcitrapa, Chenopodium album,* and *Melilotus indicus* as shown in Fig. 6.58. The reported levels of Cu in the edible parts of *Amaranthus spinosus, Amaranthus viridis, Bidens pilosa, Malva* spp., *Chenopodium album; Cichorium intybus, Nasturtium officinale, Pimpinella* spp., *Partulaca oleracea, Solanum nigrum, Sonchus asper*, and *Sonchus oleraceous* exhibited the average levels in the range of 0.250–60.00 mg/kg (Bahadur et al. 2011; Jimoh et al. 2011; Seal 2011; Aberoumand 2009; Hussain et al. 2009; Hameed et al. 2008; Akubugwo et al. 2007; Imran et al. 2007; Odhav et al. 2007; Ozcan and Akbulut 2007; Luo et al. 2011; Khan et al. 2010).

Average concentrations of Mn in edible parts of the wild edible vegetables (on dry weight basis) are depicted in Fig. 6.59, which showed significant levels in the leaves of *Persicaria amplexicaulis,* followed by *Partulaca oleracea, P. quadrifida,* and *Oxalis corniculata;* however, relatively lower concentrations were noted in *Bombax malabaracum, Bauhenia variegate var. variegate, Veronica arvensis,* and *Chenopodium album*. According to some published data around the world (Bahadur et al. 2011; Jimoh et al. 2011; Seal 2011; Aberoumand 2009; Hussain et al. 2009; Hameed et al. 2008; Akubugwo et al. 2007; Imran et al. 2007; Odhav et al. 2007), Mn levels in *Amaranthus spinosus, Amaranthus*

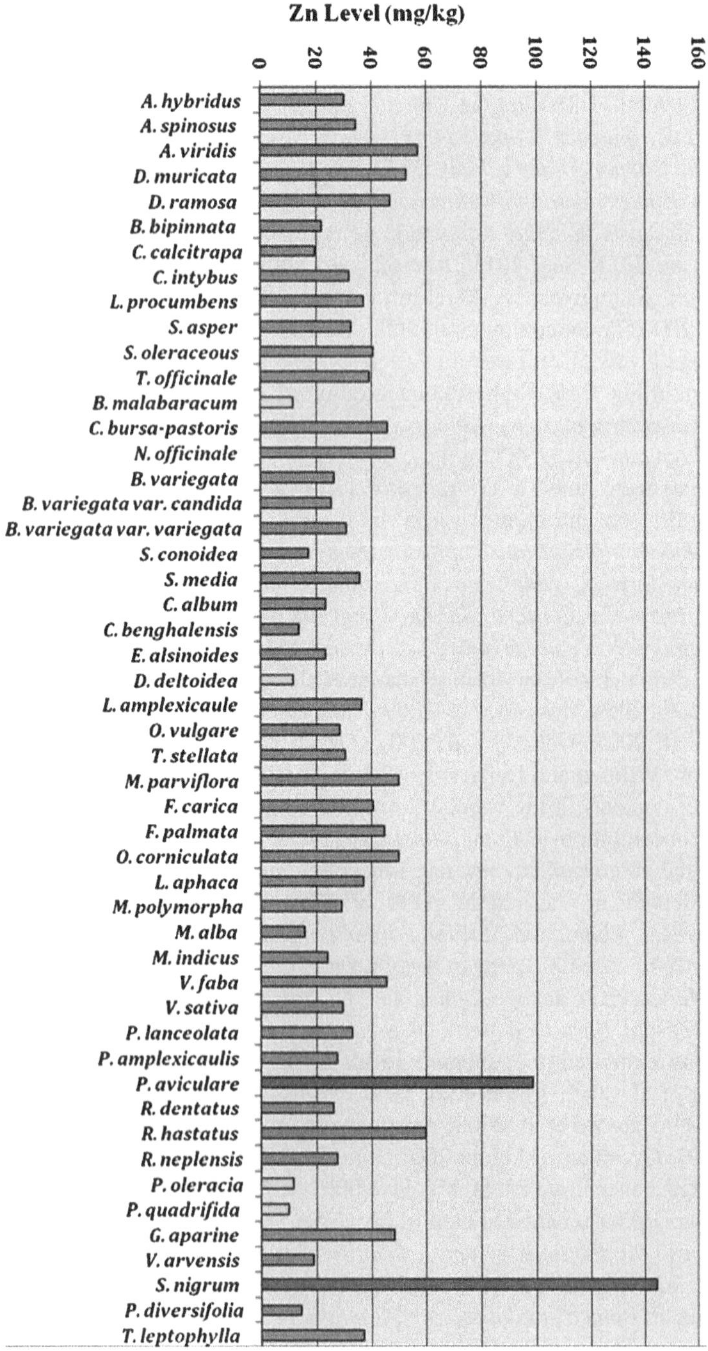

Fig. 6.57 Zinc level (mg/kg) in wild edible vegetables

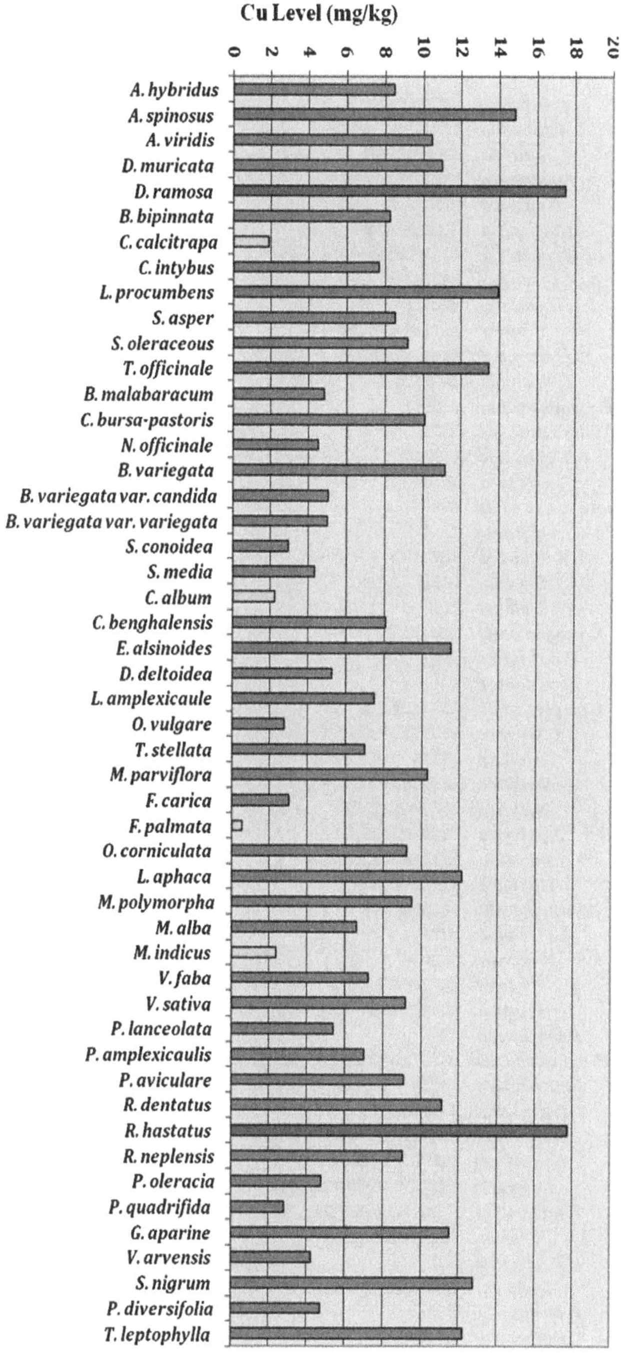

Fig. 6.58 Copper level (mg/kg) in wild edible vegetables

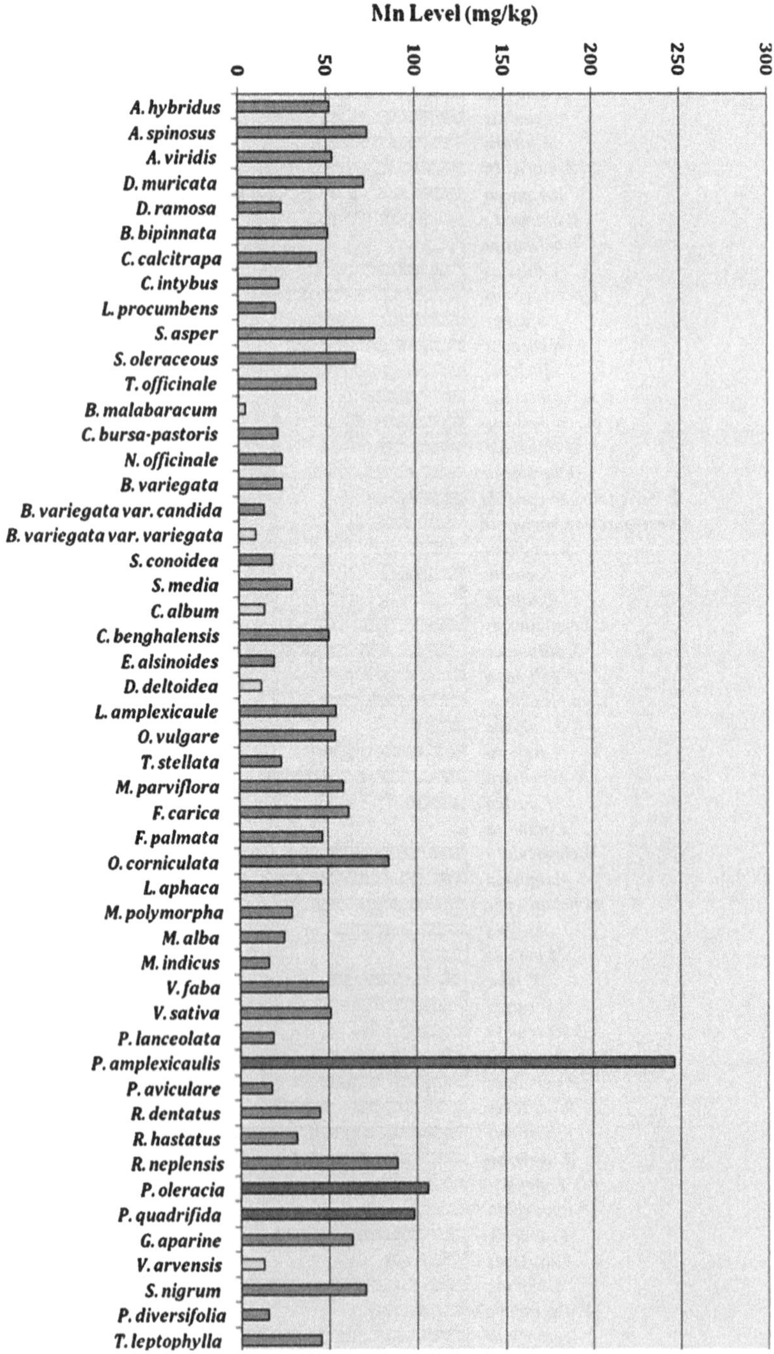

Fig. 6.59 Manganese level (mg/kg) in wild edible vegetables

viridis, Cichorium intybus, Medicago denticulata, Nasturtium officinale, Partulaca oleracea, Solanum nigrum, Sonchus arvensis, Sonchus asper, and *Sonchus oleraceous* were found in the range of 0.200–80.0 mg/kg which were in good agreement with the present results for most of the cases.

Measured levels of Li in the vegetables samples revealed highest contribution of Li in *Evolvulus alsinoides* (10.69 mg/kg), followed by *Amaranthus spinosus, Dryopteris ramosa, Capsella bursa,* and *Vicia sativa* as illustrated in Fig. 6.60. Comparatively lower concentrations were observed in the edible parts of *Silene conoidea, Bidens bipinnata, Lathyrus aphaca,* and *Persicaria amplexicaulis.*

Cobalt is an important element for the plants having ability to fix nitrogen in the root tubercles. Vitamin B_{12} is the main source of Co in animal food. The recommended daily intake of vitamin B_{12} for adults is 3 μg (0.13 μg Co). It played important role in thyroid metabolism in humans (Thunus and Lejeune 1994). Average measured levels of Co in different parts of the wild vegetables are shown in Fig. 6.61, which indicated its highest concentration in the leaves of *Cichorium intybus* (35.66 mg/kg), followed by *Solanum nigrum, Pimpinella diversifolia,* and *Bauhenia variegate.* However, the lowest metal contents were recorded in the leaves of *Launaea procumbens* (0.050 mg/kg). On the average basis, 3.800–9.800 mg/kg of Co levels were reported in *Amaranthus viridis* and *Chenopodium album,* respectively (Hussain et al. 2009).

Chromium is considered as an essential metal since it is a cofactor for insulin metabolism (Anderson 2000). The recommended daily intake for Cr extends from 50 to 250 μg. It has been reported that in some plant species Cr proved to be toxic at 5 mg/L (Adriano 1986). In the present study, highest Cr concentration was found *Rumex dentatus* (33.60 mg/kg), followed by *Amaranthus spinosu* (32.21 mg/kg) and *Dioscorea deltoidea* (31.97 mg/kg). However, the lowest level of Cr was observed in *Chenopodium album* (0.538 mg/kg), along with *Origanum vulgare* and *Vicia faba* as shown in Fig. 6.62. In previous studies, concentration of Cr in *Amaranthus viridis, Chenopodium album, Nasturtium officinale,* and *Pimpinella* spp. was reported in the range of 0.030–24.68 mg/kg (Hussain et al. 2009; Imran et al. 2007; Ozcan and Akbulut 2007).

Average levels of Cd in different edible parts of the vegetables measured during the present study are given in Fig. 6.63. Mean concentration of Cd ranged from lowest value of 0.185 mg/kg found in *Sonchus asper* to the highest value of 6.990 mg/kg noted in *Melilotus indicus.* However, in few vegetable samples, Cd levels were noted below the detection limit. Elevated levels of Cd are considered as main health risk because they might be associated with carcinogenic and non-carcinogenic health hazards. Reported levels of this metal were found in the range of 0.008–4.290 mg/kg in *Amaranthus viridis, Bidens* spp., *Malva* spp., *Chenopodium album, Cichorium intybus, Pimpinella* spp., and *Partulaca oleracea* (Hussain et al. 2009; Ozcan and Akbulut 2007; Luo et al. 2011; Khan et al. 2010).

Lead is considered as one of the major environmental pollutants and its elevated levels are generally associated with numerous health problems, especially in the children. Mean concentrations of Pb in the wild vegetables during this study were found ranging from the lowest value of 1.021 mg/kg in *Bidens bipinnata* to the highest value of 87.60 mg/kg in *Rumex dentatus* as shown in Fig. 6.64. In some

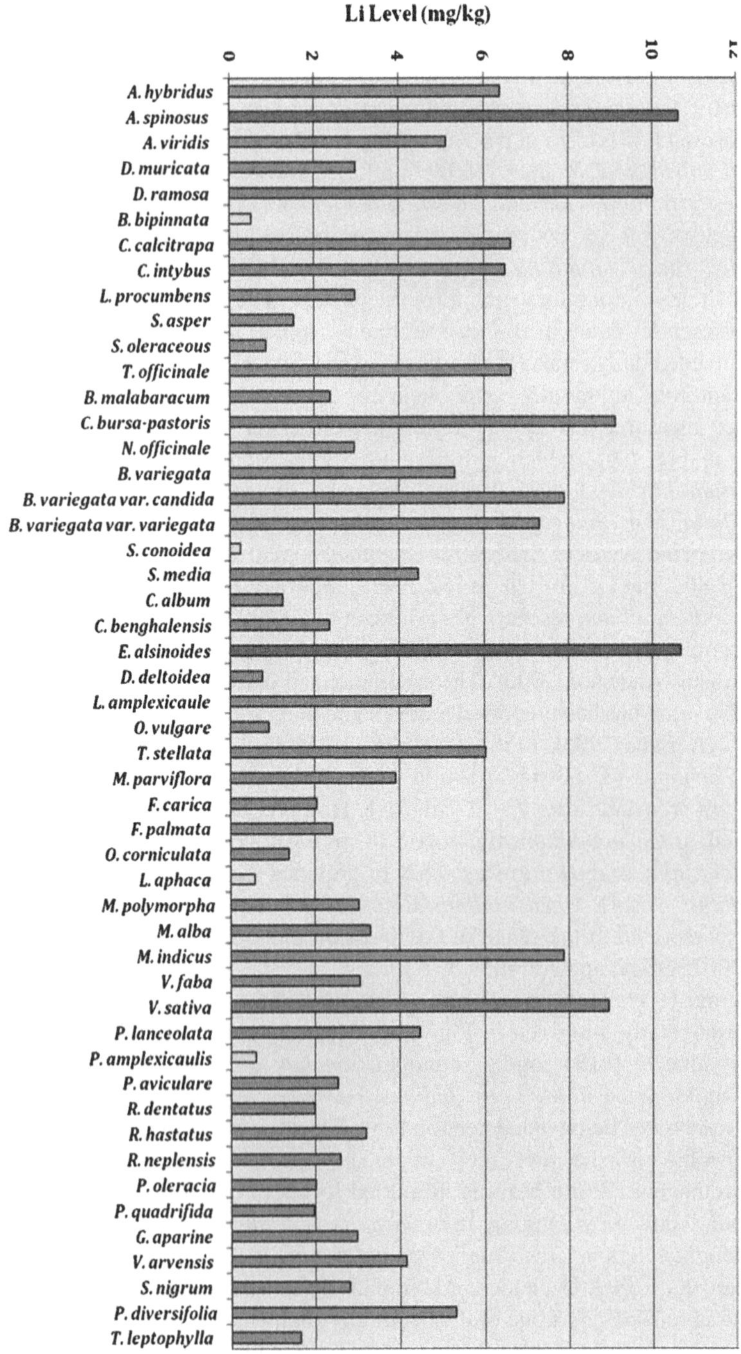

Fig. 6.60 Lithium level (mg/kg) in wild edible vegetables

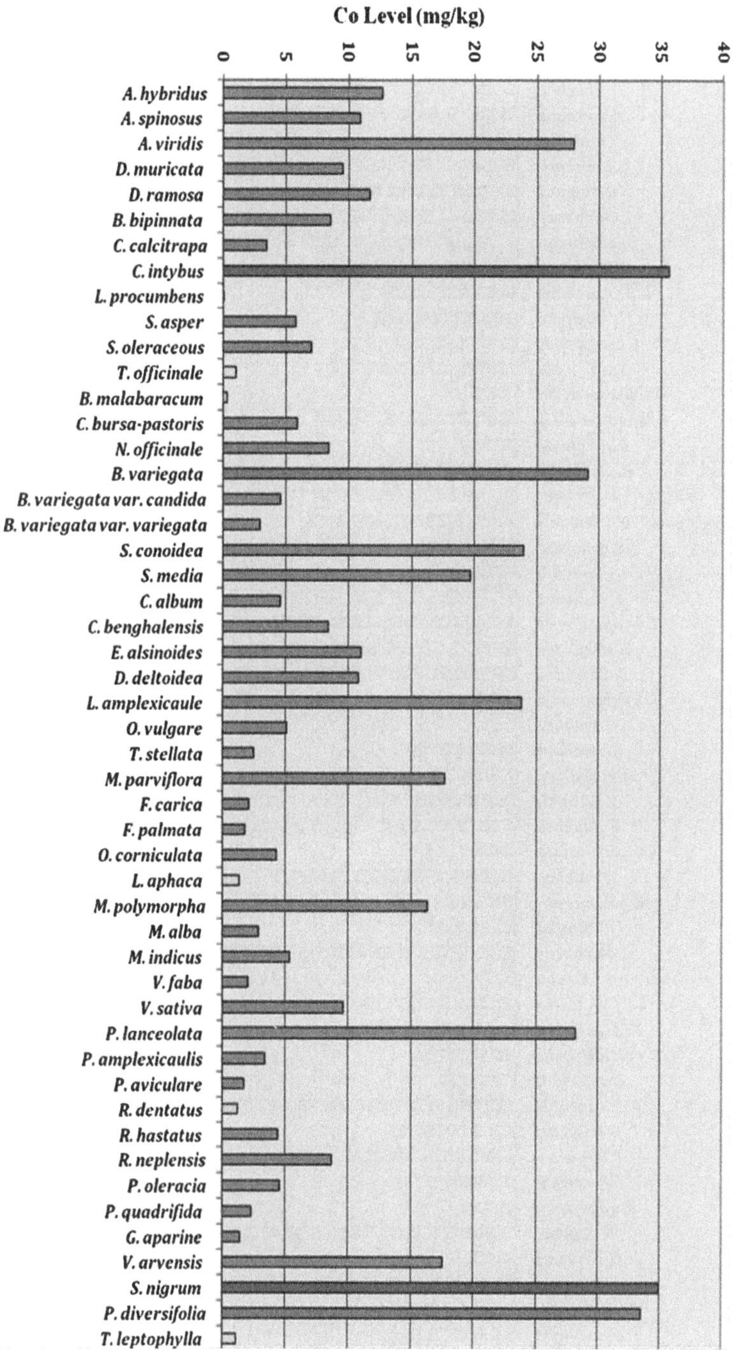

Fig. 6.61 Cobalt level (mg/kg) in wild edible vegetables

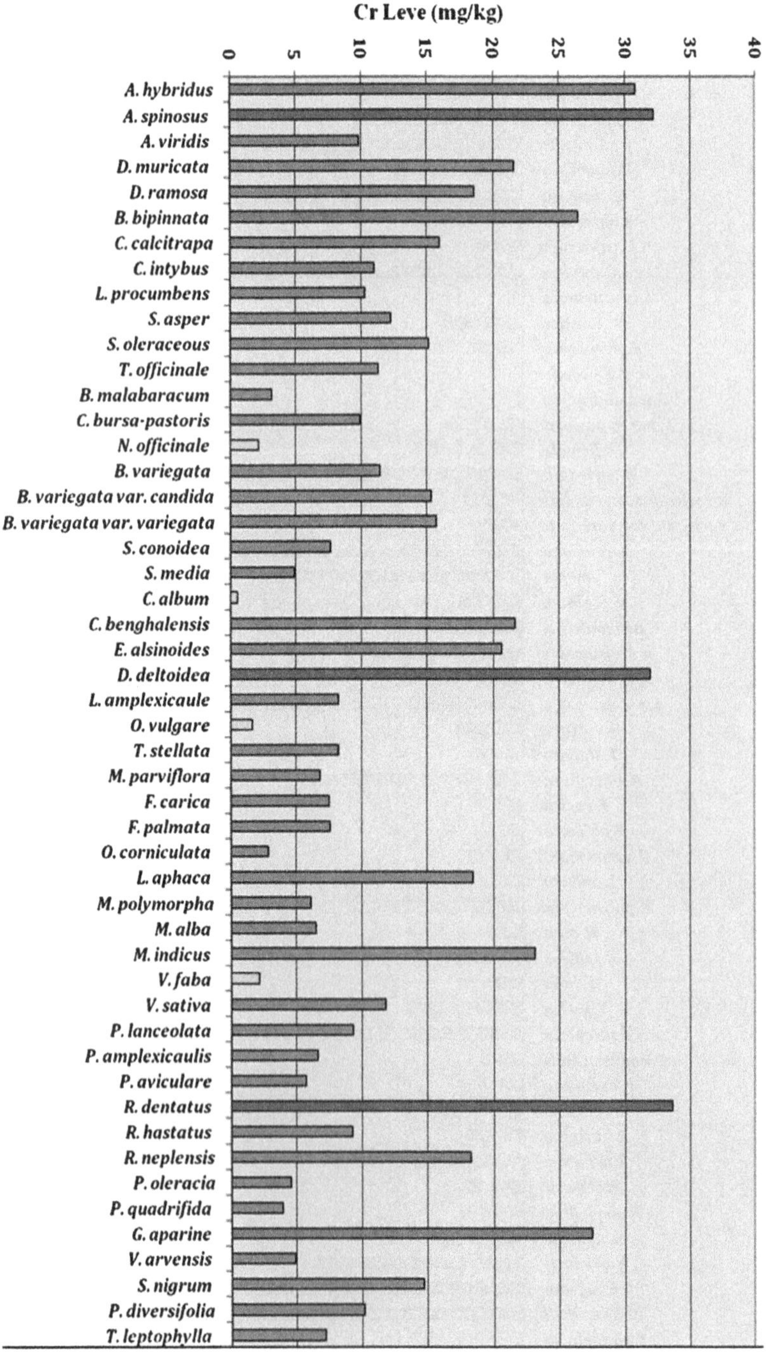

Fig. 6.62 Chromium level (mg/kg) in wild edible vegetables

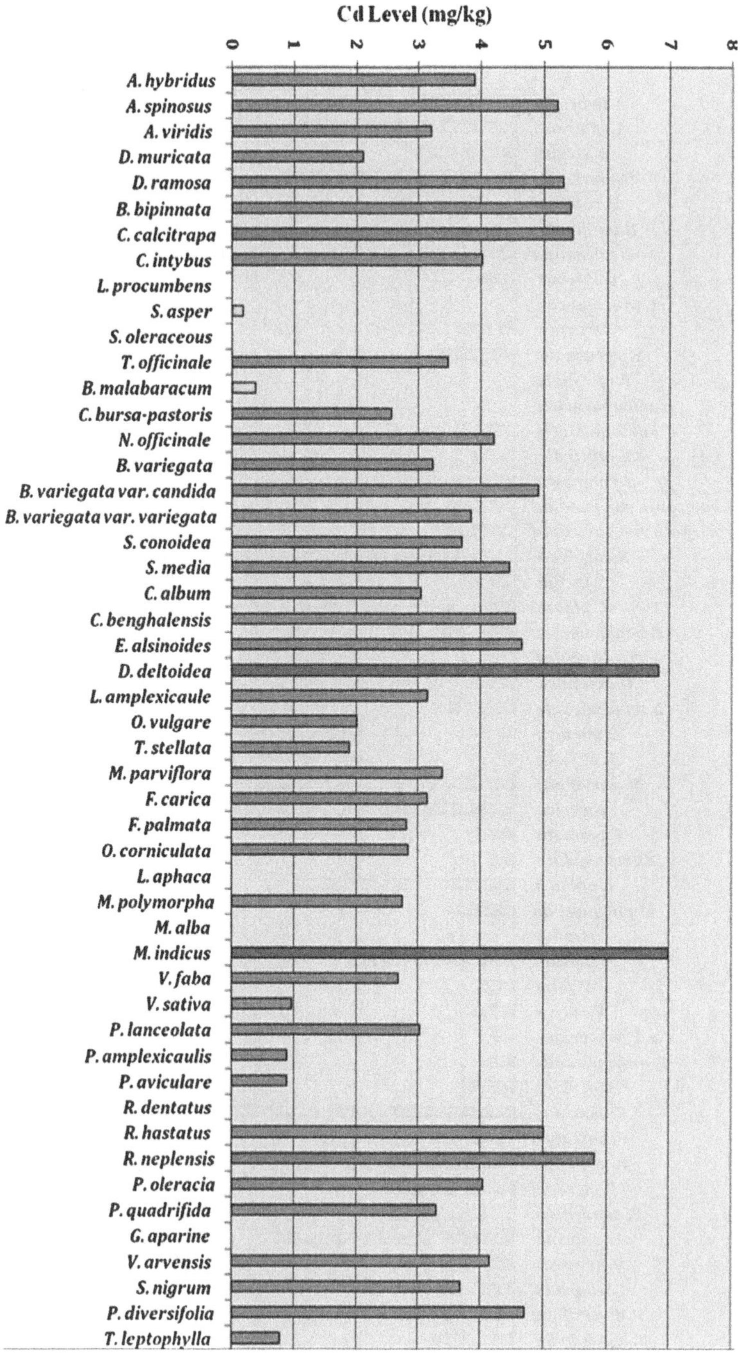

Fig. 6.63 Cadmium level (mg/kg) in wild edible vegetables

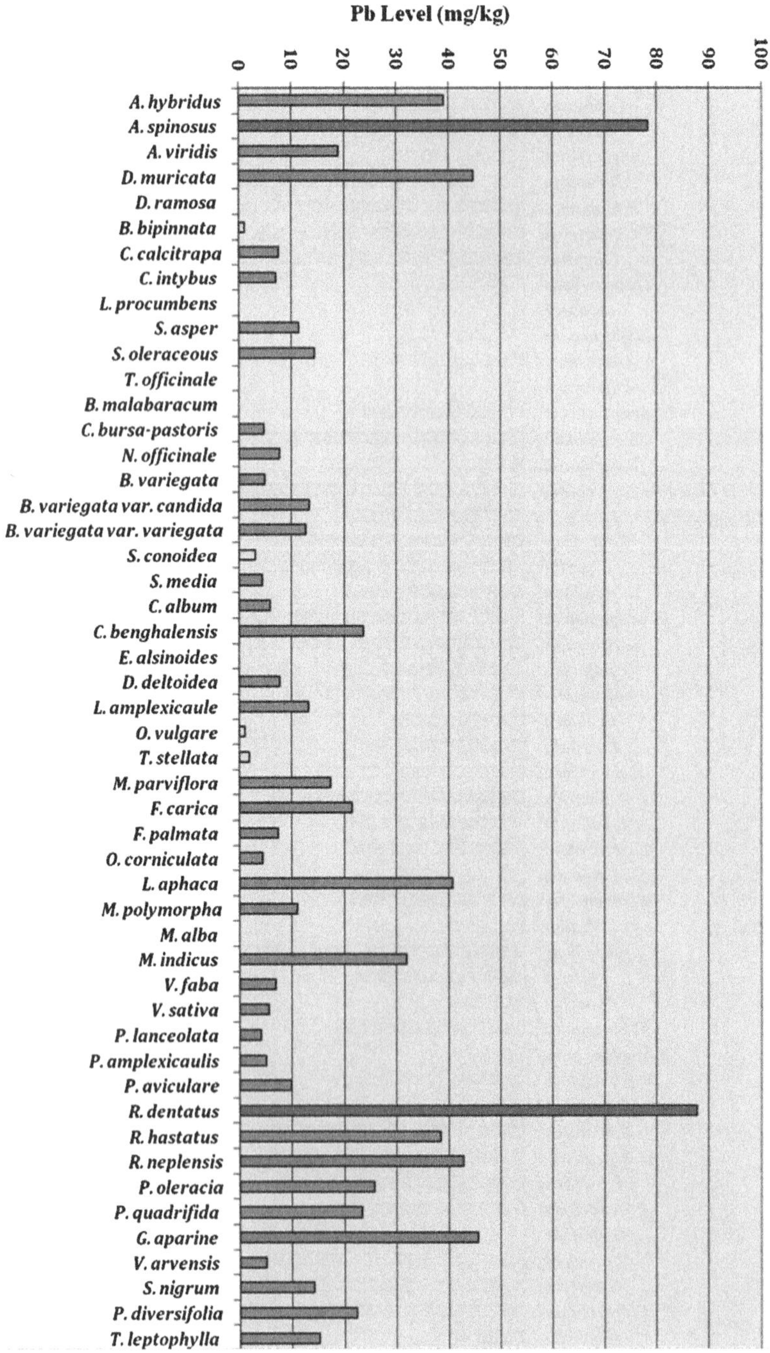

Fig. 6.64 Lead level (mg/kg) in wild edible vegetables

vegetable samples, Pb levels were noted to be lower than the instrumental detection limit. Some of the earlier studies reporting Pb levels in *Amaranthus viridis, Malva sylvestris, Bidens pilosa, Chenopodium album, Cichorium intybus,* and *Partulaca oleracea* exhibited large spread in metal concentration ranging from 0.150 to 44 mg/kg (Hussain et al. 2009; Luo et al. 2011; Khan et al. 2010). In some of the wild edible vegetable species under present investigation, Pb contents were comparatively higher which might be of major concern to the consumers using these vegetables as food supplements.

6.5 Statistical Distribution of the Metal Levels

Basic statistical parameters related to the distribution of selected metal levels (mg/kg, dry weight) in wild edible vegetables of Lesser Himalayas are given in Tables 6.2. Among studied samples of wild edible vegetables, dominant mean levels were shown by Ca (18,275 mg/kg), followed by K (5,340 mg/kg), Mg (979 mg/kg), Na (826 mg/kg), and Fe (546 mg/kg). Relatively lower average levels were noted for Sr (96.75 mg/kg), Mn (46.49 mg/kg), Zn (35.74 mg/kg), Pb (18.44 mg/kg), Cr (12.58 mg/kg), Co (10.06 mg/kg), Cu (8.006 mg/kg), Li (4.005 mg/kg), and Cd (3.485 mg/kg). Overall the decreasing trend of average metal levels in wild edible vegetables showed following order; Ca > K > Mg > Na > Fe > Sr > Mn > Zn > Pb > Cr > Co > Cu > Li > Cd. Most of the metals exhibited random and asymmetrical distribution in the vegetables as manifested by elevated SD and SE values as well as considerably higher kurtosis and skewness values. Comparatively higher asymmetry in the distribution was

Table 6.2 Statistical distribution parameters for selected metals (mg/kg dry weight) in wild edible vegetables

	Min	Max	Mean	Median	SD	SE	Kurtosis	Skew
Ca	1,394	67,486	18,257	14,076	17,278	2,444	0.352	0.980
Mg	360	3,294	979	838	570	80.66	8.350	2.799
Na	8.762	3,982	826	514	893	126	3.006	1.818
K	1,672	8,600	5,340	4,768	1,552	219	−0.188	0.631
Sr	2.520	277.9	96.75	80.70	79.35	11.22	−0.781	0.591
Fe	5.429	1,503	546	505	370	52.42	0.812	0.961
Zn	9.982	144.6	35.74	31.65	22.25	3.147	11.80	2.836
Cu	0.534	17.75	8.006	8.172	4.048	0.572	−0.299	0.337
Mn	3.619	246.5	46.49	45.15	38.01	5.376	14.95	3.123
Li	0.242	10.69	4.005	3.020	2.832	0.401	−0.170	0.838
Co	0.050	35.66	10.06	5.857	9.983	1.412	0.622	1.288
Cr	0.538	33.60	12.58	10.07	8.730	1.235	0.054	0.926
Cd	0.185	6.990	3.485	3.424	1.615	0.243	−0.137	−0.104
Pb	1.021	87.60	18.44	12.07	19.19	2.892	4.370	1.991

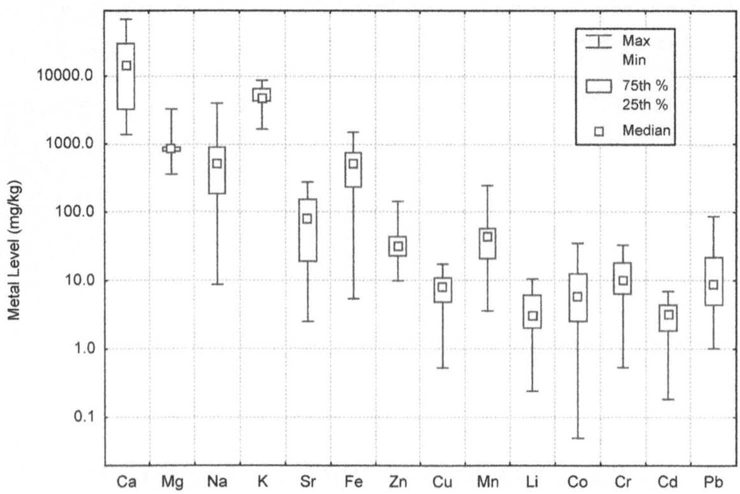

Fig. 6.65 Quartile distribution of selected metals in the wild edible vegetables

noted for Mn, Zn, Pb, Mg, and Na, while relatively lower dispersion was observed for Cu, Li, and Cd in the vegetables.

Quartile distribution of the metals in wild edible vegetables investigated in the present study is shown in Fig. 6.65 as a box and whisker plot. Most of the metals revealed broad distribution and significant variations in their quartile levels. Predominantly broad distribution was observed for Ca, Na, Sr, Fe, and Co; nonetheless, narrow distribution was shown by Mg and K in the vegetables. Relatively symmetrical variations were noted for Zn, Pb, Ca, Na, and Mn which evidenced their consistently varying levels in the vegetables. However, markedly asymmetrical spread was observed for Cd, Co, Cr, Li, Cu, and Fe which exhibited their randomness in the distribution.

6.6 Correlation Study of the Metals

Pearson correlation coefficients (r) among selected metal pairs were calculated to examine their interrelationships in the wild vegetables samples as shown in Table 6.3, wherein the significant r-values are shown in bold at $p < 0.01$. Strong positive correlations were observed for following pairs; Sr–Cr ($r = 0.741$), Ca–Sr ($r = 0.701$), Sr–Pb ($r = 0.699$), K–Co ($r = 0.683$), Cr–Pb ($r = 0.652$), Mg–Zn ($r = 0.589$), Mg–K ($r = 0.560$), and Cr–Cd ($r = 0.556$).

The correlation study thus pointed out significant mutual associations among these metals in the vegetables. Nevertheless, significant negative correlations were also observed between few metal-pairs; K–Sr ($r = -0.574$), Ca–K ($r = -0.462$), and Ca–Co ($r = -0.402$), suggesting their opposing variations in the vegetables.

Table 6.3 Correlation coefficient[a] matrix for selected metals in wild edible vegetables

	Ca	Mg	Na	K	Sr	Fe	Zn	Cu	Mn	Li	Co	Cr	Cd	Pb
Ca	1.000													
Mg	−0.230	1.000												
Na	0.318	0.290	1.000											
K	**−0.462**	**0.560**	−0.024	1.000										
Sr	**0.701**	−0.293	0.327	**−0.574**	1.000									
Fe	0.094	−0.145	0.046	−0.128	0.186	1.000								
Zn	−0.115	**0.589**	0.441	0.256	−0.021	−0.014	1.000							
Cu	0.213	0.200	0.264	0.002	0.456	0.169	0.420	1.000						
Mn	0.290	0.215	0.139	−0.108	0.077	0.343	0.061	0.064	1.000					
Li	−0.090	−0.135	0.184	−0.066	0.256	0.142	−0.018	0.241	−0.306	1.000				
Co	**−0.402**	0.393	0.000	**0.683**	−0.398	0.057	0.188	0.030	−0.154	0.178	1.000			
Cr	0.320	−0.129	0.197	−0.407	**0.741**	0.179	−0.065	0.340	−0.023	0.256	−0.025	1.000		
Cd	−0.039	−0.104	0.223	−0.075	0.171	0.076	−0.187	−0.016	−0.251	0.280	0.198	**0.556**	1.000	
Pb	0.341	−0.055	0.249	−0.300	**0.699**	0.373	−0.001	0.471	0.130	0.125	−0.194	**0.652**	0.329	1.000

[a]Bold r-values are significant at $p < 0.01$

All other metals exhibited weak positive or negative relationships. It was also noted that Na showed positive correlations with all other metals except K. Similarly Fe and Cu also showed positive correlations with all metals except Zn and Cd, respectively. However, some of the metals (K, Ca, Mg, Zn, etc.) also showed negative correlations with most of other metals. The correlation study thus pointed out some mutual associations among the metals in the vegetables which might be contributed by some common sources in the study area. This aspect of the study would be further explored by multivariate statistical methods in ensuing sections.

6.7 Correlation Study of the Metals and Macronutrients

Correlation coefficients among selected metals and macronutrients were also computed to examine their inter-relationships in the wild vegetables samples as shown in Table 6.4, wherein the significant r-values are shown in bold at $p < 0.01$. Some significant correlations in the vegetable samples were found between ash contents and Ca ($r = 0.336$), ash contents and Fe ($r = 0.291$), energy value and Mg ($r = 0.269$), moisture contents and Na ($r = 0.236$), dry matter and Mg ($r = 0.225$), moisture contents and Ca ($r = 0.212$), Ash contents and Sr ($r = 0.208$). These correlations indicated direct relationships of ash contents with Ca, Fe, and Sr, whereas moisture contents exhibited preferences for Na and Ca. Some inverse relationships among the metals and macronutrients were also noted; e.g., crude fats and Cr ($r = -0.392$), crude fats and Sr ($r = -0.351$), crude fibers and Mg ($r = -0.349$), and energy value and Fe ($r = -0.327$). The correlation study thus indicated that the fats and fiber

Table 6.4 Correlation coefficient[a] matrix for selected metals and macro-nutrients in wild edible vegetables

	MC	DM	CP	CF	Carb	Fib	Ash	EV
Ca	**0.212**	−0.212	−0.066	−0.181	−0.047	−0.173	**0.336**	−0.247
Mg	−0.225	**0.225**	−0.025	0.007	0.172	**−0.349**	−0.109	**0.269**
Na	**0.236**	−0.236	−0.021	−0.015	−0.006	−0.185	0.155	−0.043
K	−0.166	0.166	0.042	0.076	0.031	−0.202	−0.023	0.141
Sr	0.002	−0.002	−0.078	**−0.351**	0.022	−0.089	**0.208**	−0.225
Fe	0.079	−0.079	0.131	−0.120	−0.250	0.097	**0.291**	**−0.327**
Zn	−0.003	0.003	−0.078	0.073	0.038	0.016	0.006	0.007
Cu	−0.115	0.115	−0.262	−0.033	0.129	0.173	0.020	−0.112
Mn	−0.038	0.038	−0.067	0.031	−0.043	0.000	0.175	−0.141
Li	−0.075	0.075	0.139	−0.034	−0.104	0.102	−0.039	−0.026
Co	−0.145	0.145	0.023	−0.098	0.067	−0.225	−0.012	0.097
Cr	−0.095	0.095	−0.056	**−0.392**	0.152	−0.200	−0.013	0.007
Cd	0.056	−0.056	0.145	−0.130	0.003	−0.263	−0.041	0.149
Pb	0.092	−0.092	0.043	−0.193	−0.023	−0.123	0.090	−0.069

[a]Bold r-values are significant at $p < 0.01$

contents of the vegetables were not interacting with the selected metals, which were mostly contributed as inorganic compounds as evidenced by their mutual associations with the ash contents. All other correlations between the metals contents and the macronutrients were found either feebly positive or negative and hence insignificant.

6.8 Multivariate Analyses of the Metals in Vegetables

Multivariate methods have been used for the apportionment and source identification of the pollutants in various environmental segments. Two most commonly used multivariate statistical methods were employed in this study for source identification and apportionment of the metals in wild edible vegetables, namely, Principal Component Analysis (PCA) and Cluster Analysis (CA).

6.8.1 Principal Component Analysis

PCA is a method for projecting the cloud of variables onto subspaces with fewer dimensions, while maintaining the distance between variables as much as possible. The most widely used form of PCA with rotation is varimax PCA. This is based on the principle of maximizing, for each component, the sum of squares of the correlation coefficients, with the result that each component is strongly correlated with some variable and weakly correlated with the others. Thus, some variables have a high contribution to each axis, while the others have very low contributions (Toffery 2011). PCA attempts to decompose the initial data to reduce the number of columns so that it is much less than the number of initial measurements. It means that the different columns/measurements are strongly correlated and that the data structure is best described not by total measurements, but by a much smaller number. Thus, with information about the independent variables, the data can be greatly reduced (Koel and Kaliurand 2010).

The PC scores contain information on the entire chemical measurements combined into a single number, with the loadings indicating the relative contribution each element makes to that score. The software "STATISTICA" normalizes the loadings to the eigen value. An analysis of principal components often reveals relationships among original variables that were not previously suspected and thereby allows a new insight to the data. Again in replacing the original correlated variables by a fewer number of independent variables, some information has to be necessarily sacrificed. PCA analysis is optimum in the sense that the amount of lost information is kept at minimum among all the similar processes (Mukhopadhyay 2009). This method is widely used for the apportionment and source identification in environmental studies.

In this study PCA was carried out for the apportionment of selected metal levels in wild edible vegetables samples employing varimax normalized rotation on the dataset as shown in Table 6.5. Overall, five principal components (PC) were

Table 6.5 Principal component loadings of the metals in wild edible vegetables

	PC 1	PC 2	PC 3	PC 4	PC 5
Eigen value	3.754	2.522	1.958	1.269	1.029
Total variance (%)	26.81	18.02	13.98	9.064	7.346
Cumulative variance (%)	26.81	44.83	58.81	67.88	75.23
Ca	0.794	0.083	−0.185	0.127	0.024
Mg	−0.387	0.767	−0.223	−0.006	0.054
Na	0.360	0.671	−0.027	0.001	0.424
K	0.783	−0.363	0.037	0.056	−0.070
Sr	0.838	0.117	0.305	0.342	−0.007
Fe	−0.040	−0.130	−0.068	0.849	−0.040
Zn	−0.117	0.837	0.037	−0.030	−0.179
Cu	0.204	0.557	0.420	0.353	−0.368
Mn	−0.148	−0.155	0.722	−0.459	0.018
Li	0.001	0.006	0.742	0.142	0.252
Co	0.717	−0.248	−0.211	−0.202	−0.313
Cr	0.510	0.069	0.412	0.459	0.265
Cd	−0.129	−0.081	0.260	0.027	0.866
Pb	0.427	0.179	0.067	0.611	0.098

extracted, together explaining more than 75 % of the variance of data. First PC showed elevated loadings for Ca, K, Sr, Co, and Cr accounting for 26.81 % variance of the data. These metals were mostly contributed by the agricultural activities in the adjoining areas. The vegetable samples were collected mostly around the agricultural fields; thus the accumulation of these metals was mostly governed by the application of fertilizers and agricultural sprays. Second PC exhibited significant loadings for Mg, Na, Zn, and Cu and their contribution towards overall variance was 18.02 %. Third PC showed higher loadings in favor of Mn and Li. The metals in these two PCs were mostly derived from natural weathering processes and hence lithogenic in origin. Fourth PC with 9.064 % of the variance revealed elevated loadings of Fe and Pb which were mainly contributed by the transportation sources and automobile emissions in the study area. Last PC (7.346 % variance) showed elevated loading for Cd and the major sources contributing this PC were the industrial emissions and incinerators. Hence, significant anthropogenic contributions of the metals were noted in the wild edible vegetables.

6.8.2 *Cluster Analysis*

Clustering is the statistical operation of grouping objects (individuals or variables) into a limited number of groups known as clusters or segments, which have two properties. On the one hand, they are not defined in advance by the analyst, but are

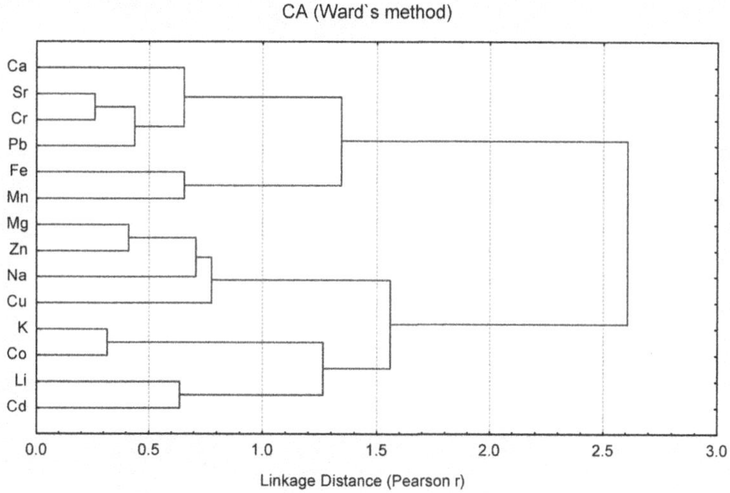

Fig. 6.66 Cluster analysis of the metals in wild edible vegetables

discovered during the operation, and on the other hand, the clusters are combination of objects having similar characteristics, which are separated from objects having different characteristics, thus resulting in internal homogeneity and external heterogeneity. The cluster to which each object belongs is not known in advance. Even the number of clusters is not fixed in advance. This is because there is no dependent variable. Hence, clustering is descriptive not predictive (Toffery 2011). The CA is typically used for non-supervised classification of observations. It searches for patterns in a data-set by grouping the observations into cluster. In the other words, it is helpful in finding the natural grouping of data. There are two common approaches to classify the observations, hierarchical clustering and partitioning. Hierarchical clustering starts with the number of clusters equal to the number of observations and ends with one cluster. In partitioning this process is reversed. One can choose from a large collection of methods of hierarchical clustering. One of the most popular is Ward's method. This very efficient, hierarchical method uses analysis of variance approach to evaluate the distance between clusters and attempts to minimize the sum of the squares of any two clusters that can be formed at each step. An important step in any clustering is to select a distance measures which determine how the similarity of two observations is calculated (Chwiej 2010). Typical output is a dendrogram where the objects are arranged in a hierarchy reflecting their similarity. Objects with similar patterns of concentrations are fused to a cluster on a low level in the dendrogram. Among the different methods of the hierarchical clustering, Ward's method normally leads to well-structured dendrogram (Gottelt et al. 1997).

Cluster analysis of the metal levels in wild edible vegetables using the Ward's method is given in Fig. 6.66, which revealed two major clusters each having two

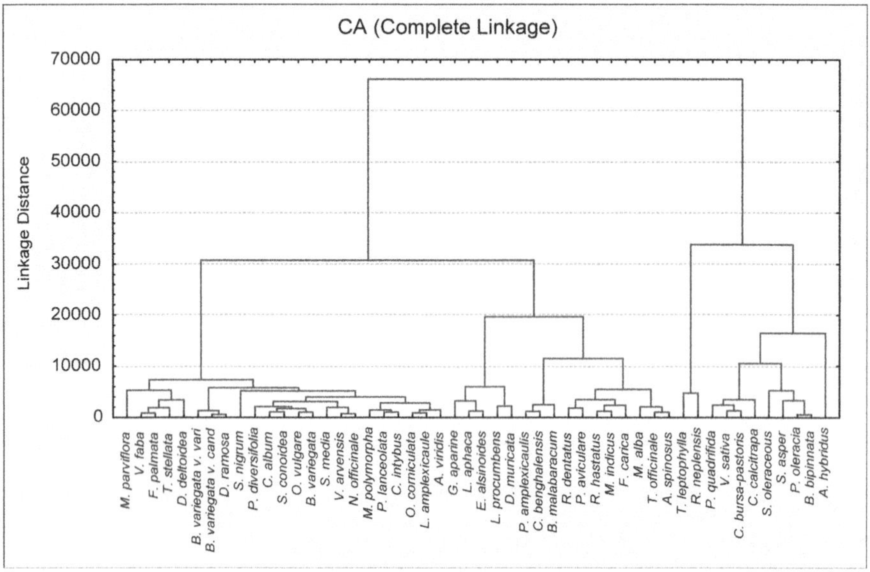

Fig. 6.67 Cluster analysis of the vegetables based on metal contents

subgroups. First major cluster comprised of Ca, Sr, Cr, Pb, Fe, and Mn, which was mostly contributed by anthropogenic activities. Second major cluster was further divided into two subgroups; Mg, Zn, Na, Cu, which were mostly lithogenic in origin, and K, Co, Li, Cd, which were mostly contributed by mixed sources including natural and anthropogenic processes.

The CA for various wild edible vegetables based on their elemental composition is shown in Fig. 6.67. In this case, three prominent clusters were observed and every group indicated closed association to each other for their metallic constituents. First cluster comprised of *Vicia faba, Ficus palmata, Dioscorea deltoidea, Tulipa stellata, Dyropteris ramosa, Pimpinella diversifolia, Chenopodium album, Silene conoidea, Origanum vulgara, Bauhinia variegata, Bauhinia variegata* var. *variegata, Bauhinia variegata* var. *candida, Stellaria media, Veronica arvensis, Nasturtium officinale, Solanum nigrum, Malva parviflora, Medicago polymorpha, Plantago lanceolata, Cichorium intybus, Oxalis corniculata, Lamium amplexicaule,* and *Amaranthus viridis.* Similarly, second cluster consisted of *Persicaria amplexicaulis, Commelina benghalensis, Galium aparine, Lathyrus aphaca, Evolvulus alsinoides, Rumex dentatus, Polygonum aviculare, Rumex hastatus, Melilotus indicus, Ficus carica, Melilotus alba, Taraxacum officinale,* and *Amaranthus spinosus,* whereas, the third cluster showed common groupings of *Portulaca quadrifida, Portulaca oleracia, R. neplensis, Vicia sativa, Capsella bursa-pastoris, Launaea procumbens, Digeria muricata, Torilis leptophylla, Sonchus oleraceous, Sonchus asper, Bidens bipinnata,* and *Amaranthus hybridus.* The third cluster was found to be significantly diverse compared to the first two clusters, while the strongest grouping was noted for the first cluster. Consequently, most of the vegetables revealed almost equivalent nutritional implication with respect to the selected metals contents.

6.9 Health Risk Assessment

Major concerns related to the metals pollution in the environment are mostly due to their toxic effects in living organisms and their non-biodegradability in the environment (Jang et al. 2006; Zhuang et al. 2009). Soil plays a critical role in the natural cycle of metals as it is the main source of the metals in the various environmental segments (Cao et al. 2010). Toxic metals present in the soil are one of the major threats to the environment and can damage human health through various absorption pathways, such as, direct ingestion, dermal contact, diet, through the soil, food chain, inhalation, and oral intake (Luo et al. 2011; Komárek et al. 2008; Park et al. 2004). Vegetables are monetary crops and play significant role in our daily diet. However, various human activities including mining, smelting, metallurgy, incineration, excavation, pesticides, automobile exhausts, and fertilizers are the major cause behind elevated metal concentrations in the environment (Zhuang et al. 2009; Cao et al. 2010; Shi et al. 2011). Vegetables can absorb the metals from contaminated soil and surface adsorption from polluted environment and their continuous consumption may result in harmful effects on beings (Wang et al. 2005; Lai et al. 2010; John and Andrew 2011). Some of the metals may cause neurologic involvement, headache, and liver disease, beyond their safe threshold values (Park et al. 2004; USEPA 2000). Use of environmentally contaminated food was related with an increased risk of postmenopausal breast cancer (Hiroaki et al. 2013). Acute and chronic exposure to toxic metals causes dermal, respiratory, cardiovascular, gastrointestinal, hematological, hepatic, renal, neurological, developmental, reproductive, immunological, genotoxic, mutagenetic, and liver cancer (Kapaj et al. 2006; Lin et al. 2013).

Vegetables are considered as an important part of food both in cooked and raw form. The people of Pakistan, particularly those of middle and low-income groups, consume significant quantities of wild edible vegetables. The required amount of the vegetables in our daily diet is about 300–350 g per person (Anonymous 1998), but only 80–90 g of vegetable per person on the average is being used in daily diet by the Pakistani population (Butt and Haq 1993). Vegetables contain large quantities of minerals, vitamins, carbohydrates, essential amino acid, and dietary fibers, which are necessary for usual performance of human metabolic activates. They are also significant to counteract the acid produced during digestion (Hashmi et al. 2007). Health risk assessment (HRA) or health risk appraisal is one of the most significant and extensively used screening tools in the field of health promotion and is often the first step in multi-component health promotion programs. Human health risk assessment is the estimation of the nature and possibility of adverse health effects in humans who may be exposed to chemicals in contaminated environmental media. Human health risk assessment addresses the types of health problems caused by environmental stressors, like chemicals/radiation; health problems in people when exposed to different levels of chemicals/radiations in their surrounding environment; types of environmental stressors, their levels and period; people which are more at risk to environmental stressors because of age, genetics, preexisting health conditions, ethnic practices, gender, work place, food, etc.

Concern over the environmental pollutants, particularly the toxic metals, has increased immensely in Pakistan during last couple of decades in the wake of population explosion, industrialization, urbanization, and other human activities (Manzoor et al. 2006; Shah and Shaheen 2007). Fossil fuel combustion, industrial emissions, automobile exhausts, mining, and quarrying are considered as the major pollution sources (Iqbal and Shah 2011; Shaheen et al. 2005). Mostly, the industrial developments have gone unplanned, without any environmental protection for the disposal of potentially toxic solid wastes, effluents, and emissions (Manzoor et al. 2006; Shaheen et al. 2005). In Pakistan, wild vegetables are commonly used by sub-urban and rural population, where these are considered as safe and healthy food by most of the consumers (Abbasi et al. 2013).

Numerous methods have been used for health risk assessment, but health risk index (HRI) is most frequently used. It is based on the concentration of the metals in edible parts of the vegetables in comparison with the reference dose of the metal and intake/body weight of the consumers (Cui et al. 2004; Pandey et al. 2012). Similarly, non-carcinogenic health risk has been evaluated through target hazard quotient (THQ) and hazard index (HI), while carcinogenic health risk is assessed by calculating the target cancer risk (TCR) (USEPA 2006, 2010; Yang et al. 2011). In order to determine health risk assessment a filed survey was conducted first in local population and different information including growth season of the vegetables, the consumption duration, intake rate, the number of family members, and the source of the vegetables were acquired in this study. Based on this survey, the average daily intake of each vegetable by an adult in a year was computed.

6.9.1 Health Risk Index

HRI is a commonly utilized method to assess the health risk to human by selected metals in food. The index depends on daily intake of metals through consumption of the vegetables and then comparing them to the prescribed reference oral dose (Luo et al. 2011; Cui et al. 2004; Pandey et al. 2012; Li et al. 2012; Singh et al. 2010). The index was calculated using following relationship:

$$\mathrm{HRI} = \frac{\sum_n (C_n \times \mathrm{D}_n)}{\mathrm{R}f\mathrm{D} \times \mathrm{B}_w}$$

where C_n = Metal level in a specific vegetable on fresh weight basis (mg/kg), D_n = Average daily intake of specific vegetable in a whole year (255 g), R*f*D = Safe

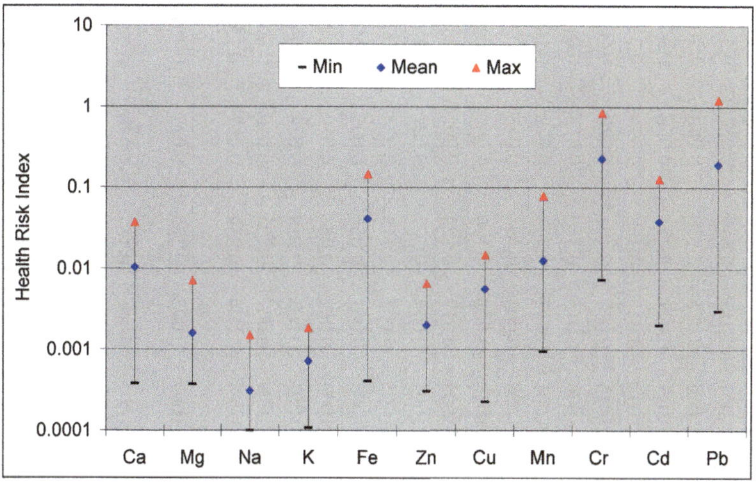

Fig. 6.68 Summary of health risk index (HRI) of the metals in wild edible vegetables

level of exposure by oral for life time, and B_w = Average body weight (65 kg for adult).

In this study, dietary reference intakes (DRI) of the metals are taken as R*f*D (FNB 2004), except Cd and Pb, for which maximum allowed levels (ML) were considered (EC 2006).

Generally, the index value less than unity (1.0) shows safe levels while the values higher than 1 are associated with adverse health effects. Minimum, maximum, and average values of HRI for various metals in the vegetables are shown in Fig. 6.68. The index values were found to considerably lower than 1 except Pb which revealed its maximum value slightly higher than 1, although the average value was well below unity. Among rest of the metals, maximum value of HRI for Cr was nearly approaching 1, while for most of the metals average index values were noted to be less than 0.01 as shown in Table 6.6. Consequently, the consumption of these wild vegetables was considered to be safe for human health.

6.9.2 Target Hazard Quotient

THQ is used to determine the non-carcinogenic health risks for humans associated with the consumption of these vegetables (Yang et al. 2011; EC, 2006). The method to estimate THQ is provided in USEPA Region III Risk-Based Concentration Table (USEPA 2006):

$$\mathrm{THQ} = \frac{\mathrm{C} \times \mathrm{I} \times 10^{-3} \times \mathrm{EF}_r \times \mathrm{ED}_{\mathrm{tot}}}{\mathrm{R}f\mathrm{D} \times \mathrm{BW}_a \times \mathrm{AT}_n}$$

Table 6.6 Health risk index for selected metals in the vegetables

	Ca	Mg	Na	K	Fe	Zn	Cu	Mn	Cr	Cd	Pb
Amaranthus hybridus	0.037	0.002	<0.001	0.001	0.100	0.002	0.007	0.017	0.649	0.057	0.576
Amaranthus spinosus	0.014	0.002	0.001	0.001	0.073	0.002	0.013	0.025	0.710	0.080	1.209
Amaranthus viridis	0.003	0.007	<0.001	0.002	0.059	0.005	0.011	0.021	0.257	0.058	0.345
Digeria muricata	0.026	0.002	<0.001	0.001	0.079	0.004	0.010	0.025	0.498	0.034	0.719
Dryopteris ramosa	0.002	0.001	<0.001	0.001	0.043	0.003	0.015	0.008	0.401	0.080	<0.001
Bidens bipinnata	0.028	0.001	<0.001	0.001	0.019	0.001	0.006	0.015	0.510	0.073	0.014
Centaurea calcitrapa	0.024	0.001	0.001	0.001	0.058	0.001	0.001	0.013	0.304	0.072	0.100
Cichorium intybus	0.002	0.001	<0.001	0.001	0.049	0.002	0.005	0.005	0.166	0.043	0.073
Launaea procumbens	0.020	0.001	<0.001	0.001	0.010	0.002	0.010	0.006	0.196	<0.001	<0.001
Sonchus asper	0.020	0.001	0.001	0.001	0.022	0.002	0.005	0.017	0.184	0.002	0.118
Sonchus oleraceous	0.021	0.001	0.001	0.001	0.033	0.002	0.005	0.014	0.207	<0.001	0.137
Taraxacum officinale	0.007	0.001	<0.001	<0.001	0.065	0.001	0.005	0.007	0.113	0.024	<0.001
Bombax malabaracum	0.007	0.001	<0.001	0.001	<0.001	0.001	0.003	0.001	0.056	0.005	<0.001
Capsella bursa-pastoris	0.008	<0.001	<0.001	<0.001	0.024	0.001	0.003	0.002	0.064	0.012	0.021
Nasturtium officinale	<0.001	<0.001	<0.001	<0.001	0.019	0.001	0.001	0.003	0.018	0.024	0.044
Bauhinia variegata	0.001	0.002	<0.001	0.001	0.011	0.002	0.008	0.007	0.205	0.040	0.061
Bauhinia variegata var. *candida*	0.002	0.001	<0.001	0.001	0.017	0.001	0.003	0.004	0.254	0.057	0.155
Bauhinia variegata var. *variegata*	0.001	0.001	<0.001	0.001	0.031	0.002	0.003	0.003	0.276	0.047	0.157
Silene conoidea	0.002	0.001	<0.001	0.001	0.052	0.001	0.002	0.006	0.167	0.056	0.046
Stellaria media	0.001	0.001	<0.001	0.001	0.045	0.002	0.002	0.006	0.065	0.041	0.040
Chenopodium album	0.001	0.001	<0.001	0.001	0.029	0.001	0.001	0.003	0.007	0.029	0.057
Commelina benghalensis	0.006	0.001	<0.001	<0.001	0.021	0.001	0.004	0.010	0.279	0.041	0.213
Evolvulus alsinoides	0.031	0.001	<0.001	0.001	0.100	0.002	0.015	0.010	0.681	0.107	<0.001
Dioscorea deltoidea	0.006	0.002	<0.001	0.001	0.014	0.001	0.005	0.005	0.841	0.126	0.142

Melilotus alba	0.003	0.003	<0.001	0.001	0.060	0.002	0.006	0.017	0.172	0.046	0.193
Melilotus indicus	0.003	0.004	<0.001	0.002	0.014	0.003	0.004	0.027	0.054	0.046	0.025
Lamium amplexicaule	0.003	0.002	<0.001	0.001	0.087	0.003	0.007	0.010	0.224	0.036	0.036
Origanum vulgare	0.008	0.006	0.001	0.002	0.083	0.005	0.013	0.028	0.217	0.075	0.385
Tulipa stellata	0.011	0.001	<0.001	<0.001	0.012	0.002	0.002	0.012	0.097	0.028	0.192
Malva parviflora	0.002	0.001	<0.001	<0.001	0.019	0.002	<0.001	0.008	0.083	0.021	0.055
Ficus carica	0.003	0.005	0.001	0.002	0.037	0.005	0.010	0.037	0.082	0.057	0.085
Ficus palmata	0.024	0.002	<0.001	0.001	0.058	0.003	0.012	0.019	0.489	<0.001	0.752
Oxalis corniculata	0.002	0.001	<0.001	0.001	0.010	0.001	0.004	0.005	0.073	0.023	0.091
Lathyrus aphaca	0.008	0.001	<0.001	<0.001	0.021	0.001	0.003	0.005	0.077	<0.001	<0.001
Medicago polymorpha	0.009	0.001	0.001	<0.001	0.003	0.001	0.001	0.003	0.260	0.055	0.249
Vicia faba	0.004	0.001	<0.001	0.001	0.042	0.003	0.006	0.016	0.044	0.038	0.099
Vicia sativa	0.023	0.001	<0.001	0.001	0.004	0.002	0.007	0.015	0.229	0.013	0.075
Plantago lanceolata	0.002	0.001	<0.001	0.001	0.019	0.002	0.004	0.005	0.168	0.038	0.050
Persicaria amplexicaulis	0.010	0.002	<0.001	0.001	0.133	0.002	0.006	0.078	0.139	0.013	0.073
Polygonum aviculare	0.015	0.002	<0.001	0.001	0.046	0.007	0.007	0.006	0.120	0.013	0.143
Rumex dentatus	0.010	0.001	<0.001	<0.001	0.057	0.001	0.006	0.009	0.436	<0.001	0.795
Rumex hastatus	0.009	0.001	0.001	<0.001	0.037	0.002	0.008	0.006	0.113	0.042	0.325
Rumex neplensis	0.032	0.001	0.001	<0.001	0.050	0.001	0.005	0.018	0.249	0.055	0.408
Portulaca olerecea	0.012	0.001	<0.001	<0.001	0.021	<0.001	0.002	0.013	0.037	0.023	0.146
Portulaca quadrifida	0.016	0.001	<0.001	<0.001	0.028	<0.001	0.001	0.020	0.052	0.030	0.213
Galium aparine	0.008	0.001	<0.001	<0.001	0.049	0.001	0.004	0.008	0.237	<0.001	0.275
Veronica arvensis	<0.001	<0.001	<0.001	<0.001	0.021	<0.001	0.001	0.002	0.039	0.023	0.028
Solanum nigrum	0.001	0.004	0.001	0.001	0.015	0.006	0.007	0.015	0.200	0.035	0.136
Pimpinella diversifolia	0.002	0.001	<0.001	0.001	0.146	0.001	0.004	0.005	0.227	0.074	0.348
Torilis leptophylla	0.026	0.001	<0.001	<0.001	0.009	0.001	0.006	0.008	0.085	0.006	0.125
DRI/ML	1,000	400	1,500	4,700	8.00	11.0	0.900	2.30	0.035	0.050	0.050

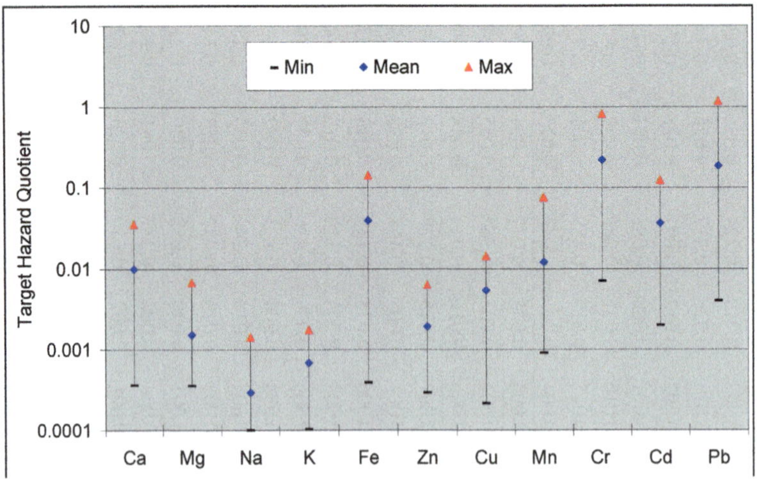

Fig. 6.69 Target hazard quotient (THQ) of the metals in wild edible vegetables

where C = Mean metal level in a specific vegetable on fresh weight basis (mg/kg), I = Ingestion rate (255 g/day/person), EF_r = Exposure frequency (350 days/year), ED_{tot} = Total exposure duration (30 years), BW_a = Average body weight, adult (65 kg), and AT_n = Averaging time, non-carcinogens ($ED_{tot} \times 365$ day/year).

The minimum to maximum range of THQ along with mean values for various metal levels in the vegetables is shown in Fig. 6.69. The calculated THQ values for the metals were found to be within the safe levels (<1) except the maximum value for Pb which slightly exceeded the limit. In most of the cases average THQ values were found to be less than 0.01 manifesting no significant non-carcinogenic health hazard associated with selected metals and the consumption of the wild vegetables was considered safe. Hence, these wild vegetables may be considered as alternate and healthy food resources in the study area.

6.9.3 *Hazard Index*

Hazard index (HI) is also used to determine the non-carcinogenic health risks for humans associated with the consumption of these vegetables (Yang et al. 2011; EC, 2006). The HI can be expressed as the sum of the hazard quotients for all selected metals (USEPA 2006):

$$HI = THQ_1 + THQ_2 + \cdots + THQ_n$$

where, THQ_{1-n} = THQs for 1–n metals.

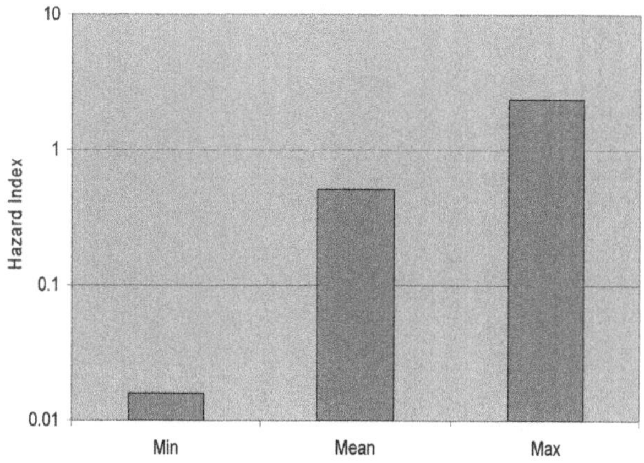

Fig. 6.70 Hazard index (HI) of the metals in wild edible vegetables

The mean value of HI along with the lowest and highest values is shown in Fig. 6.70. The average HI value was found to be significantly lower than the limit (1.0), indicating no considerable health hazard associated with the metal levels in wild vegetables. However, highest HI value exceeded the safety limit suggesting some remedial measures to be adopted in order to safeguard the consumers. Consequently, it may be considered the right time to launch an awareness campaign to control the anthropogenic pollutants in the study area.

6.9.4 Target Cancer Risk

TCR is used to assess the carcinogenic risks to the consumers and its acceptable limit is one out of a million (1×10^{-6}). The method to estimate TCR is also provided in USEPA Region III Risk-Based Concentration Table (USEPA 2006):

$$\text{TCR} = \frac{\text{C} \times \text{I} \times 10^{-3} \times \text{CPS}_o \times \text{EF}_r \times \text{ED}_{\text{tot}}}{\text{BW}_a \times \text{AT}_c}$$

where, C = Mean metal level in a specific vegetable on fresh weight basis (mg/kg), I = Ingestion rate (255 g/day/person), CPS_o = carcinogenic potency slope, oral $(\text{mg/kg/day})^{-1}$, EF_r = Exposure frequency (350 days/year), ED_{tot} = Total exposure duration (30 years), BW_a = Average body weight, adult (65 kg), and AT_c = Averaging time, carcinogens (70×365 days).

Since CPS_o values were not available for all selected metals, so the TCR of Cr, Cd, and Pb were computed in this study (Fig. 6.71) to indicate the life time carcinogenic risk to the consumers (USEPA 2010). Mean TCR values calculated for Cr and Cd were found to be higher than the acceptable risk limit (1×10^{-6}),

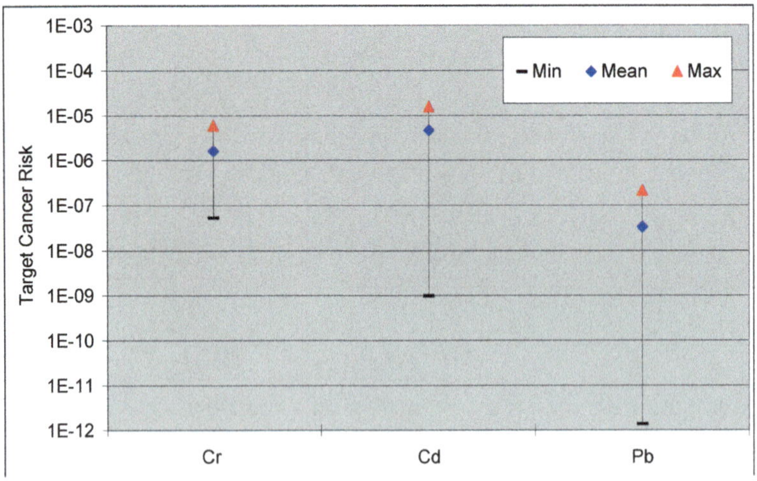

Fig. 6.71 Target cancer risk (TCR) of selected metals in wild edible vegetables

indicating that the consumers were exposed to lifetime carcinogenic risk with respect to Cr and Cd levels; however, no significant risk was associated with respect to Pb contents in the vegetables.

References

Abbasi, A. M., Iqbal, J., Khan, M. A., & Shah, M. H. (2013). Health risk assessment and multivariate apportionment of trace metals in wild leafy vegetables from Lesser Himalayas, Pakistan. *Ecotoxicology and Environmental Safety, 92*, 237–244.

Aberoumand, A. (2009). Nutritional evaluation of edible *Portulaca oleracia* as plant food. *Food Analytical Methods, 2*, 204–207.

Adriano, D. C. (1986). *Trace elements in the terrestrial environment* (p. 99). New York: Springer.

Akubugwo, I. E., Obasi, N. A., Chinyere, G. C., & Ugbogu, A. E. (2007). Nutritional and chemical value of *Amaranthus hybridus* L. leaves from Afikpo, Nigeria. *African Journal of Biotechnology, 6*(24), 2833–2839.

Anderson, R. A. (2000). Chromium in the prevention and control of diabetes. *Diabetes and Metabolism, 26*(1), 22–27.

Anonymous. (1998). *Health guideline for the use of wastewater in agriculture and aquaculture*. Tech. Rep. Sci. 778, 10, Report of WHO Science Group. Geneva: World Health Organization.

Arora, M., Kiran, B., Rani, S., Rani, A., Kaur, B., & Mittal, N. (2008). Heavy metal accumulation in vegetables irrigated with water from different sources. *Food Chemistry, 111*, 811–815.

Bahadur, A., Chaudhry, Z., Jan, G., Danish, M., Rehman, A. U., Ahmad, R., Khan, A., Khalid, S., Ullah, I., Shah, Z., Ali, F., Mushtaq, T., & Jan, F. G. (2011). Nutritional and elemental analysis of some selected fodder species used in traditional medicine. *African Journal of Pharmacy and Pharmacology, 5*(8), 1157–1161.

Birch, N. J., & Padgham, C. (1994). Potassium. In H. G. Seiler, A. Sigel, & H. Sigel (Eds.), *Handbook on metals in clinical and analytical chemistry* (p. 531). New York: Marcel Dekker.

Butt, S. J., & Haq, M. I. A. (1993). Vegetable cultivation-secret of manifold prosperity's science. *Technology and Development, 12*(3), 46–51.

Cao, Y., Chen, A., Jones, R. L., Radcliffe, J., Caldwell, K. L., & Dietrich, K. N. (2010). Does background post–natal methyl mercury exposure in toddlers affects cognition and behavior? *Neurotoxicology, 31*, 1–9.

Chwiej, J. (2010). The use of cluster and discriminant analysis in the investigations of the role of trace metals in the pathogenesis of Parkinson's disease. *Journal of Trace Elements in Medicine and Biology, 24*, 78–88.

Cui, Y. J., Zhu, Y. G., Zhai, R. H., Chen, D. Y., Huang, Y. Z., Qiu, Y., & Liang, J. Z. (2004). Transfer of metals from soil to vegetables in an area near a smelter in Nanning, China. *Environment International, 30*, 785–791.

EC. (2006). Commission regulation no 1881/2006: Setting maximum levels for certain contaminants in food stuffs. European Commission (EC). Retrieved July 17, 2011, from http://www.eurlex.europa.eu/LexUriServ/site/en/oj/2006/l_364/l36420061220en00050024.pdfS

FNB. (2004). Dietary reference intakes (DRIs): Recommended intakes for individuals. Food and Nutrition Board, Institute of Medicine, National Academies. Retrieved July 15, 2011, from http://www.iom.edu/Global/NewsAnnouncements/~/media/Files/ActivityFiles/Nutrition/DRIs/DRISummaryListing2.ashx

Gottelt, U., Henrion, G., & Henrion, R. (1997). Multivariate data analysis of trace element content in animal organs measured by simultaneous atomic absorption spectrometry. *Fresenius Journal of Analytical Chemistry, 358*, 781–784.

Hameed, I., Dasatgir, G., & Hussain, F. (2008). Nutritional and elemental analysis of some selected medicinal plants of the family Polygonaceae. *Pakistan Journal of Botany, 40*(6), 2493–2502.

Hashmi, D. R., Ismail, S., & Shaikh, G. H. (2007). Assessment of the level of trace metals in commonly edible vegetables locally available in the markets of Karachi city. *Pakistan Journal of Botany, 39*(3), 747–751.

Hiroaki, I., Motoki, I., Norie, S., Ribeka, T., Yoshio, K., Shiro, Y. (2013). Dietary cadmium intake and breast cancer risk in Japanese women: A case control study. *Journal of Hazardous Materials,* 2013. http://dx.doi.org/10.1016/j.jhazmat.2012.12.049.

Hussain, J., Khan, A. L., Rehman, N. U., Zainullah, K. F., Hussain, S. T., & Shinwari, Z. K. (2009). Proximate and nutrient investigations of selected medicinal plants species of Pakistan. *Pakistan Journal of Nutrition, 8*(5), 620–624.

Hussain, J., Rehman, N. U., Khan, A. L., Hussain, H., Harrasi, A. A., Ali, L., Sami, F., & Shinwari, Z. K. (2011). Determination of macro and micronutrients and nutritional prospects of six vegetable species of Mardan, Pakistan. *Pakistan Journal of Botany, 43*(6), 2829–2833.

Imran, M., Khan, H., Shah, M., Khan, R., & Khan, F. (2010). Chemical composition and antioxidant activity of certain *Morus* species. *Journal of Zhejiang University-Science B, 11* (12), 973–980.

Imran, M., Talpur, F. N., Jan, M. I., Khan, A., & Khan, I. (2007). Analysis of nutritional components of some wild edible plants. *Journal of Chemical Society of Pakistan, 29*(5), 500–508.

Iqbal, J., & Shah, M. H. (2011). Distribution, correlation and risk assessment of selected metals in urban soils from Islamabad, Pakistan. *Journal of Hazardous Materials, 192*, 887–898.

Jan, G., Khan, M. A., Ahmad, M., Iqbal, Z., Afzal, A., Afzal, M., Shah, G. M., Majid, A., Fiaz, M., Zafar, M., Waheed, A., & Gul, F. (2011). Nutritional analysis, micronutrients and chlorophyll contents of *Cichorium intybus* L. *Journal of Medicinal Plants Research, 5*(12), 2452–2456.

Jang, C. S., Liu, C. W., Lin, K. H., & Wang, S. W. (2006). Spatial analysis of potential carcinogenic risks associated with ingesting arsenic in aquacultural tilapia (Oreochromis mossambicus) in black foot disease hyperendemic areas. *Environmental Science and Technology, 40*, 1707–1713.

Jimoh, F. O., Adedapo, A. A., & Afolayan, A. J. (2011). Comparison of the nutritive value, antioxidant and antibacterial activities of *Sonchus asper* and *Sonchus oleraceous*. *Records of Natural Products, 5*(1), 29–42.

John, G. F., & Andrew, B. (2011). A lead isotopic study of the human bioaccessibility of lead in urban soils from Glasgow, Scotland. *Science of Total Environment, 409*, 4958–4965.

Kapaj, S., Peterson, H., Liber, K., & Bhattacharya, P. (2006). Human health effects from chronic arsenic poisoning a review. *Journal of Environment Science and Health, 41*, 2399–2428.

Khan, S., Rehman, S., Khan, A. Z., Khan, M. A., & Shah, M. T. (2010). Soil and vegetables enrichment with heavy metals from geological sources in Gilgit, northern Pakistan. *Ecotoxicology and Environmental Safety, 73*, 1820–1827.

Koel, M., & Kaliurand, M. R. (2010). *Green analytical chemistry* (p. 332). Cambridge, England: Royal Society of Chemistry.

Komárek, M., Chrastný, V., & Mihaljevič, M. (2008). Lead isotopes in environmental sciences: A review. *Environmental International, 34*, 562–577.

Lai, H. Y., Hseu, Z. Y., Chen, T. C., Chen, B. C., Guo, H. Y., & Chen, Z. S. (2010). Health risk-based assessment and management of heavy metals-contaminated soil sites in Taiwan. *International Journal of Environmental Research and Public Health, 7*, 3595–3614.

Li, Q. S., Chen, Y., Fu, H. B., Cui, Z. H., Shi, L., Wang, L. L., & Liu, Z. F. (2012). Health risk of heavy metals in food crops grown on reclaimed tidal flat soil in the Pearl River Estuary, China. *Journal of Hazardous Materials, 227*, 148–154.

Lin, H. J., Sunge, T., Cheng, C. Y., & Guo, H. R. (2013). Arsenic levels in drinking water and mortality of liver cancer in Taiwan. *Journal of Hazardous Materials, 262*, 1132–1138. http://dx.doi.org/10.1016/j.jhazmat.2012.12.049.

Luo, C., Liu, C., Wang, Y., Liu, X., Li, F., Zhang, G., & Li, X. (2011). Heavy metal contamination in soils and vegetables near an e-waste processing site, south China. *Journal of Hazardous Materials, 186*, 481–490.

Manzoor, S., Shah, M. H., Shaheen, N., Khalique, A., & Jaffar, M. (2006). Multivariate analysis of trace metals in textile effluents in relation to soil and groundwater. *Journal of Hazardous Materials, 137*, 31–37.

Morel, F. M., & Price, N. M. (2003). The biogeochemical cycles of trace metals in the oceans. *Science, 300*(5621), 944–947.

Mukhopadhyay, P. (2009). *Multivariate statistical analysis* (p. 549). Singapore: World Scientific Publishing.

Odhav, B., Beekrum, S., Akula, U., & Baijnath, H. (2007). Preliminary assessment of nutritional value of traditionally leafy vegetables in KwaZulu-Natal, South Africa. *Journal of Food Composition and Analysis, 20*, 430–435.

Ozcan, M. M., & Akbulut, M. (2007). Estimation of minerals, nitrate and nitrite contents of medicinal and aromatic plants used as spices, condiments and herbal tea. *Analytical, Nutritional and Clinical Methods, 106*, 852–858.

Pandey, R., Shubhashish, K., & Pandey, J. (2012). Dietary intake of pollutant aerosols via vegetables influenced by atmospheric deposition and wastewater irrigation. *Ecotoxicology and Environmental Safety, 76*, 200–208.

Park, R. M., Bena, J. F., Stayner, L. T., Smith, R. J., Gibb, H. J., & Lees, P. S. (2004). Hexavalent chromium and lung cancer in the chromate industry: A quantitative risk assessment. *Risk Analysis, 24*, 1099–1108.

Reddy, M.B., Chidambaram, M.V. & Bates, G.W. (1987). Iron Transport in Microbes, Plants and Animals. In: Winkelmann, G., Helm, Van der D. and Neilands, J.B. Eds., VCH, New York.

Schwarz, K. (1977). Essentiality versus toxicity of metals. Pp. 3-22 in Clinical Chemistry and Chemical Toxicology of Metals, S.S. Brown, ed. Amsterdam: Elsevier.

Seal, T. (2011). Determination of nutritive value, mineral conmtents and antioxidant activity of some edible plants from Meghalaya state, India. *Asian Journal of Applied Sciences, 4*(3), 238–246.

Shah, M. H., & Shaheen, N. (2007). Annual TSP and trace metal distribution in urban atmosphere of Islamabad in comparison with mega-cities of the world. *Human Ecological Risk Assessment, 13*, 884–899.

Shaheen, N., Shah, M. H., & Jaffar, M. (2005). A study of airborne selected metals and particle size distribution in relation to climatic variables and their source identification. *Water, Air, and Soil Pollution, 164*, 275–294.

Shi, J. C., Yu, X. L., Zhang, M. K., Lu, S. G., Wu, W. H., & Wu, J. J. (2011). Potential risks of copper, zinc, and cadmium pollution due to pig manure application in a soil–rice system under intensive farming: A case study of Nanhu, China. *Journal of Environmental Quality, 40*, 1695–1704.

Singh, A., Sharma, R. K., Agrawal, M., & Marshall, F. M. (2010). Health risk assessment of heavy metals via dietary intake of foodstuffs from the wastewater irrigated site of a dry tropical area of India. *Food and Chemical Toxicology, 48*, 611–619.

Thunus, L., & Lejeune, R. (1994). Cobalt, zinc. In H. G. Seiler, A. Sigel, & H. Sigel (Eds.), *Handbook on metals in clinical and analytical chemistry* (pp. 333–667). New York: Marcel Dekker.

Tiwari, K. K., Singh, N. K., Patel, M. P., Tiwari, M. R., & Rai, U. N. (2011). Metal contamination of soil and translocation in vegetables growing under industrial wastewater irrigated agricultural field of Vadodara, Gujarat, India. *Ecotoxicology and Environmental Safety, 74*, 1670–1677.

Toffery, S. (2011). *Data mining and statistics for decision making* (p. 716). West Sussex, England: Wiley.

Trichopoulou, A., Vasilopoulou, E., Hollman, P., Chamalides, C., Foufa, E., Kaloudis, T., Kromhout, D., Miskaki, P., Petrochilou, I., Poulima, E., Stafilakis, K., & Theophilou, D. (2000). Nutritional composition and flavonoids content of wild edible green and green pies: A potential rich source of antioxidant nutrients in the Mediterranean diet. *Food Chemistry, 70*, 319–323.

USEPA. (2000). *Risk based concentration table.*. Washington, DC: United States Environmental Protection Agency.

USEPA. (2006). *USEPA region III risk-based concentration table: Technical background information*. Washington, DC: Unites States Environmental Protection Agency.

USEPA. (2010). Risk-based Concentration Table. United States Environmental Protection Agency. Retrieved August 12, 2011, from http://www.epa.gov/reg3hwmd/risk/human/index.htmS

Wang, X. L., Sato, T., Xing, B. S., & Tao, S. (2005). Health risks of heavy metals to the general public in Tianjin, China via consumption of vegetables and fish. *Science of the Total Environment, 350*, 28–37.

Wilman, D., & Derrick, D. (1994). Concentration and availability to sheep of N, P, K, Ca, Mg and Na in chickweed, dandelion, dock, ribwort and spurrey, compared with perennial rye grass. *Journal of Agriculture Science, 122*, 217–223.

Yang, Q. W., Xu, Y., Liu, S. J., He, J. F., & Long, F. Y. (2011). Concentration and potential health risk of heavy metals in market vegetables in Chongqing, China. *Ecotoxicology and Environmental Safety, 74*, 1664–1669.

Zafar, M., Khan, M. A., Ahmad, M., Jan, G., Sultana, S., Ullah, K., Marwat, S. K., Ahmad, F., Jabeen, A., Nazir, A., Abbasi, A. M., Rehman, Z. U., & Ullah, Z. (2010). Elemental analysis of some medicinal plants used in traditional medicine by atomic absorption spectrophotometer (AAS). *Journal of Medicinal Plant Research, 4*(19), 1987–1990.

Zahoor, A., Jaffar, M., & Saqib, M. (2003). Elemental distribution in summer fruits of Pakistan. *Nutrition and Food Science, 33*(5), 203–207.

Zhuang, P., Mcbridge, B. B., Xia, H. P., Li, N. Y., & Li, Z. A. (2009). Health risk from heavy metals via consumption of food crops in the vicinity of Dabaoshan mine, South China. *Science of Total Environment, 407*, 1551–1561.

Chapter 7
Phytochemical Contents of Wild Edible Vegetables

7.1 Secondary Metabolites/Phytochemicals

Plant-based secondary metabolites and bioactive compounds play significant role in the prevention of different human diseases. Phenolic compounds, flavonoids, and vitamins are the important sources of plant-based natural antioxidants in human diet (Kim et al. 2003). These substances are capable of scavenging free super-oxide radicals, reducing the risk of cancer and protecting biological systems against the harmful effects of oxidative processes on macromolecules, such as carbohydrates, proteins, lipids, and DNA (Halliwell and Gutteridge 1990). Phenolic compounds confer unique taste, flavor, and health-promoting properties found in vegetables and fruits (Tomas-Barberan and Espin 2001). Though plant base phenolics and flavonoids have various biological activities, like antiviral, anticancer, antidiabetic, etc., they also contribute along with antioxidant vitamins and enzymes, to support the total antioxidant defense system of the human body (Mandal et al. 2010). In addition phenolic compounds and their functions have been the subject of a great number of agricultural, biological, chemical, and medical studies. In recent years, the importance of antioxidant activities of phenolic compounds and their potential usage in processed foods as a natural antioxidant compound has reached a new level (Gryglewski et al. 1987).

Like as phenolic, flavonoids are also widely distributed secondary metabolites in plants (Manach et al. 2004). Tea, apple, tomato, cherry, onion, parsley, soya-beans, grape fruit, legumes, orange, lemon, and neem are the major dietary sources of flavonoids in the form of flavonols, flavones, isoflavones, and flavonones (Beecher 2003). In recent years flavonoids have gained much attention because of their broad biological and pharmacological activities including antimicrobial, cyto-toxicity, anti-inflammatory, anti-tumor, and the best one, their capacity to act as powerful antioxidants, which can protect the human body from free radicals and reactive oxygen species (Bors et al. 1990). The antioxidant potential of flavonoids mainly depends upon their molecular structure, the position of hydroxyl groups (Formica

A.M. Abbasi et al., *Wild Edible Vegetables of Lesser Himalayas*,
DOI 10.1007/978-3-319-09543-1_7

and Regelson 1995). The recognized dietary antioxidants are vitamin C, vitamin E, selenium, and carotenoids. However, recent studies have confirmed that flavonoids found in fruits and vegetables are a potential source of natural antioxidants (Sies et al. 1992). It was observed that flavonoids such as luteolin and cathechins are better antioxidants than the nutrients antioxidants such as vitamin C, vitamin E, and β-carotene (Herrmann 1976).

Flavonoids provide attractive colors to flowers which are helpful to plant pollinators (Harborne 1976). In leaves, these compounds promote physiological survival of the plant, protecting it from fungal pathogens and UV-radiations (Harborne 1993). In addition, flavonoids are involved in photosensitization, energy transfer, the actions of plant growth hormones and growth regulators, control of respiration and photosynthesis, morphogenesis, and sex determination (Harborne and Baxter 1999). On the other hand, phenolic compounds are associated with diverse functions in plants, including nutrient uptake, protein synthesis, enzyme activity, photosynthesis, structural components, and allelopathy (Einhellig et al. 1986).

Flavonoids and phenolics are well-known for their antioxidant activity (Kukic et al. 2006). Plants are potential sources of invaluable antioxidants. Results of various studies verified that the antioxidants of plant origin with free radical scavenging properties could have enormous importance as therapeutic agents in diseases caused due to oxidative stress (Ramchoun et al. 2009). Because of nutritional difference requirement for antioxidants in Indo-Pak conditions differs from that of industrialized western countries. There are different institutes and laboratories from China, Pakistan, and India working on the antioxidant effect of plant compounds, mainly derived from natural sources that are capable of protecting against cell damage and different diseases.

Vitamin C or ascorbic acid is a water-soluble vitamin that is necessary for normal growth and development. It forms an important protein used to make skin, tendons, ligaments, and blood vessels. It heals wounds and form scar tissue, repair and maintain cartilage, bones, and teeth. Ascorbic acid is a weak acid which is structurally similar to glucose. Ascorbic acid is well-known for its antioxidant activity, acting as a reducing agent to reverse oxidation in liquids. Individuals experiencing oxidative stress have ascorbate blood levels lower than, 45 μmol/L, compared to healthy individual who range between 61 and 80 μmol/L (Schorah et al. 1996). Vitamin C may contribute to decreased risk of cardiovascular disease and strokes through reduction in systolic blood pressure (Fotherby et al. 2000).

In this section we have focused on phytochemical contents, i.e., total phenolics, flavonoids, flavonols, and ascorbic acid in wild edible vegetables used by the local communities of Lesser Himalayas-Pakistan. Though different techniques are used to estimate the photochemical contents in vegetables and fruits, we determine total phenolics, flavonoids, flavonols, and ascorbic acid contents in water and acetone extracts of wild edible vegetables on fresh weight basis by using UV-spectrophotometer.

7.2 Methodology

7.2.1 Extraction

The freshly collected vegetable samples were rinsed under tap and deionized water, then cut into small pieces and dried at 70 °C for 48 h in an electric oven. The dried samples were ground into fine powder.

A two-step extraction process (Ji et al. 2011) was adopted to assess the hydrophilic (water) and hydrophobic (acetone) contributions. First the samples were extracted with water (hydrophilic part) and then with acetone (hydrophobic part). About 1 g of dried powered sample was mixed with 10 mL of distilled water and this homogenate was centrifuged at 6,000 rpm for 15 min. This extraction procedure is repeated three times with the solid residue and is further extracted thrice in acetone (1:10 w/v).

7.2.2 Determination of Total Phenolic Contents

The total phenolic contents of wild edible vegetables in water and acetone extracts were determined according to the method reported by Lin et al. (2011). Aliquots of 1.0 mL of water or acetone extracts were mixed with 5 mL of tenfold diluted Folin-Ciocalteu reagent and 4 mL of 7.5 % sodium carbonate. The mixture was allowed to stand for 90 min at room temperature before the absorbance was measured at 760 nm spectrophotometrically. Gallic acid was used as standard and the final results were expressed as gallic acid equivalents (GAE) on fresh weight basis.

7.2.3 Determination of Total Flavonoid Contents

The flavonoid contents in water and acetone extracts of wild edible vegetables were measured using a colorimetric method described by Lin et al. (2011). A volume of 5 mL of water or acetone extract was transferred to the test tube, mixed with the 0.3 mL of 5 % sodium nitrite for 5 min. Then 0.3 mL of 10 % aluminium chloride was added. After 6 min, reaction was stopped by addition of 2 mL sodium hydroxide. The mixture was further diluted with distilled water up to 10 mL. The absorbance of the mixture was immediately measured at 510 nm by using UV spectrophotometer. Rutin was used as standard and the flavonoid contents were calculated and expressed as rutin equivalents (Rt) on fresh weight basis.

7.2.4 Determination of Total Flavonol Contents

Total flavonol contents in the water and acetone extracts were deliberate using the method of Kumaran and Karunakaran (2006). About 2.0 mL of sample was taken and 2.0 mL of 2 % $AlCl_3$ and 3 mL sodium acetate (50 g/L) solutions were added into it. The absorption was measured at 440 nm by UV spectrophotometer after 2.5 h at 20 °C. Total flavonols contents were calculated as Rutin (mg/g) on fresh weight basis.

7.2.5 Determination of Ascorbic Acid Contents

Ascorbic acid contents in wild edible vegetables were determined according to the method of Klein and Perry (1982). Water and acetone extracts were re-extracted with meta-phosphoric acid (1 %, 10 mL) for 45 min at room temperature and filtered through Whatman No 4 filter paper. The filtrate (1.0 mL) was mixed with 9 mL of 2,6-dichloroindophenol (0.8 g/1,000 mL) and the absorbance was measured within 30 min at 515 nm by using UV spectrophotometer. Ascorbic acid contents were calculated on the basis of calibration curve of L-ascorbic acid (0.006–0.1 mg/mL; $y = 3.006x + 0.007$; $R^2 = 0.999$), and results were expressed as ascorbic acid equivalents on fresh weight basis.

7.3 Phytochemical Contents of Wild Vegetables

7.3.1 *Amaranthus hybridus* L.

In the leaves of *A. hybridus*, total phenolic and flavonoid contents on fresh weight basis expressed as mg GAE/100 g, FW and mg Rt/100 g, FW, respectively, were estimated higher as compared to the acetone extracts. However, flavonols contents expressed as Rt/100 g, FW were found more in acetone extract than water extract. Similarly high concentration of ascorbic acid contents was noted in acetone extract expressed as mg AA/100 g, FW than water extract (Fig. 7.1).

7.3.2 *Amaranthus spinosus* L.

Phytochemical contents deliberated in the leaves of *A. spinosus* are shown in Fig. 7.2, which indicate that water extract showed higher concentration of total phenolic contents (mg GAE/100 g, FW) as compared to acetone extract. On the other hand acetone extracts showed higher concentrations of flavonols and

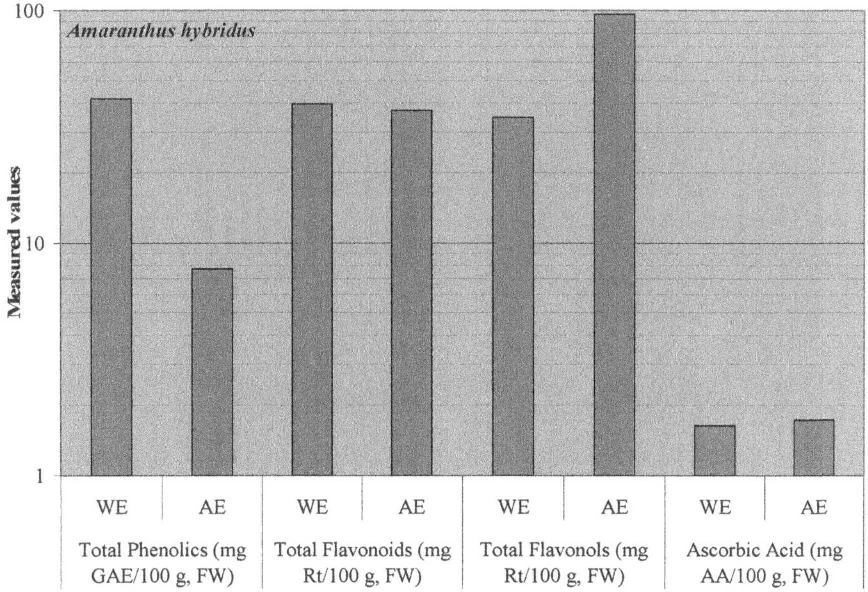

Fig. 7.1 Phytochemical contents in *Amaranthus hybridus*

flavonoids contents (mg Rt/100 g, FW) than water extracts. Relatively higher levels of ascorbic acid contents were estimated in water extract (mg AA/100 g, FW) as compared to the acetone extract.

7.3.3 *Amaranthus viridis L.*

Water extracts showed significant higher value of total phenolic contents (GAE/100 g, FW) as compared to acetone extracts. However, flavonols and flavonoids contents (mg Rt/100 g, FW) were higher in acetone extracts than water extracts. Ascorbic acid contents (mg AA/100 g, FW) were found to be almost equal in both water and acetone extracts (Fig. 7.3).

7.3.4 *Digera muricata (L.) Mart.*

On fresh weight basis phytochemical investigation revealed that young leaves of *D. muricata* showed higher total phenolics and flavonoids contents in water extracts expressed as mg GAE/100 g, FW and Rt/100 g, FW, respectively, as compared to acetone extracts (Fig. 7.4). Flavonols contents (Rt/100 g, FW) were found relatively

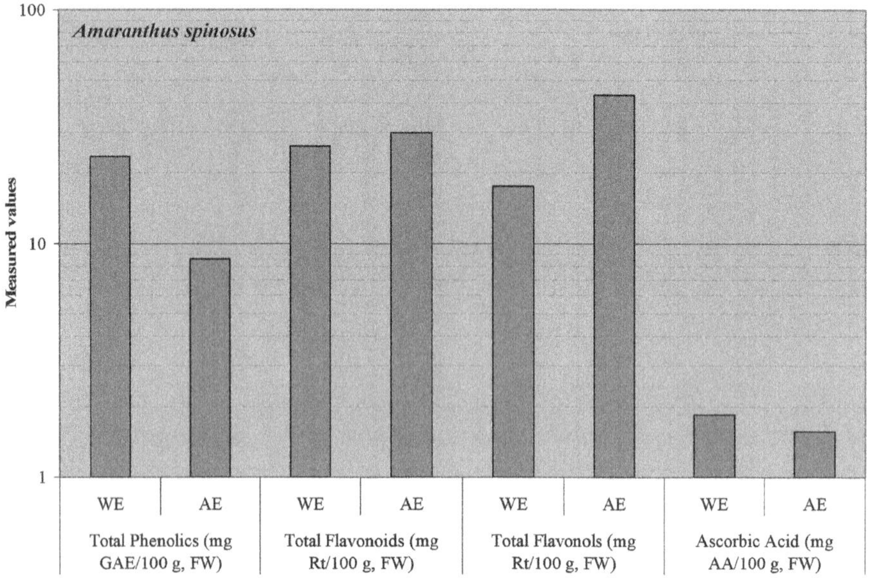

Fig. 7.2 Phytochemical contents in *Amaranthus spinosus*

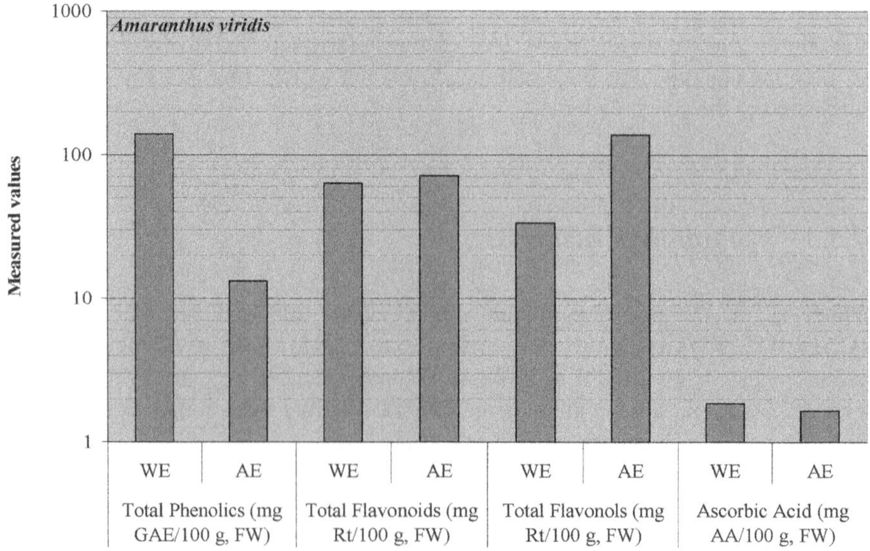

Fig. 7.3 Phytochemical contents in *Amaranthus viridis*

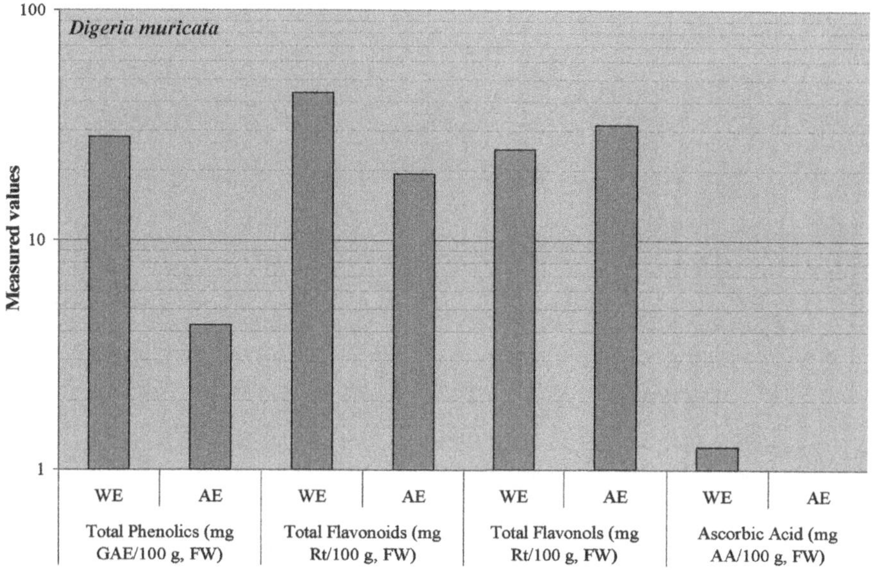

Fig. 7.4 Phytochemical contents in *Digera muricata*

higher in acetone extract than water extract. Comparatively higher ascorbic acid contents (AA/100 g, FW) were recorded in water extract than acetone extract.

7.3.5 Dryopteris ramosa (Hope) C. Chr.

Phytochemical contents of the leaves of *D. ramosa* as mentioned (Fig. 7.5) showed that water extracts exhibited higher values for total phenolics (GAE/100 g, FW) and flavonoids contents (mg Rt/100 g, FW) as compared to acetone extracts. However, flavonols (mg Rt/100 g, FW) and ascorbic acid contents (mg AA/100 g, FW) were estimated more in acetone extracts than water extracts.

7.3.6 Bidens bipinnata L.

Phytochemical investigation of *B. bipinnata* presented in Fig. 7.6 revealed that water extracts showed higher concentrations for total phenolics (mg GAE/100 g, FW) and flavonoids contents (mg Rt/100 g, FW), respectively, as compared to acetone extracts. Significantly higher flavonols contents (mg Rt/100 g, FW) were observed in acetone extract as compared to water extract, whereas ascorbic acid value was recorded higher in water extract (mg AA/100 g, FW) than acetone extract.

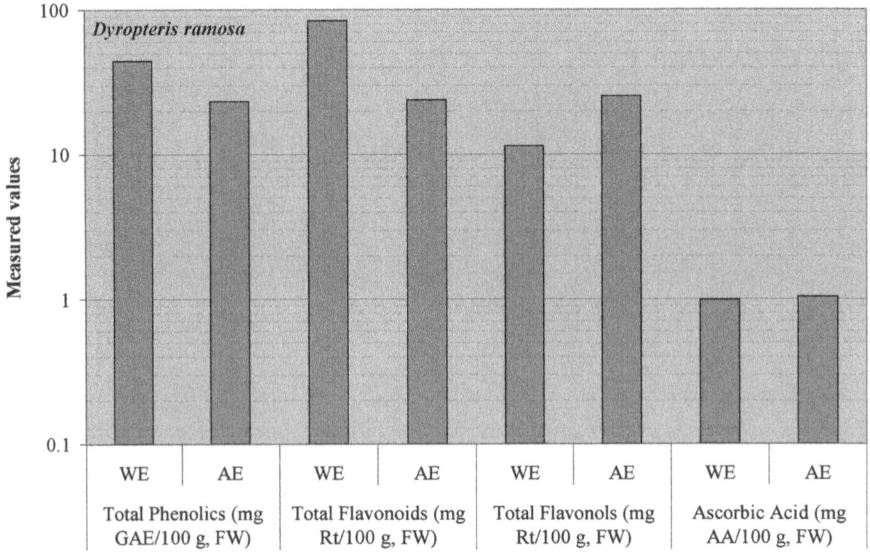

Fig. 7.5 Phytochemical contents in *Dryopteris ramosa*

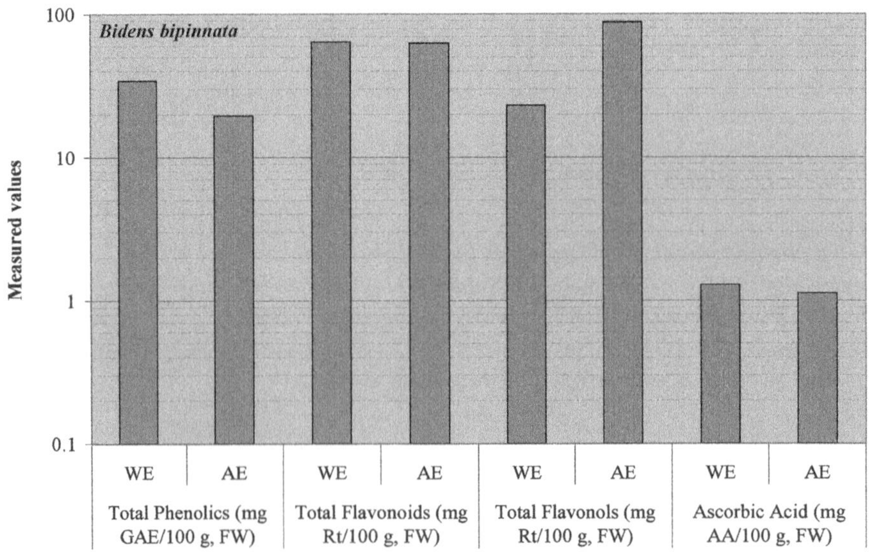

Fig. 7.6 Phytochemical contents in *Bidens bipinnata*

7.3.7 *Centaurea calcitrapa* L.

Phytochemical contents of *C. calcitrapa* are given in Fig. 7.7, which shows that water extract of fresh edible leaves exhibited higher concentration of total phenolic contents (mg GAE/100 g, FW) as compared to acetone extract. Similarly flavonoid

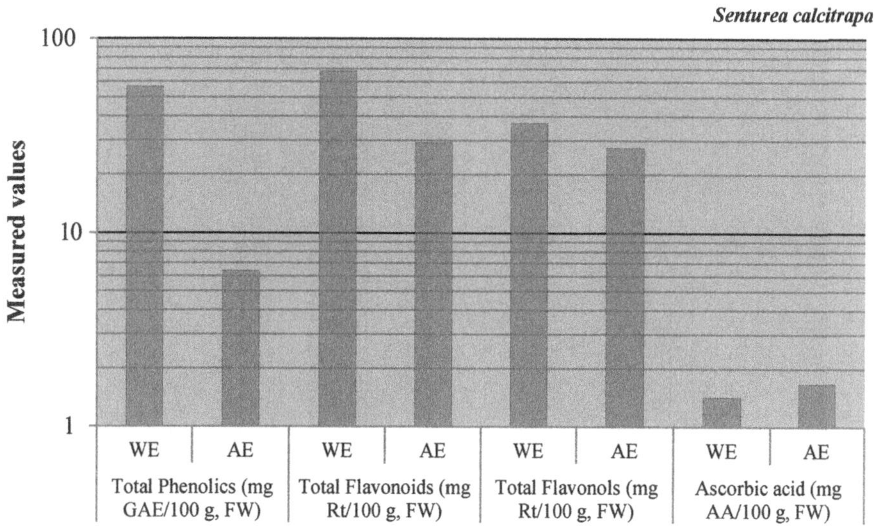

Fig. 7.7 Phytochemical contents in *Centaurea calcitrapa*

and flavonol contents (mg Rt/100 g, FW) were also measured higher in water extracts as compared to acetone extracts on fresh weight basis. However, acetone extract showed moderately higher value of ascorbic acid contents (mg AA/100 g, FW) as compared to the water extract on fresh weight basis.

7.3.8 Cichorium intybus L.

Phytochemical assessment of the edible leaves of *C. intybus* indicates that the concentrations of total phenolic (mg GAE/100 g, FW) and flavonoid contents (mg Rt/100 g, FW) are comparatively higher than acetone extracts, whereas flavonols contents (mg Rt/100 g, FW) were found higher in acetone extract as compared to water extract. However, water extract showed more ascorbic acid contents (mg AA/100 g, FW) than acetone extract (Fig. 7.8).

7.3.9 Launaea procumbens (Roxb.)

Phytochemical profile of *L. procumbens* demonstrated that water extracts exhibited higher phenolics (mg GAE/100 g, FW) and flavonoids contents (mg Rt/100 g, FW) as compared to acetone extracts. However, flavonols contents (mg Rt/100 g, FW) and ascorbic acid concentration (mg AA/100 g, FW) were estimated comparatively higher in acetone extract than water extract (Fig. 7.9).

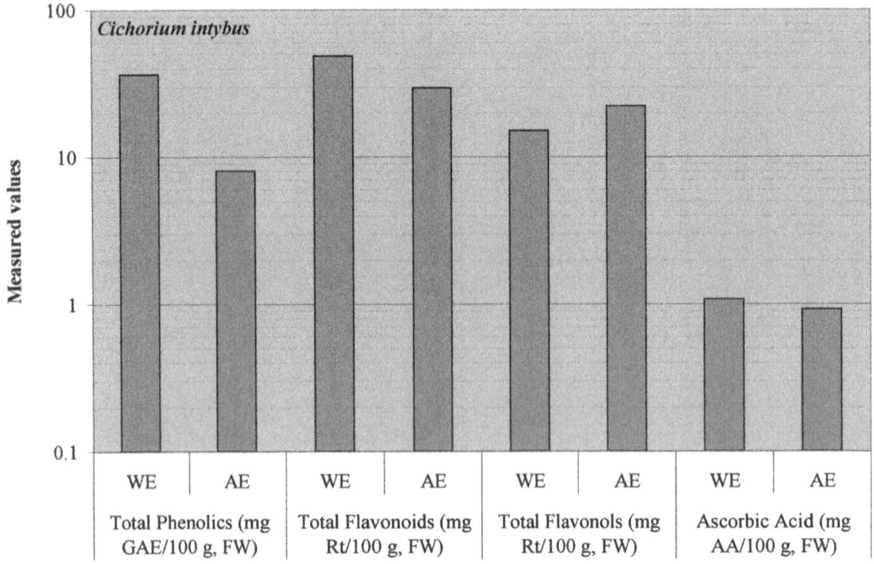

Fig. 7.8 Phytochemical contents in *Cichorium intybus*

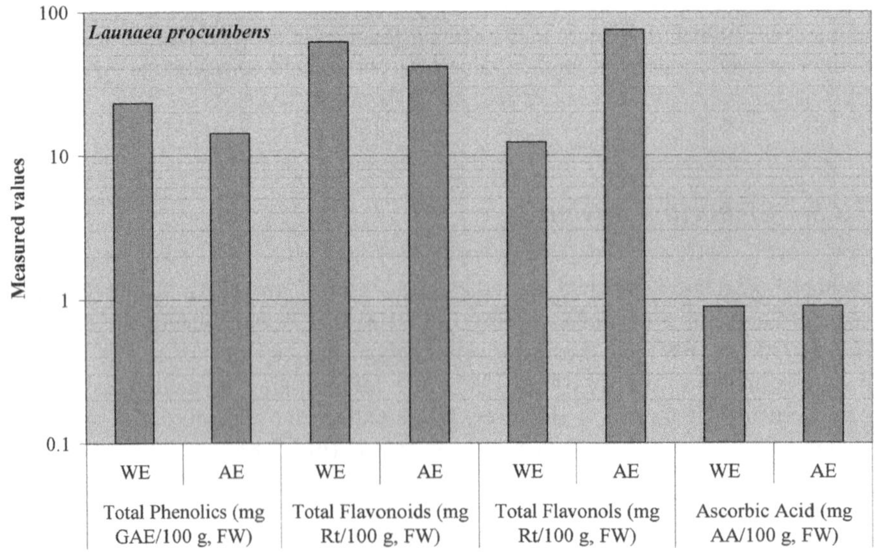

Fig. 7.9 Phytochemical contents in *Launaea procumbens*

7.3.10 *Sonchus asper* L.

Phytochemical contents investigation of *S. asper* revealed moderately higher concentration of total phenolics (mg GAE/100 g, FW) in water extract as compared to acetone extract (Fig. 7.10). However, flavonoids and flavonols contents

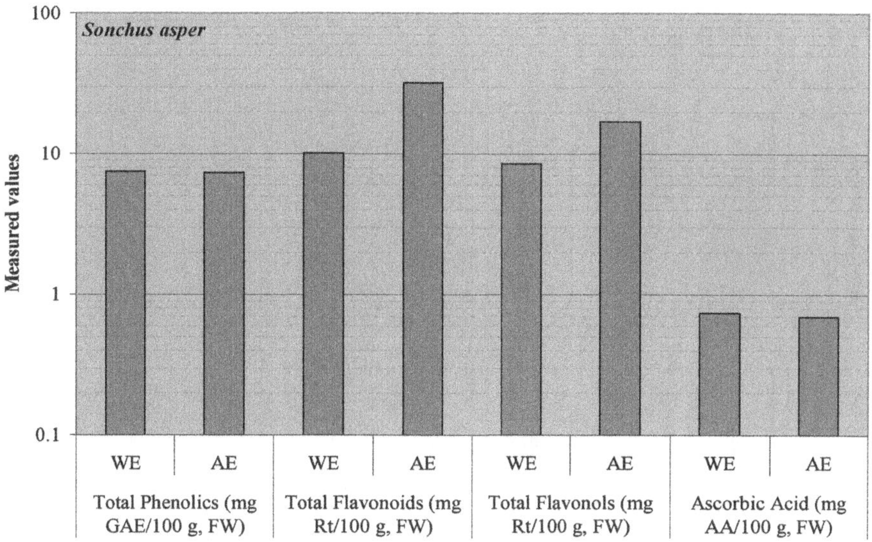

Fig. 7.10 Phytochemical contents in *Sonchus asper*

(mg Rt/100 g, FW) were estimated higher in acetone extract than water extract. Ascorbic acid contents were found relatively higher in water extract (mg AA/100 g, FW) as compared to acetone extract.

7.3.11 Sonchus oleraceous L.

Results of the phytochemicals contents in water and acetone extracts of *S. oleraceous* are mentioned in Fig. 7.11, which revealed that water extract exhibited higher total phenolic contents (mg GAE/100 g, FW) than acetone extract on fresh weight basis. On the other hand, flavonoid and flavonol contents (mg Rt/100 g, FW) were measured higher in acetone extract as compared to water extract. However, ascorbic acid contents (mg AA/100 g, FW) were found relatively higher in water extract than acetone extract.

7.3.12 Taraxacum officinale L.

Comparative contributions of phytochemical constituents in acetone and water extracts of *T. officinale* leaves are also given in Fig. 7.12. Total phenolic contents (mg GAE/100 g, FW) were found higher in water extract than acetone extract. However, acetone extracts showed higher concentrations of

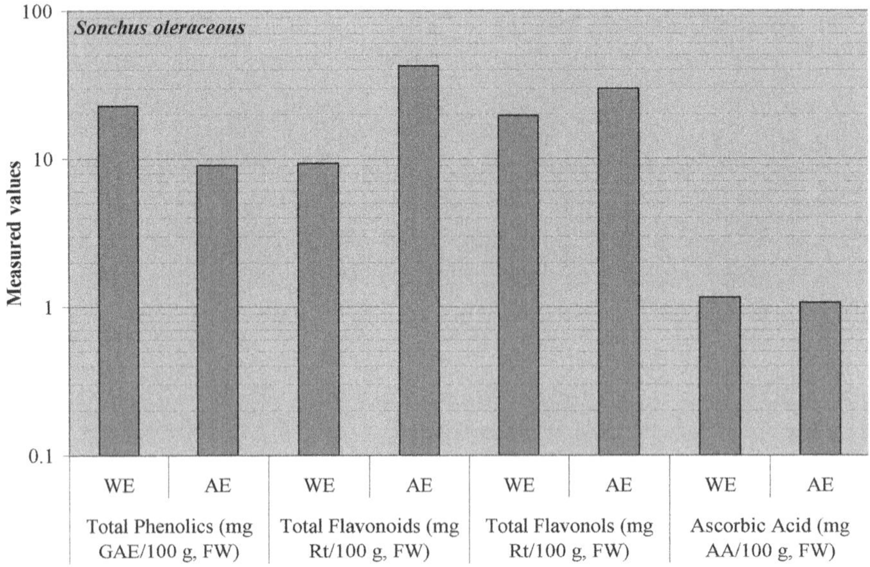

Fig. 7.11 Phytochemical contents in *Sonchus oleraceous*

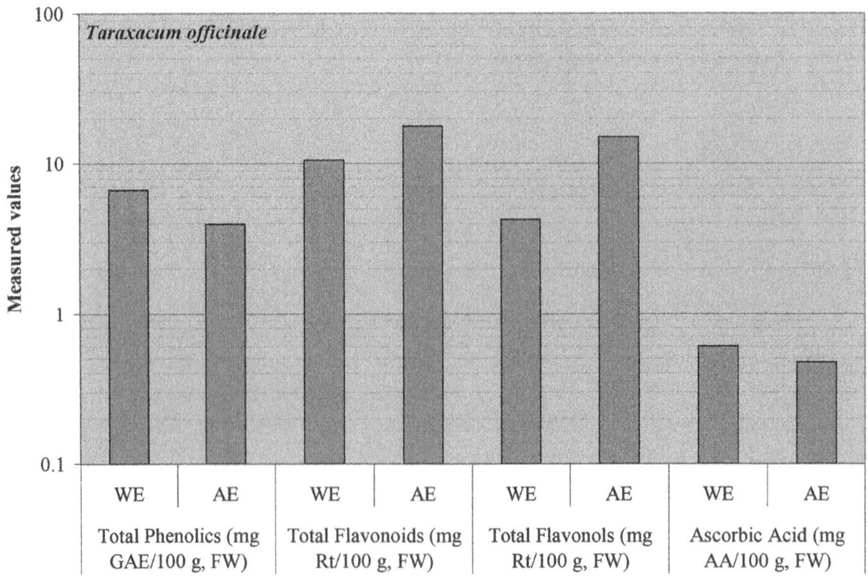

Fig. 7.12 Phytochemical contents in *Taraxacum officinale*

flavonoids and flavonols (mg Rt/100 g, FW) as compared to water extracts. Comparatively higher value of ascorbic acid contents (mg AA/100 g, FW) were found in water extract (mg AA/100 g, FW) as compared to acetone extract (mg AA/100 g, FW)

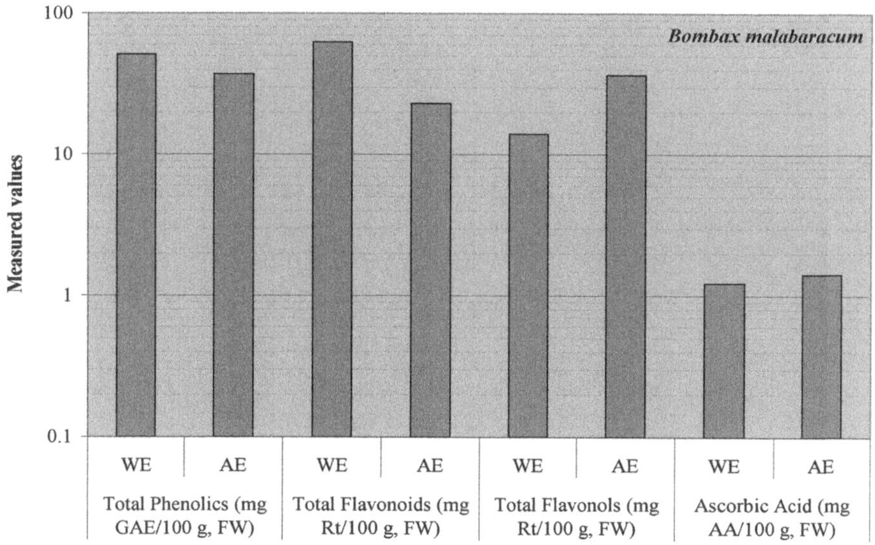

Fig. 7.13 Phytochemical contents in *Bombax malabaracum*

7.3.13 Bombax malabaracum DC.

Measured levels of phytochemical contents in both water and acetone extracts of *B. malabaracum* are presented in Fig. 7.13, which shows comparatively higher levels of total phenolics (mg GAE/100 g, FW) and flavonoids contents (mg Rt/100 g, FW) in water extracts as compared to acetone extracts. However, both flavonols contents (mg Rt/100 g, FW) and ascorbic acid contents (mg AA/100 g, FW) were calculated higher in acetone extract as compared to water extract on fresh weight basis.

7.3.14 Capsella bursa-pastoris (L.) Medic.

Phytochemical assessment of *C. bursa-pastoris* is given in Fig. 7.14, which indicates that water extract showed higher concentration of total phenolics (mg GAE/ 100 g, FW) as compared to acetone extract. Flavonoids and flavonol contents (mg Rt/100 g, FW) were recorded higher in acetone extract than water extract, while water extract showed slightly higher value of ascorbic acid contents (mg AA/100 g, FW) than acetone extract.

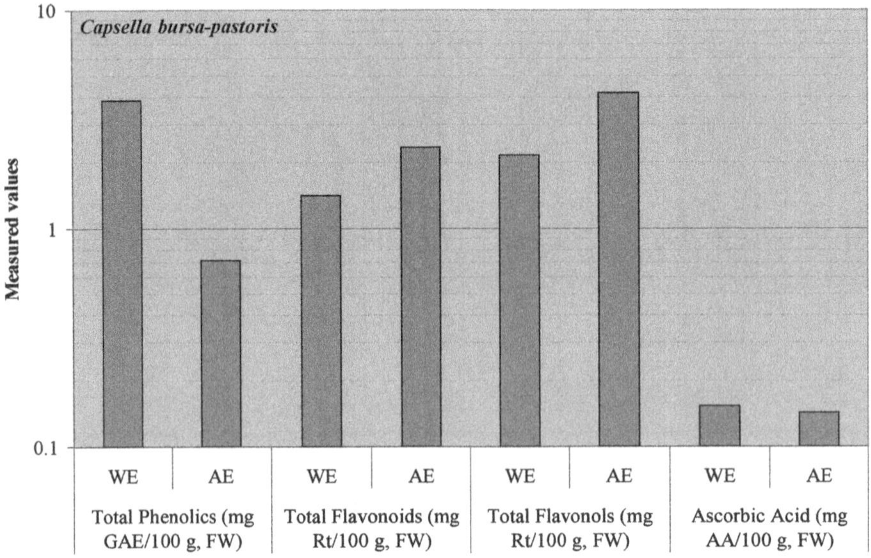

Fig. 7.14 Phytochemical contents in *Capsella bursa-pastoris*

7.3.15 *Nasturtium officinale R. Br.*

Phytochemical profile of *N. officinale* demonstrated that water extract showed higher concentration of total phenolics contents (mg GAE/100 g, FW) than acetone extract. Total flavonoid and flavonol contents were observed higher in acetone extract (mg Rt/100 g, FW) as compared to water extract on fresh weight basis (Fig. 7.15). Water extract showed slightly higher value of ascorbic acid (mg AA/100 g, FW) than acetone extract on fresh weight basis.

7.3.16 *Bauhinia variegata L.*

Phytochemical profile of *B. variegata* leaves used as vegetable is given in Fig. 7.16. It was observed that water extracts exhibited higher concentrations of phenolic contents (mg GAE/100 g, FW) and flavonoids contents (mg Rt/100 g, FW) as compared to acetone extracts. However, total flavonols (mg Rt/100 g, FW) and ascorbic acid contents (mg AA/100 g, FW) were found to be more in acetone extracts than water extract on fresh weight basis.

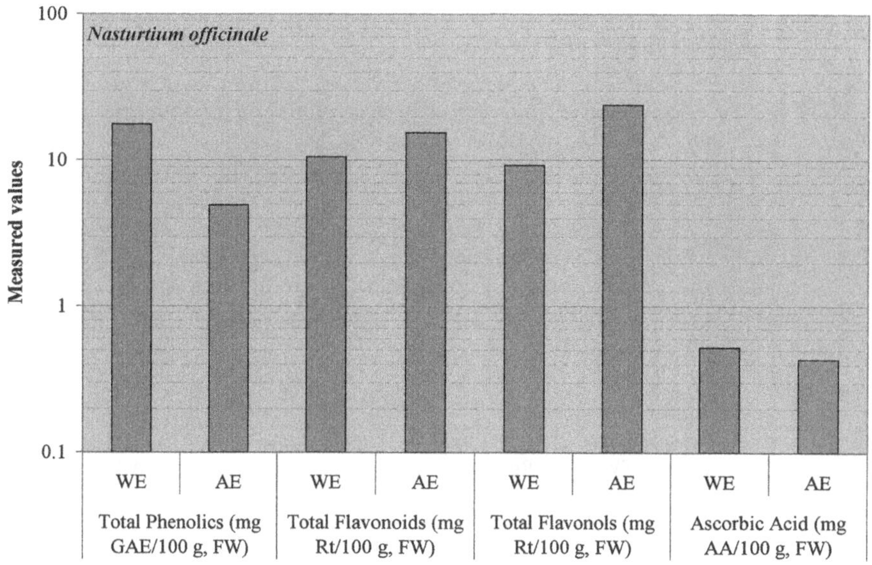

Fig. 7.15 Phytochemical contents in *Nasturtium officinale*

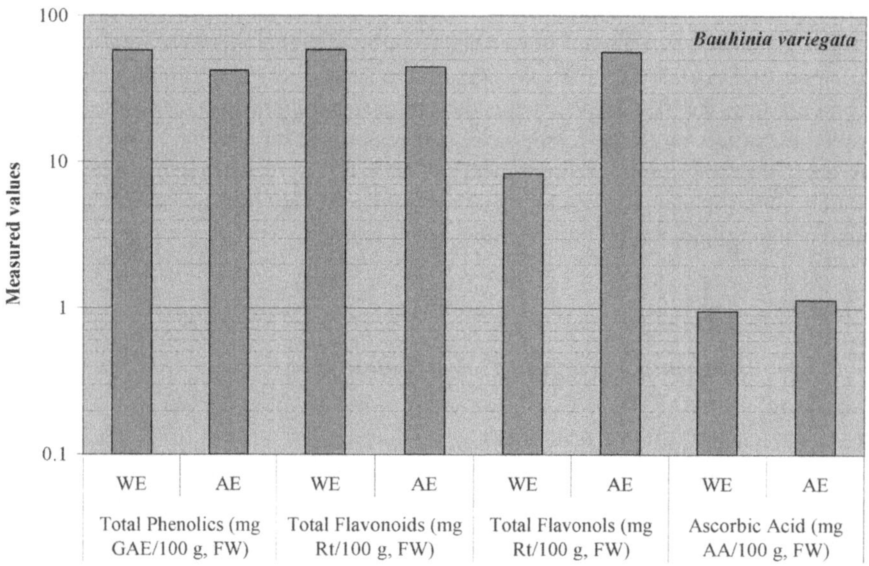

Fig. 7.16 Phytochemical contents in *Bauhinia variegata*

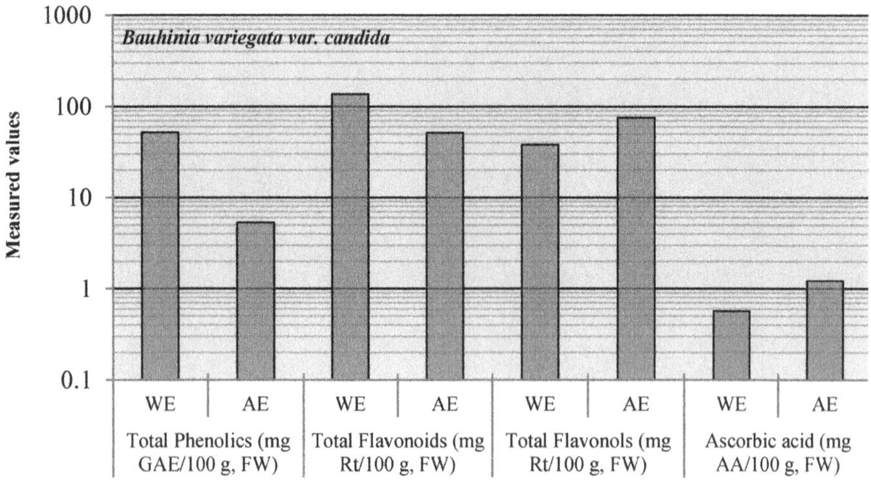

Fig. 7.17 Phytochemical contents in *B. variegata var. candida*

7.3.17 Bauhinia variegata var. candida Voigt

Phytochemical profile of *B. variegata*.var. *candida* is given in Fig. 7.17, which depicted that water extract of flower buds exhibited higher values of total phenolic contents (mg GAE/100 g, FW) as compared to acetone extract. Similarly flavonoid contents (mg Rt/100 g, FW) were also measured higher in water extracts than acetone extracts on fresh weight basis, whereas in the case of flavonols comparatively higher concentration (mg Rt/100 g, FW) was observed in acetone extract than water extract. Again acetone extract showed moderately higher value of ascorbic acid contents (mg AA/100 g, FW) than water extract.

7.3.18 Bauhinia variegata L. var. variegata

Phytochemical profile of *B. variegata*.var. *variegata* is given in Fig. 7.18, which shows that water extract of flower buds exhibited higher concentration of total phenolic contents (mg GAE/100 g, FW) as compared to acetone extract. Similarly flavonoid contents (mg Rt/100 g, FW) were also measured higher in water extracts as compared to acetone extracts on fresh weight basis, whereas in the case of flavonols comparatively higher concentration (mg Rt/100 g, FW) was observed in acetone extract than water extract. On fresh weight basis again acetone extract showed moderately higher value of ascorbic acid contents (mg AA/100 g, FW) than water extract.

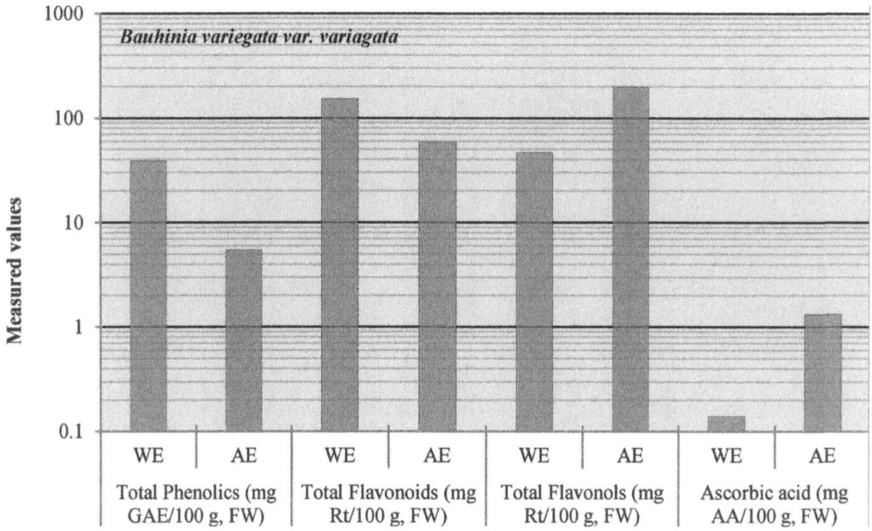

Fig. 7.18 Phytochemical contents in *B. variegata var. variegata*

7.3.19 Silene conoidea L.

Phytochemical contents measured in the leaves of *S. conoidea* are depicted in Fig. 7.19, which indicates that water extracts contained higher phenolic (mg GAE/100 g, FW) and flavonoid contents (mg Rt/100 g, FW) as compared to acetone extracts on fresh weight basis. In the case of flavonols comparatively higher concentration (mg Rt/100 g, FW) was observed in acetone extract than water extract. On fresh weight basis water extract showed moderately higher value of ascorbic acid contents (mg AA/100 g, FW) than acetone extract.

7.3.20 Stellaria media (L.) Cyr.

Phytochemical investigation revealed that water extract showed higher phenolic contents (mg GAE/100 g, FW) than acetone extract. In the case of flavonoids and flavonols higher concentrations (mg Rt/100 g, FW) were observed in acetone extracts than water extracts. Comparatively, water extract showed higher concentration of ascorbic acid expressed as mg AA/100 g, FW than acetone extract (Fig. 7.20).

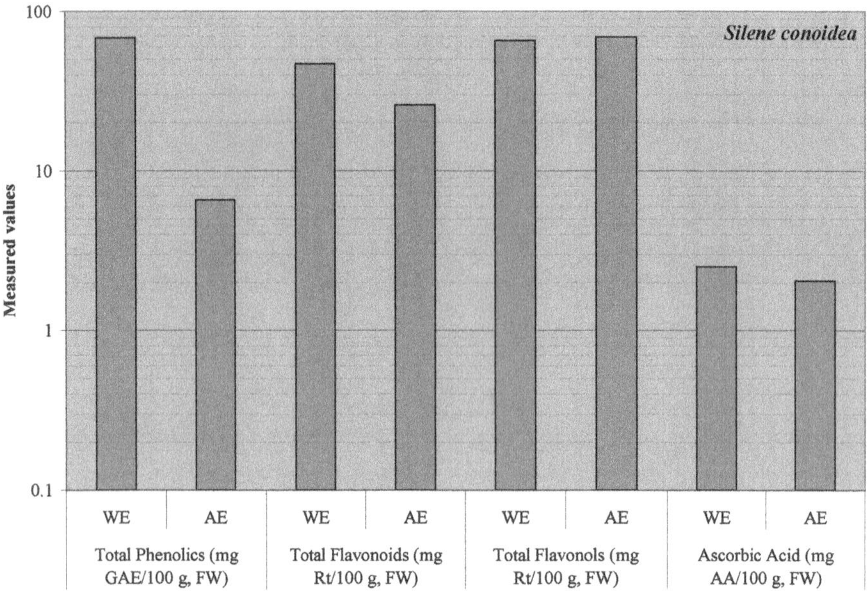

Fig. 7.19 Phytochemical contents in *Silene conoidea*

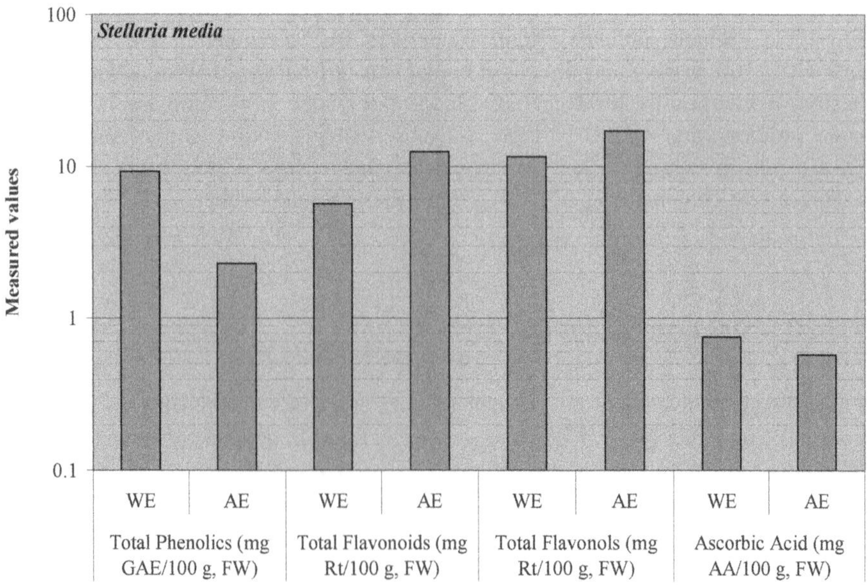

Fig. 7.20 Phytochemical contents in *Stellaria media*

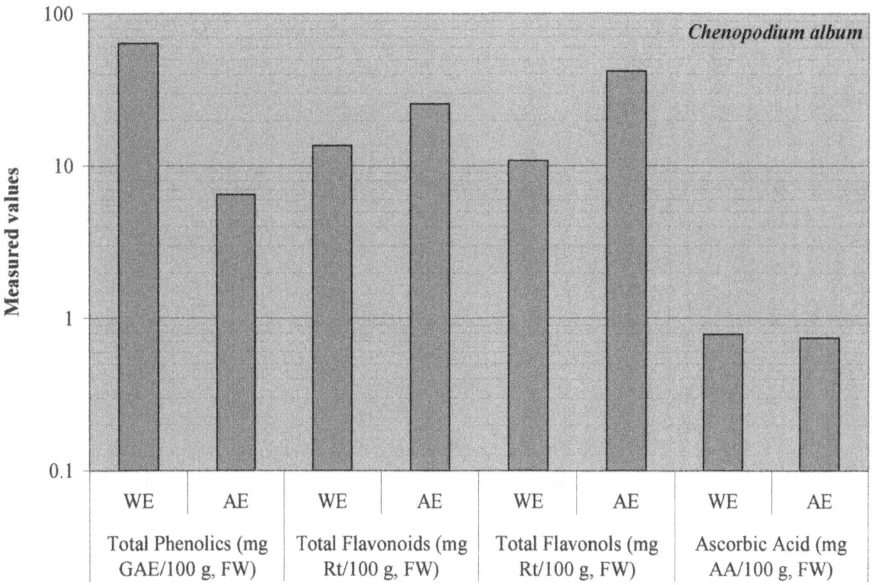

Fig. 7.21 Phytochemical contents in *Chenopodium album*

7.3.21 *Chenopodium album* L.

Phytochemical investigation of *C. album* is shown in Fig. 7.21, which demonstrated that water extract showed higher value of total phenolics contents (mg GAE/100 g, FW) than acetone extract. Flavonoid and flavonol contents were found higher in acetone extracts (mg Rt/100 g, FW) as compared to water extracts on fresh weight basis. Water extract showed slightly more concentration of ascorbic acid contents (mg AA/100 g, FW) than acetone extract on fresh weight basis.

7.3.22 *Commelina benghalensis* L.

Water extract showed significantly higher concentration of total phenolics (mg GAE/100 g, FW) as compared to acetone extract. In the case of flavonoids higher value (mg Rt/100 g, FW) was recorded in acetone extract than water extract. Similarly flavonols contents (mg Rt/100 g, FW) were also estimated higher in acetone extract as compared to water extract on fresh weight basis (Fig. 7.22). Comparatively higher concentration of ascorbic acid contents (mg AA/100 g, FW) was noted for water extract than acetone extract.

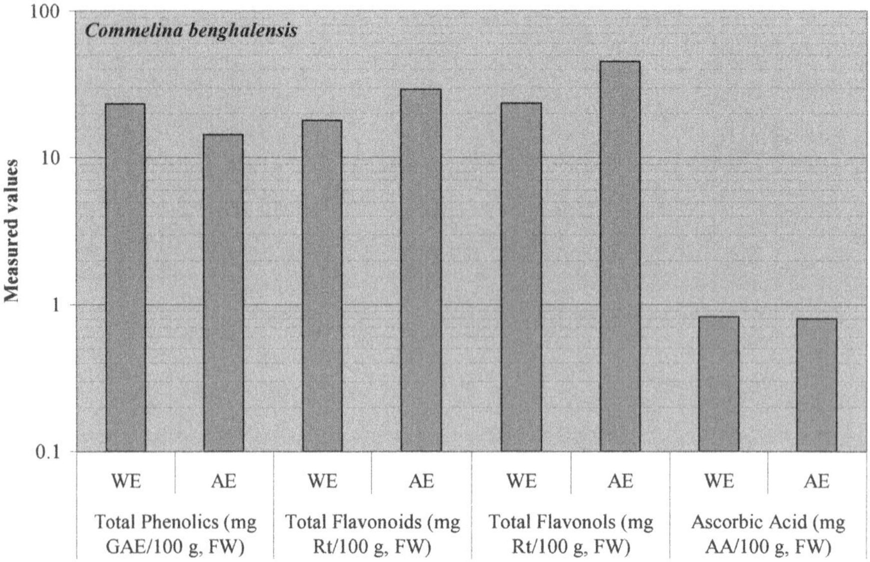

Fig. 7.22 Phytochemical contents in *Commelina benghalensis*

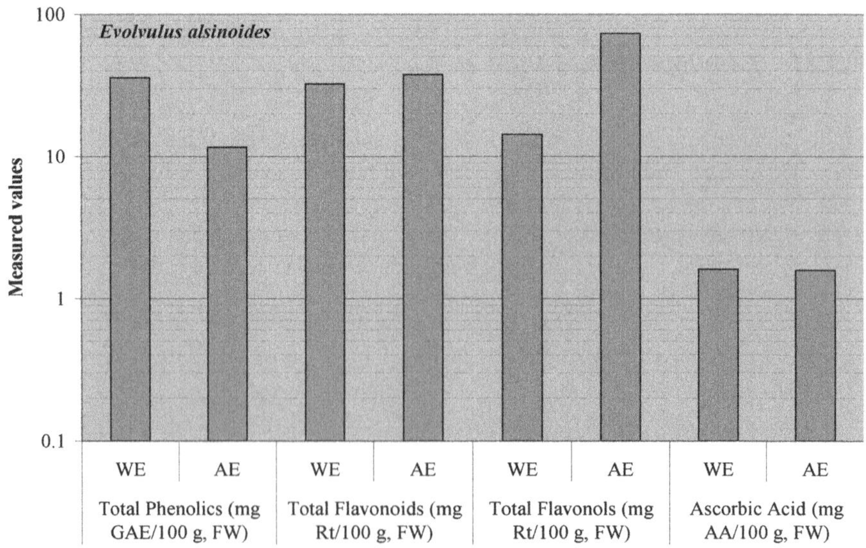

Fig. 7.23 Phytochemical contents in *Evolvulus alsinoides*

7.3.23 *Evolvulus alsinoides* L.

Phytochemical profile of *E. alsinoides* is mentioned in Fig. 7.23, which revealed higher concentration for total phenolic contents (mg GAE/100 g, FW) in water extract as compared to acetone extract. Flavonoid and flavonol contents

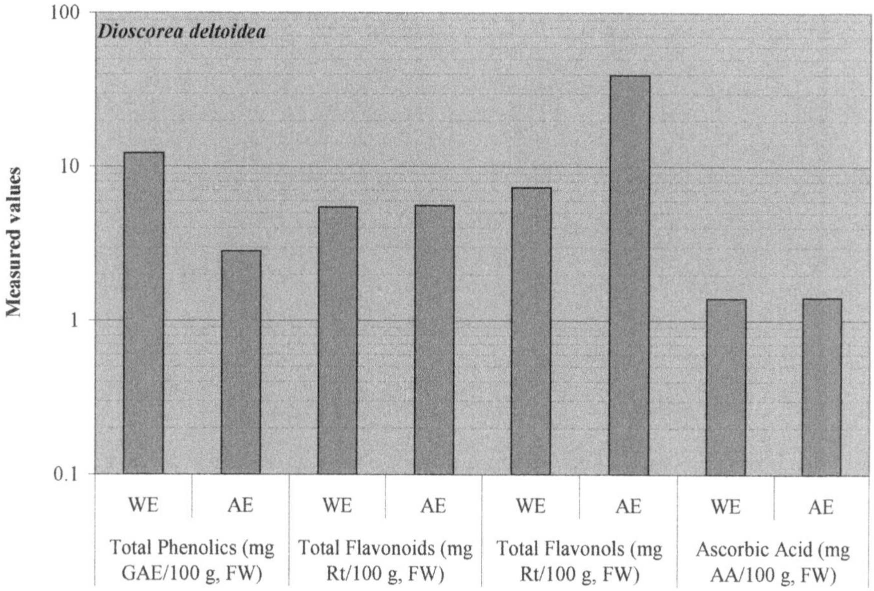

Fig. 7.24 Phytochemical contents in *Dioscorea deltoidea*

(mg Rt/100 g, FW) were noted higher in acetone extract as compared to water extract. Water extract showed comparatively higher concentration of ascorbic acid contents (mg AA/100 g, FW) than acetone extract on fresh weight basis.

7.3.24 *Dioscorea deltoidea* Wall. ex, Griseb

Phytochemical evaluation of *D. deltoidea* indicated that water extract exhibited higher concentration of total phenolic contents (mg GAE/100 g, FW) than acetone extract. Flavonoid and flavonol contents expressed as mg Rt/100 g, FW were found significantly higher in acetone extracts as compared to water extracts. Ascorbic acid contents were found comparatively higher in acetone extract (mg AA/100 g, FW) than water extract on fresh weight basis (Fig. 7.24).

7.3.25 *Lamium amplexicaule* L.

Phytochemical investigation of *L. amplexicaule* revealed that water extracts showed relatively higher concentrations of total phenolics (mg GAE/100 g, FW) and flavonoids contents (mg Rt/100 g, FW) than acetone extracts. Similarly water extracts showed higher values of flavonols (mg Rt/100 g, FW) and ascorbic acid contents (mg AA/100 g, FW) than acetone extract (Fig. 7.25).

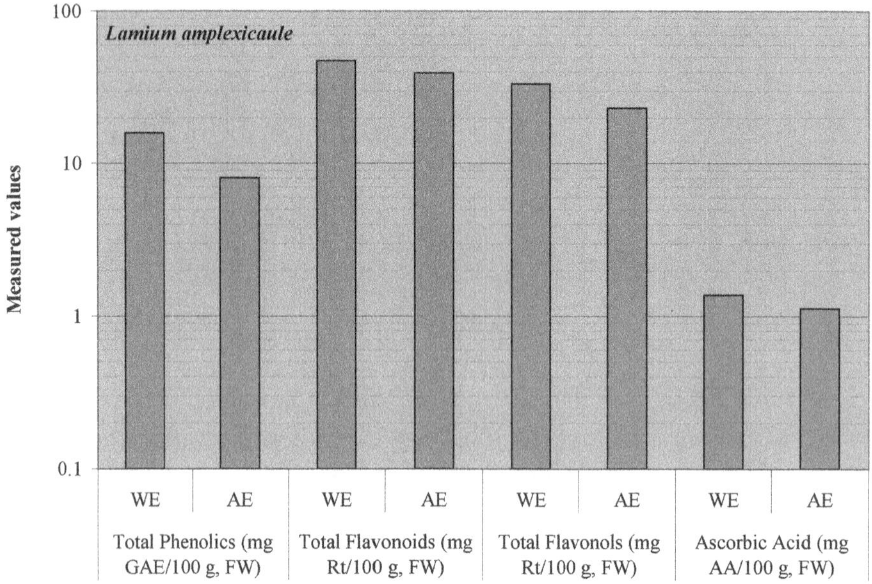

Fig. 7.25 Phytochemical contents in *Lamium amplexicaule*

7.3.26 *Origanum vulgare subsp. Hirtum (Link) Letsw*

Results of phytochemical contents in the leaves of *O. vulgare* subsp. *Hirtum* on fresh weight basis are mentioned in Fig. 7.26, which indicates significantly higher concentrations of total phenolic contents expressed as mg GAE/100 g, FW and flavonoids (mg Rt/100 g, FW) in water extracts as compared to acetone extracts. Water extracts again exhibited higher flavonols contents expressed as mg Rt/100 g, FW and ascorbic acid contents (mg AA/100 g, FW) as compared to acetone extracts on fresh weight basis.

7.3.27 *Tulipa stellata var. clusiana Hk. f.*

Among the phytochemicals constituents measured in *T. stellata* it was observed that water extracts showed higher concentrations of total phenolics (mg GAE/100 g, FW) and flavonoids contents (mg Rt/100 g, FW) as compared to acetone extracts. Flavonols contents (mg Rt/100 g, FW) were observed relatively higher in acetone extract than water extract, while water extract contained moderately higher value of ascorbic acid contents (mg AA/100 g, FW) than acetone extract (Fig. 7.27).

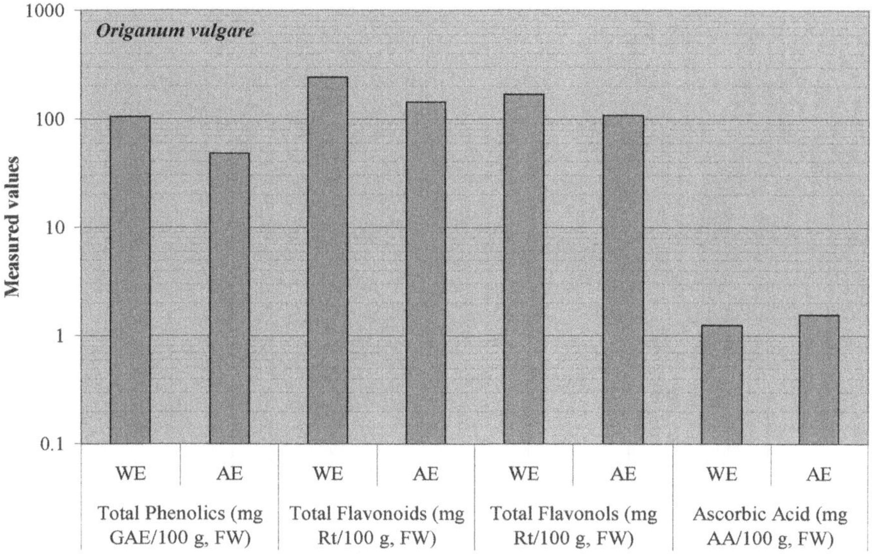

Fig. 7.26 Phytochemical contents in *O. vulgare* subsp. Hirtum

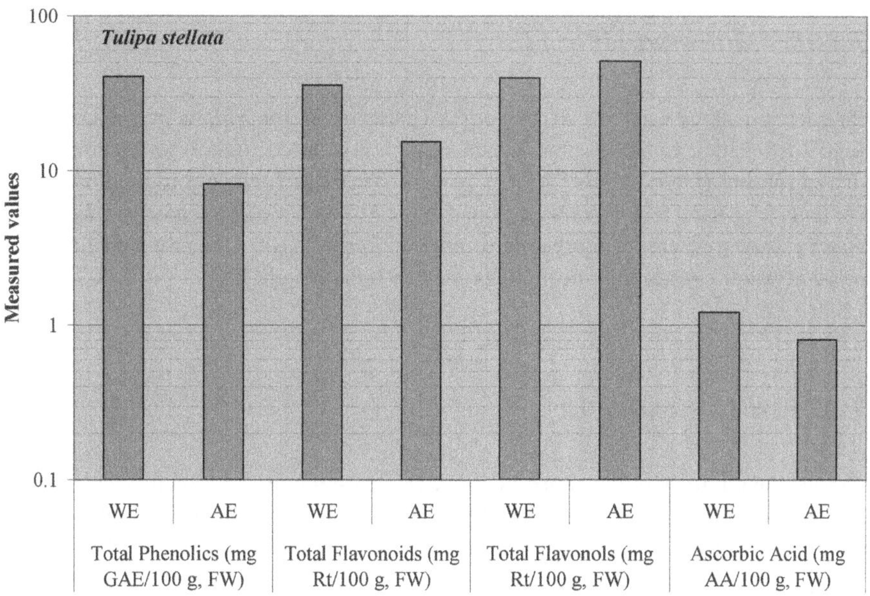

Fig. 7.27 Phytochemical contents in *T. stellata* var. clusiana

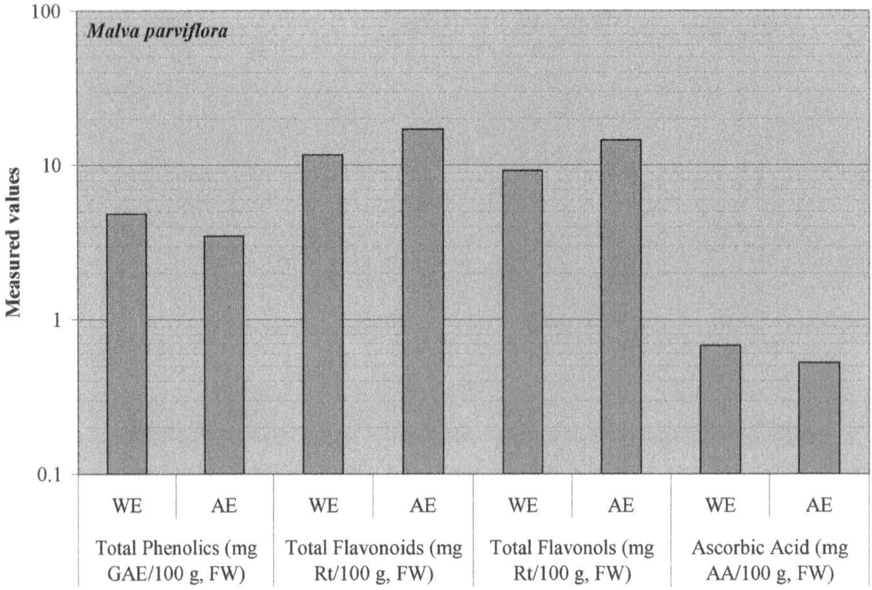

Fig. 7.28 Phytochemical contents in *Malvaparviflora*

7.3.28 *Malva parviflora* L.

Phytochemical assessment of the edible parts of *M. parviflora* is presented in Fig. 7.28, which indicates that water extracts exhibited higher concentration of total phenolics (mg GAE/100 g, FW) than acetone extract. Flavonoids and flavonols contents (mg Rt/100 g, FW) were found higher in acetone extracts as compared to water extracts, whereas ascorbic acid contents (mg AA/100 g, FW) were calculated more in water extract than acetone extract on fresh weight basis.

7.3.29 *Ficus carica* L.

Among the phytochemical contents of *F. carica* water extracts showed significantly higher concentrations of total phenolics (mg GAE/100 g, FW) and total flavonoids contents (mg Rt/100 g, FW) as compared to acetone extracts (Fig. 7.29). Significantly higher concentration of flavonols contents expressed in mg Rt/100 g, FW and ascorbic acid (mg AA/100 g, FW) were recorded in acetone extract as compared to water extract on fresh weight basis.

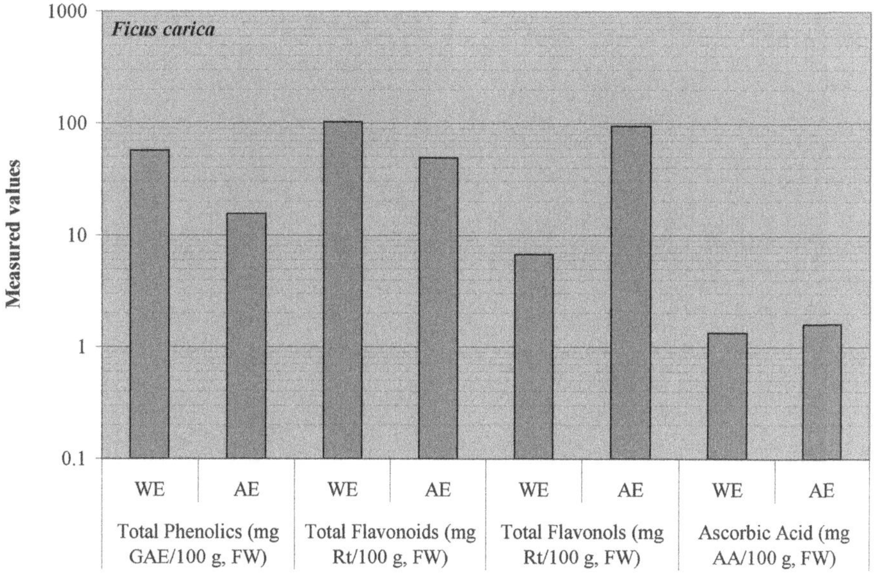

Fig. 7.29 Phytochemical contents in *Ficus palmata*

7.3.30 Ficus palmata Forssak.

Phytochemical profile of *F. palmata* is shown in Fig. 7.30, which indicates that water extracts show higher concentrations of total phenolics (mg GAE/100 g, FW) and flavonoids (mg Rt/100 g, FW) contents as compared to acetone extracts, whereas acetone extract exhibited more concentration of total flavonols (mg Rt/100 g, FW) than water extract. Comparatively higher value of ascorbic acid contents (mg AA/100 g, FW) were found in water extract as compared to acetone extract on fresh weight basis.

7.3.31 Oxalis corniculata L.

Phytochemical profile of *O. corniculata* demonstrated that water extract showed higher concentration of total phenolics (mg GAE/100 g, FW) as compared to acetone extract. Flavonoids and flavonols contents (mg Rt/100 g, FW) were recorded more in acetone extract than water extract, whereas ascorbic acid contents (mg AA/100 g, FW) were found slightly higher in water extract as compared to acetone extract on fresh weight basis (Fig. 7.31).

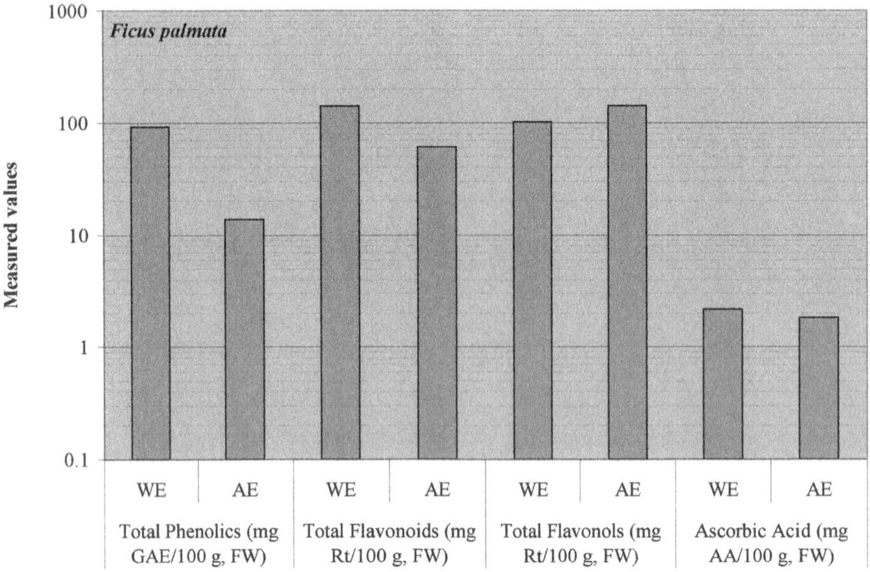

Fig. 7.30 Phytochemical contents in *Ficus palmata*

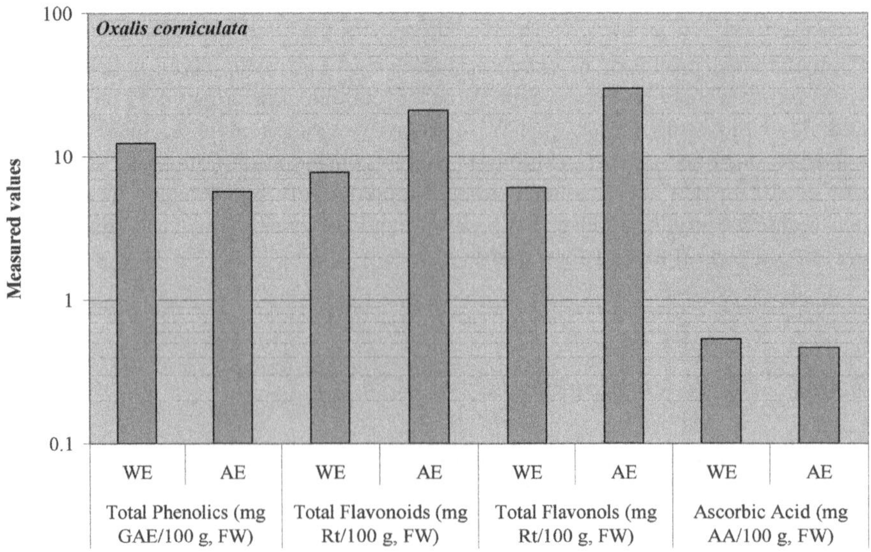

Fig. 7.31 Phytochemical contents in *Oxalis corniculata*

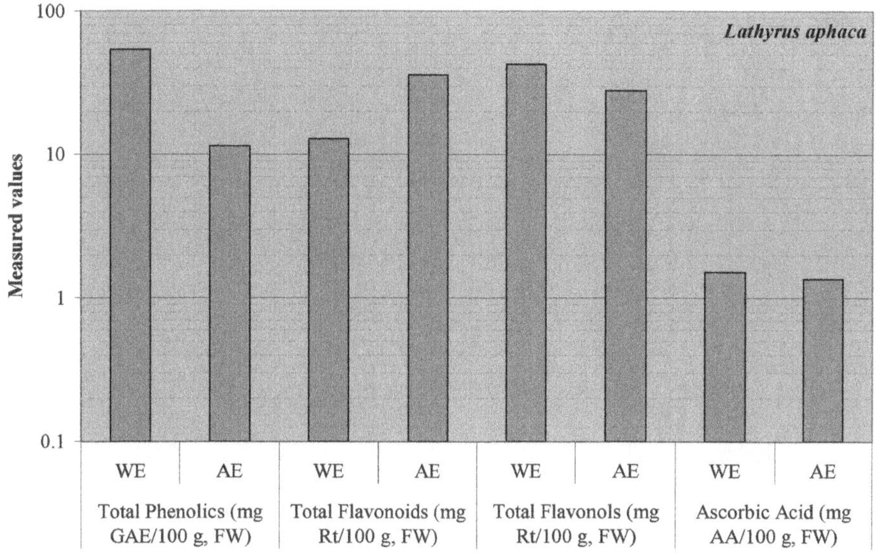

Fig. 7.32 Phytochemical contents in *Lathyrus aphaca*

7.3.32 *Lathyrus aphaca* L.

Comparative assessment of phytochemical constituents of *L. aphaca,* as shown in Fig. 7.32, revealed that water extract showed higher contribution of total phenolics (mg GAE/100 g, FW) as compared to acetone extract. Flavonoids contents (mg Rt/100 g, FW) were noted higher in acetone extract than water extract. Water extracts showed higher value of total flavonols contents (mg Rt/100 g, FW) than acetone extract. Ascorbic acid contents (AA/100 g, FW) were also calculated relatively higher in water extract than acetone extract.

7.3.33 *Medicago polymorpha* L.

Phytochemical investigation of edible parts of *M. polymorpha* indicates that water extracts showed higher concentration of total phenolics contents (mg GAE/100 g, FW) as compared to acetone extract. Flavonoids and flavonols contents (mg Rt/100 g, FW) were found higher in acetone extracts as compared to water extracts. Comparatively higher ascorbic acid contents (mg AA/100 g, FW) were found in water extract than acetone extract on fresh weight basis (Fig. 7.33).

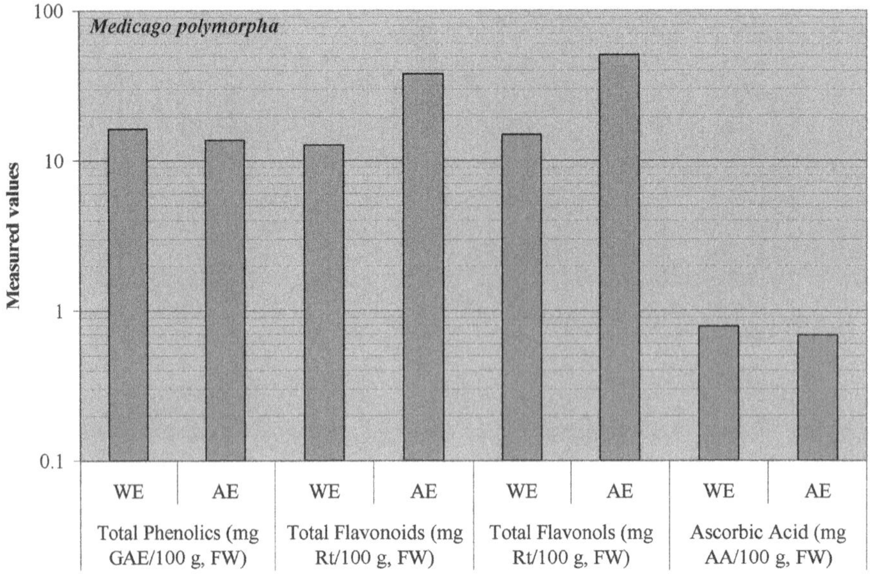

Fig. 7.33 Phytochemical contents in *Medicago polymorpha*

7.3.34 Melilotus alba Ledeb.

Comparative contributions of phytochemical constituents in acetone and water extracts of *M. alba* are given in Fig. 7.34. Water extracts exhibited higher concentrations of total phenolics (GAE/100 g, FW) and flavonoids contents (mg Rt/100 g, FW) as compared to acetone extracts. Flavonols contents articulated as mg Rt/100 g, FW were calculated higher in acetone extract than water extract, whereas ascorbic acid value was found slightly higher in water extract than acetone extract.

7.3.35 Melilotus indicus (L.) All.

Phytochemical investigation of *M. indicus* is given in Fig. 7.35, which revealed that water extracts showed more concentration compared to acetone extracts on fresh weight basis. Total phenolics (mg GAE/100 g, FW) and flavonoids contents were recorded higher in water extract than acetone extract. Comparatively higher value of flavonols contents (mg Rt/100 g, FW) was recorded in acetone extract than water extract, whereas ascorbic acid value expressed in (mg AA/100 g, FW) was estimated higher in water extract than acetone extract on fresh weight basis.

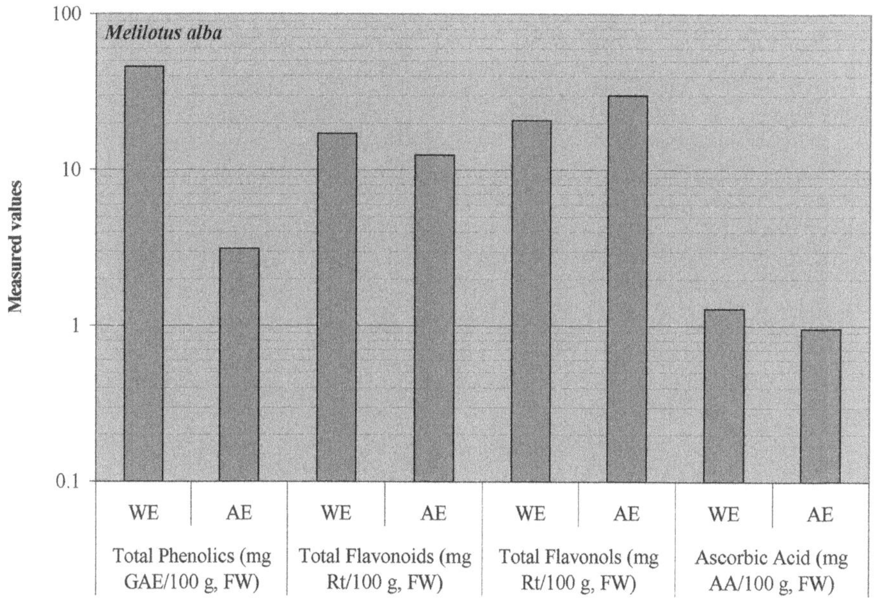

Fig. 7.34 Phytochemical contents in *Melilotus alba*

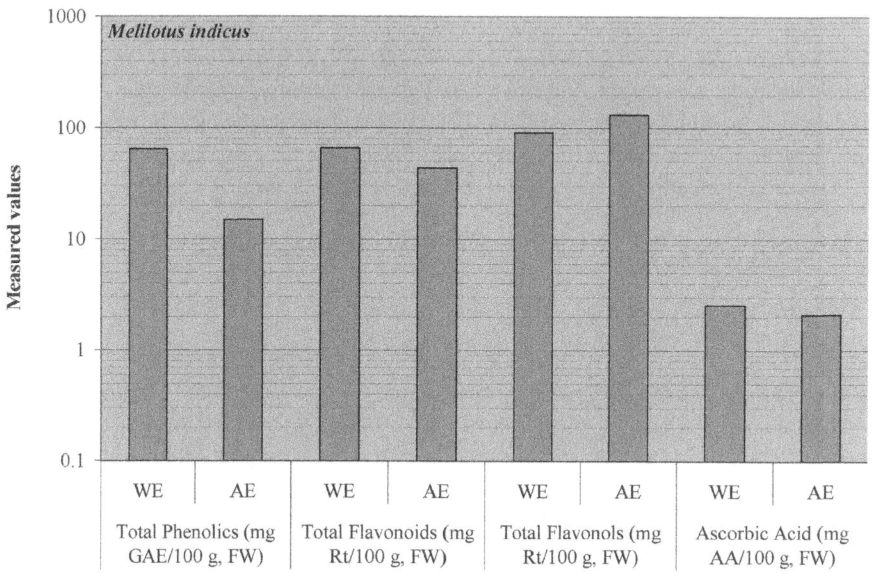

Fig. 7.35 Phytochemical contents in *Melilotus indicus*

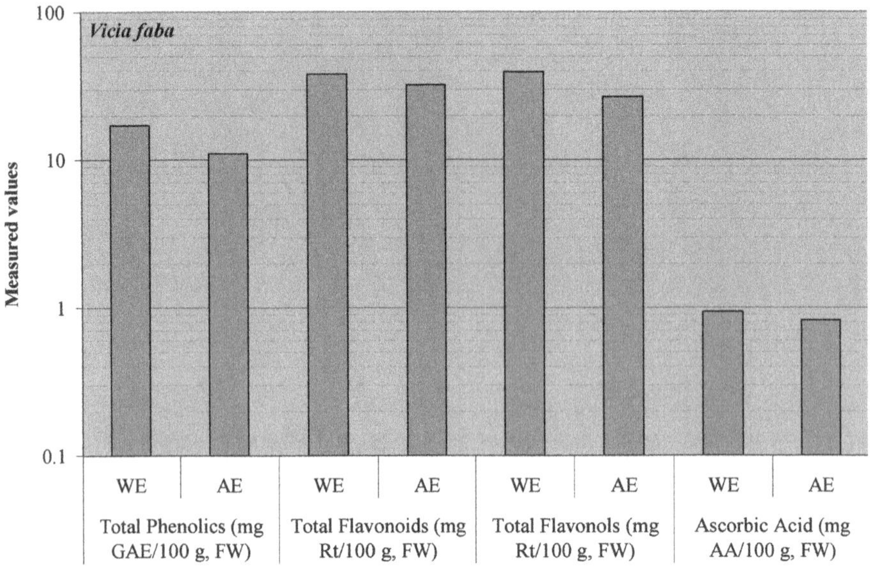

Fig. 7.36 Phytochemical contents in *Vicia faba*

7.3.36 *Vicia faba* L.

Phytochemical contents measured in the leaves of *V. faba* are depicted in Fig. 7.36. Water extracts showed higher concentrations of phenolics, flavonoids, and flavonols contents articulated as mg GAE/100 g, FW and mg Rt/100 g, FW, respectively, as compared to acetone extract. Similarly ascorbic acid contents (mg AA/100 g, FW) were found slightly higher in water extract than acetone extract.

7.3.37 *Vicia sativa* L.

Phytochemical evaluation of *V. sativa* demonstrated that water extract showed comparatively higher concentration of total phenolics (mg GAE/100 g, FW) than acetone extract. In the case of flavonoids and flavonols (mg Rt/100 g, FW), higher values were observed in acetone extracts compared to water extracts. Comparatively higher value of ascorbic acid contents (mg AA/100 g, FW) was found in water extract compared to acetone extract (Fig. 7.37).

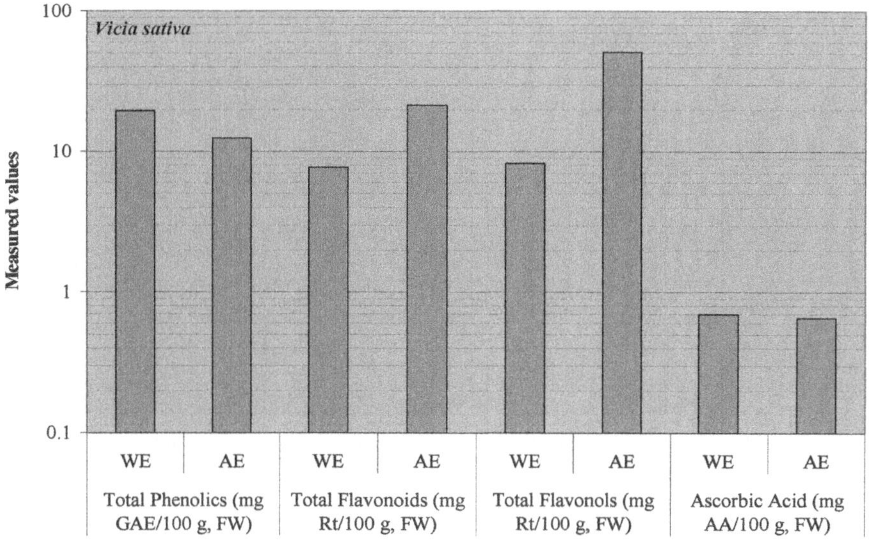

Fig. 7.37 Phytochemical contents in *Vicia sativa*

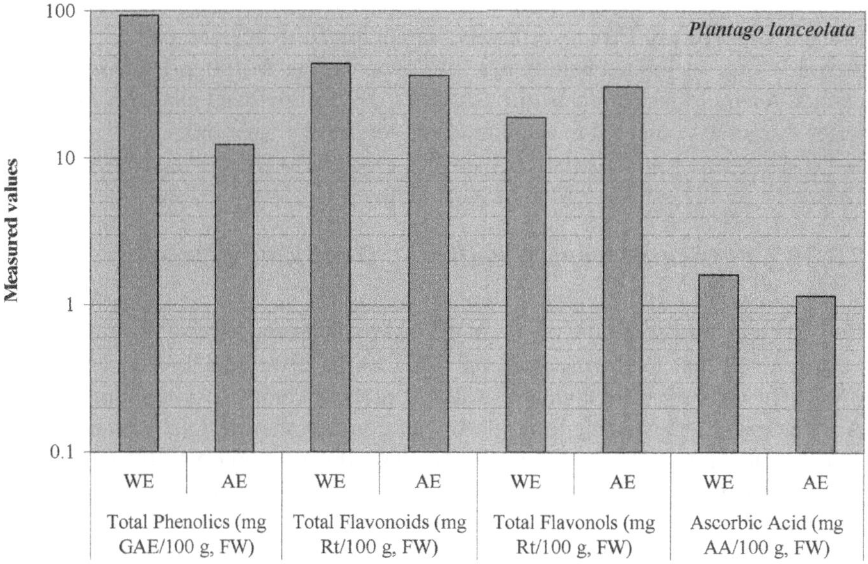

Fig. 7.38 Phytochemical contents in *Plantago lanceolata*

7.3.38 Plantago lanceolata L.

Phytochemical investigation of edible parts of *P. lanceolata* is mentioned in Fig. 7.38, which revealed that water extracts show significantly higher concentrations of total phenoloic and flavonoids contents articulated as mg GAE/100 g, FW

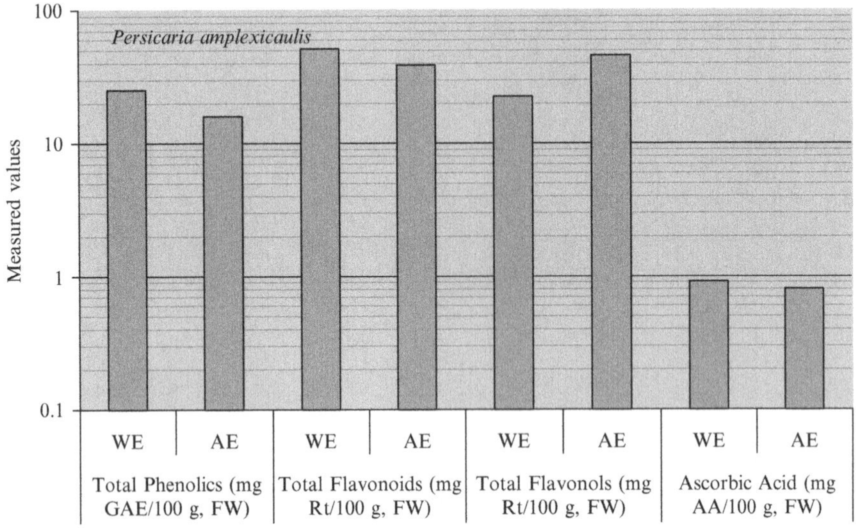

Fig. 7.39 Phytochemical contents in *Persicaria amplexicaulis*

and mg GAE/100 g, FW, respectively, as compared to acetone extract, whereas acetone extract exhibited better value of flavonols (mg Rt/100 g, FW) than water extract. Ascorbic acid contents (mg AA/100 g, FW) were found relatively higher in water extract as compared to acetone extract on fresh weight basis.

7.3.39 *Persicaria amplexicaulis (D. Don) Ronse Decr.*

Phytochemical investigation of *P. amplexicaulis* is mentioned in Fig. 7.39, which demonstrated that total phenolic (mg GAE/100 g, FW) and flavonoid contents (mg Rt/100 g, FW) were higher in water extracts as compared to acetone extract. Acetone extract showed significantly higher concentration of total flavonols contents (mg Rt/100 g, FW) than water extract. Ascorbic acid (mg AA/100 g, FW) was found comparatively higher in water extract as compared to acetone extract.

7.3.40 *Polygonum aviculare L.*

Phytochemical analysis of edible parts of *P. aviculare* revealed that total phenolic contents (mg GAE/100 g, FW) were found to be significantly higher in water extract as compared to acetone extract. On the other hand, acetone extract showed higher concentration of flavonoids and flavonols (mg Rt/100 g, FW) than water

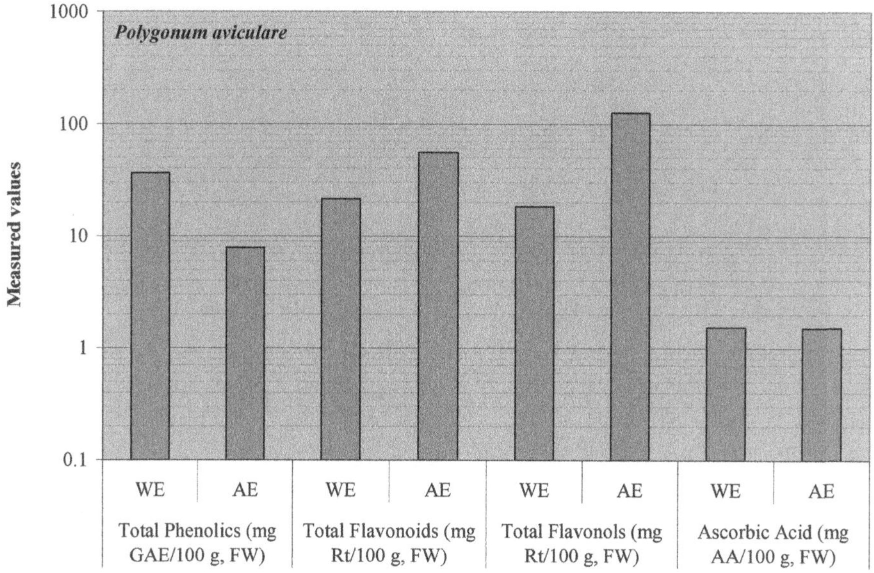

Fig. 7.40 Phytochemical contents in *Polygonum aviculare*

extract. Ascorbic acid contents (mg AA/100 g, FW) were found relatively higher in water extract than acetone extract (Fig. 7.40).

7.3.41 *Rumex dentatus* L.

Phytochemical contents in the leaves of *R. dentatus* are shown in Fig. 7.41, which exhibited elevated levels of total phenolics (mg GAE/100 g, FW) in water extract as compared to acetone extract, whereas acetone extract exhibited higher concentration of flavonoids and flavonols (mg Rt/100 g, FW) than water extract. However, ascorbic acid contents (mg AA/100 g, FW) were calculated comparatively higher in water extract than acetone extract on fresh weight basis.

7.3.42 *Rumex hastatus* D. Don, Prodor.

Phytochemical constituents of *R. hastatus* are given in Fig. 7.42, which revealed higher concentrations for acetone extracts than water extract. Total phenolics and flavonoids contents articulated as mg GAE/100 g, FW and mg Rt/100 g, FW, respectively, were recorded higher in acetone extract than water extract. Similarly flavonols contents (mg Rt/100 g, FW) and ascorbic acid contents (mg AA/100 g, FW) were also estimated higher in acetone extract than water extract on fresh weight basis.

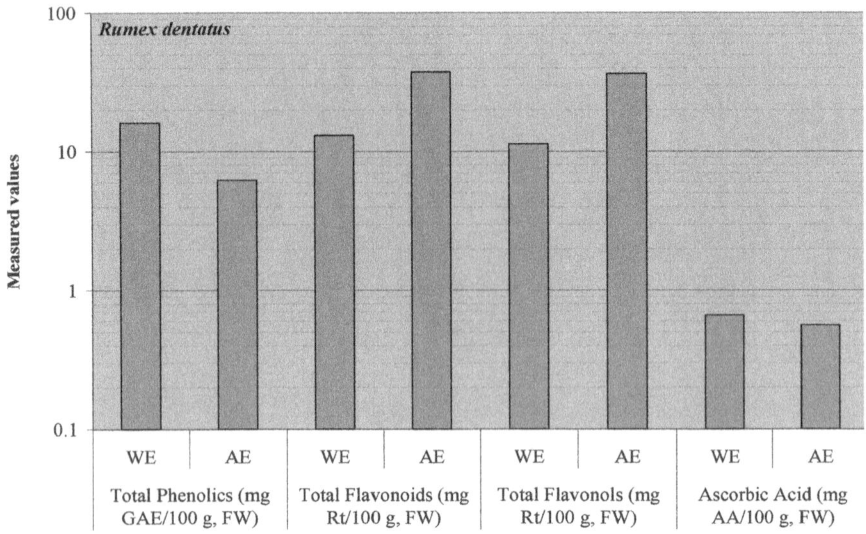

Fig. 7.41 Phytochemical contents in *Rumex dentatus*

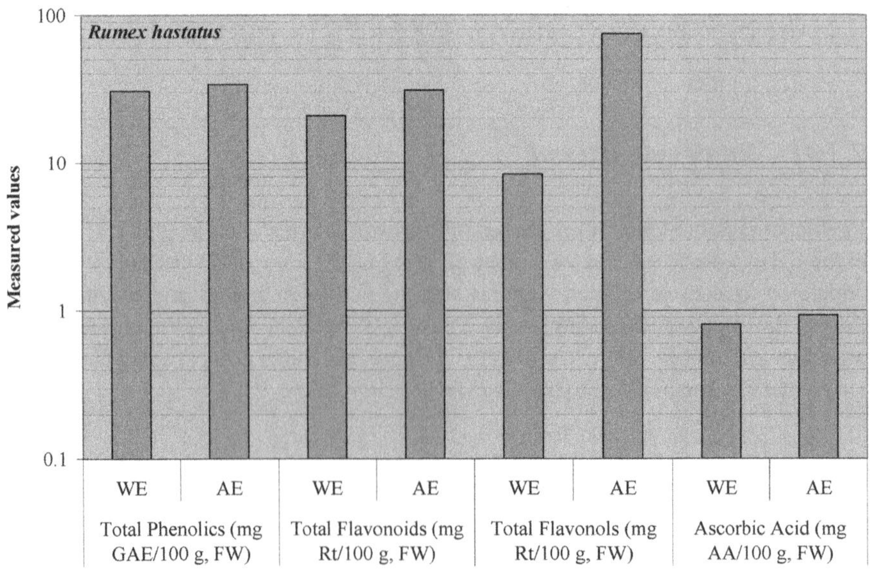

Fig. 7.42 Phytochemical contents in *Rumex hastatus*

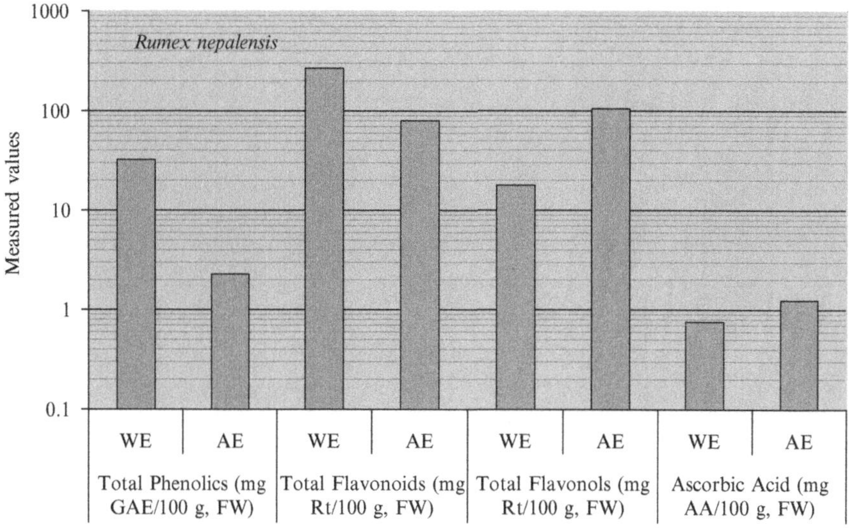

Fig. 7.43 Phytochemical contents in *Rumex nepalensis*

7.3.43 *Rumex nepalensis* Spreng.

Phytochemical profile of *R. nepalensis* is mentioned in Fig. 7.43, which revealed that on fresh weight basis water extract showed higher concentration of total phenolics contents (mg GAE/100 g, FW) than acetone extract. Total flavonoid and flavonols contents (mg Rt/100 g, FW) were estimated higher in water and acetone extracts, respectively. However, acetone extract showed slightly higher concentration of ascorbic acid (mg AA/100 g, FW) than water extract on fresh weight basis.

7.3.44 *Portulaca quadrifida* L.

Phytochemical profile of *P. quadrifida* demonstrated that water extract showed higher concentration of total phenolics contents (mg GAE/100 g, FW) than acetone extract, whereas total flavonoids and flavonols contents (mg Rt/100 g, FW) were estimated higher in acetone extract as compared to water extract. However, water extract showed slightly higher concentration of ascorbic acid (mg AA/100 g, FW) than acetone extract on fresh weight basis (Fig. 7.44).

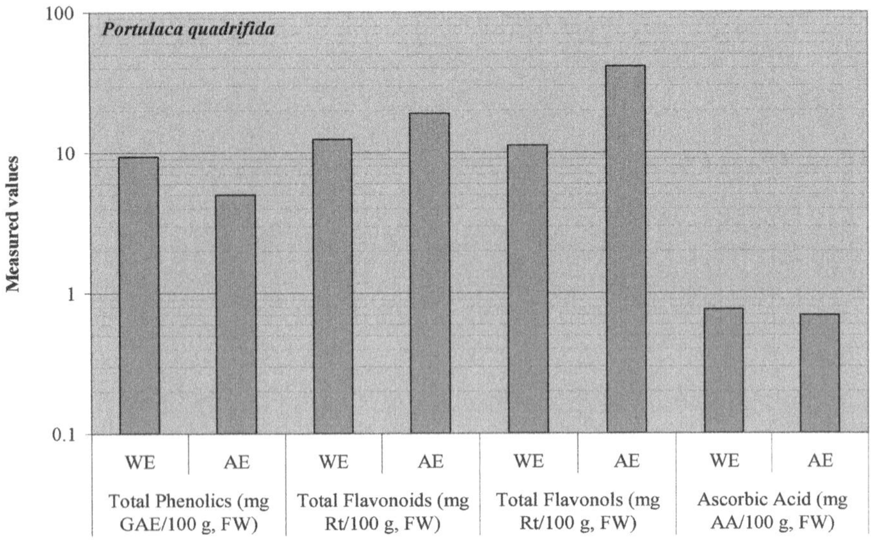

Fig. 7.44 Phytochemical contents in *Portulaca quadrifida*

7.3.45 *Portulaca oleracea* L.

Phytochemical profile of *Portulaca oleracea* is given in Fig. 7.45, which indicates that water extract exhibited higher concentration of total phenolic contents (mg GAE/100 g, FW) compared to acetone extract. Similarly flavonoid contents (mg Rt/100 g, FW) were also measured higher in water extracts as compared to acetone extracts on fresh weight basis, whereas in the case of flavonols comparatively higher concentration (mg Rt/100 g, FW) was measured in acetone extract than water extract. Acetone extract also exhibited moderately higher value of ascorbic acid contents (mg AA/100 g, FW) as compared to the water extract on fresh weight basis.

7.3.46 *Galium aparine* L.

Phytochemical assessment of *G. aparine*, as shown in Fig. 7.46, indicates that water extract exhibited significantly higher value of total phenolics contents (mg GAE/ 100 g, FW) and flavonoids (mg Rt/100 g, FW) as compared to acetone extract. In the case of flavonols (mg Rt/100 g, FW) higher concentration was recorded in acetone extract than water extract. Water extract showed comparatively higher concentration of ascorbic acid contents (mg AA/100 g, FW) than acetone extract on fresh weight basis.

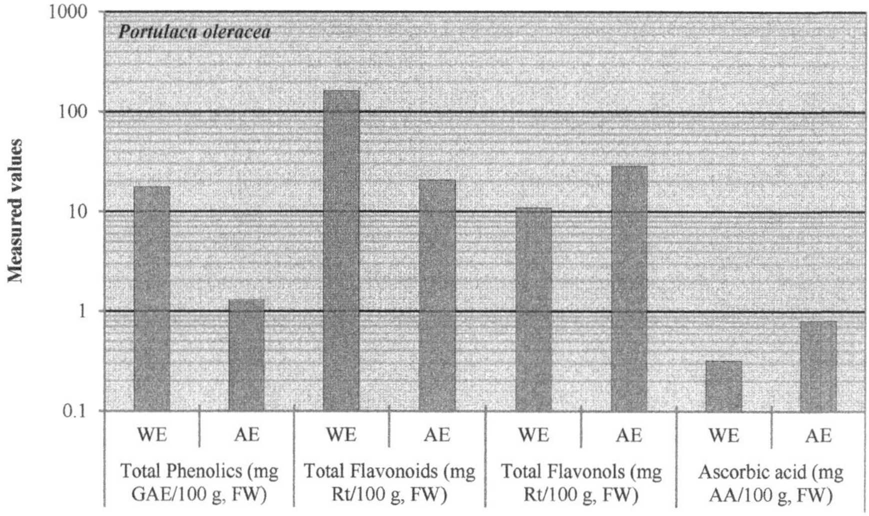

Fig. 7.45 Phytochemical contents in *Portulaca oleracea*

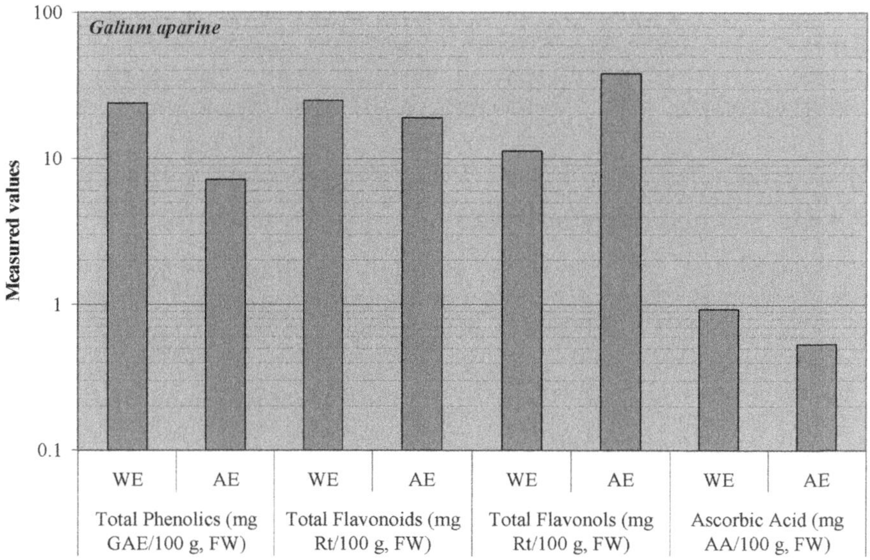

Fig. 7.46 Phytochemical contents in *Galium aparine*

7.3.47 Veronica arvensis L.

Phytochemical investigation of *V. arvensis* on fresh weight basis is shown in Fig. 7.47, which indicates that water extract showed higher concentration of total phenolics contents (mg GAE/100 g, FW) than acetone extract. In the case of

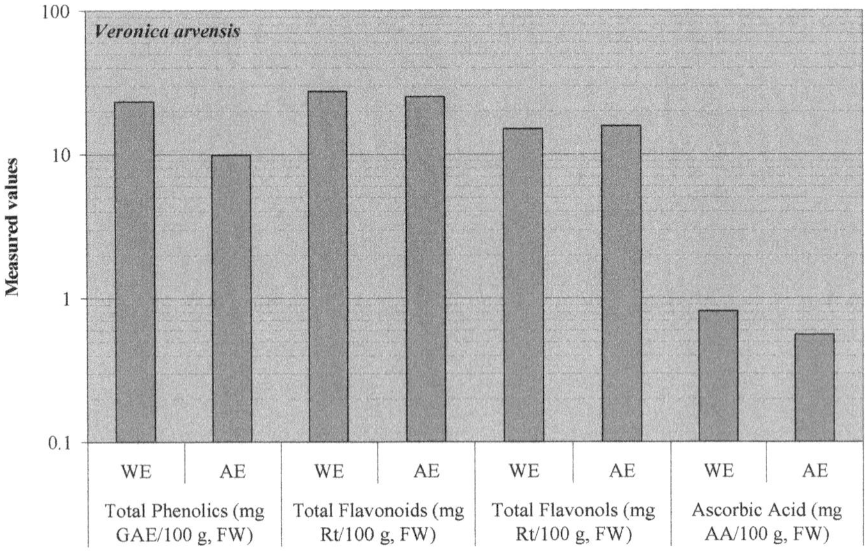

Fig. 7.47 Phytochemical contents in *Veronica arvensis*

flavonoids and flavonols almost comparable value expressed as mg Rt/100 g, FW were recorded in water extracts as compared to acetone extracts. Water extract showed higher concentration of ascorbic acid (mg AA/100 g, FW) than acetone extract.

7.3.48 Solanum nigrum auct.

Among the phytochemicals measured in the leaves of *S. nigrum* total phenolics contents (mg GAE/100 g, FW) were observed higher in water extract than acetone extract. Comparatively elevated values of flavonoid and flavonol contents (mg Rt/100 g, FW) were found in acetone extract as compared to water extract. Ascorbic acid (mg Rt/100 g, FW) was calculated slightly more in water extract than acetone extract on fresh weight basis (Fig. 7.48).

7.3.49 Pimpinella diversifolia (Wall.) DC. Prodor.

Phytochemical contents of *P. diversifolia* water extract showed higher concentration of phenolic (mg GAE/100 g, FW) and flavonoid contents (mg Rt/100 g, FW) than acetone extract on fresh weight basis (Fig. 7.49). Flavonol contents (mg Rt/100 g, FW) were also estimated higher in water extract as compared to acetone extract. However, water extract showed comparatively higher concentration of ascorbic acid (mg AA/100 g, FW) than acetone extract.

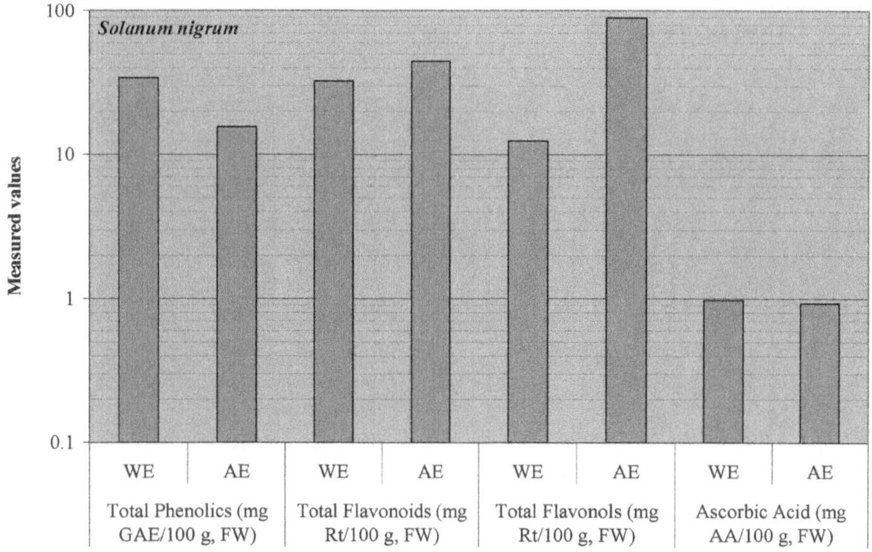

Fig. 7.48 Phytochemical contents in *Solanum nigrum*

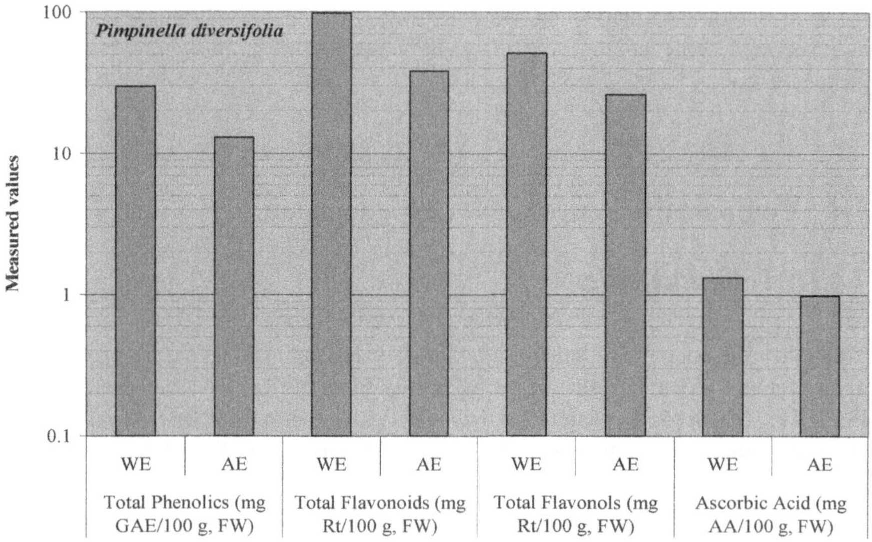

Fig. 7.49 Phytochemical contents in *Pimpinella diversifolia*

7.3.50 *Torilis leptophylla* (L.)Reichb.f.

Phytochemical profile of *T. leptophylla,* as shown in Fig. 7.50, revealed that total phenolic contents (mg GAE/100 g, FW) are higher in water extract as compared to acetone extract. Flavonoid and flavonol contents (mg Rt/100 g, FW) were estimated

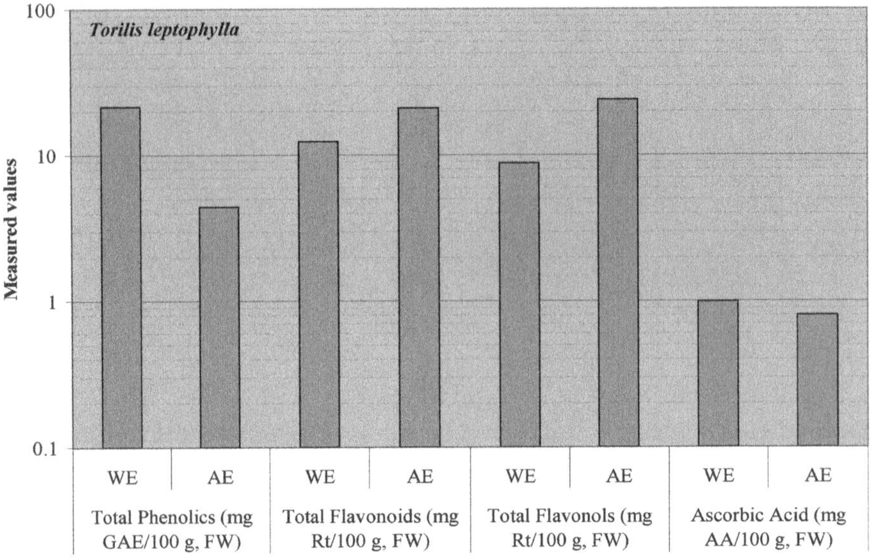

Fig. 7.50 Phytochemical contents in *Torilis leptophylla*

more in acetone extracts than water extracts. Comparatively higher value of ascorbic acid contents (mg AA/100 g, FW) was observed in water extract than acetone extract.

7.4 Comparative Assessment of Phytochemical Contents

7.4.1 Total Phenolics

Total phenolic contents of wild vegetables in water and acetone extracts expressed as milligram GAE on fresh weight basis (mg GAE/100 g, FW) are presented in Fig. 7.51. Measured values indicate that in case of water extract *Amaranthus viridis* showed highest concentration (139.9 mg GAE/100 g), followed by *Origanum vulgare*, *Plantago lanceolata* and *Ficus palmata*, whereas lowest values were noted for *Capsella bursa-pastoris*, *Malvaparviflora*, *Taraxacum officinale*, and *Sonchus asper*. In case of acetone extracts edible parts of *Origanum vulgare* showed significantly higher value (47.86 mg GAE/100 g) followed by *Bauhinia variegata*, *Bombax malabaracum*, and *Rumex hastatus*, whereas lowest concentrations of total phenolic contents were measured in *Capsella bursa-pastoris* followed by *Portulaca oleracea* and *Stellaria media*. Reported levels of phenolic contents have been observed between 33.00 and 790.0 mg GAE/100 g in acetonitrile, ethanolic, methanolic and water extracts of *Amaranthus viridis*, *Portulaca oleracea*, *Origanum vulgare*, *Rumex tuberosus*, and *Solanum nigrum* (Barros et al. 2010; Andarwulan et al. 2010; Motamed and Naghibi 2010; Lako et al. 2007).

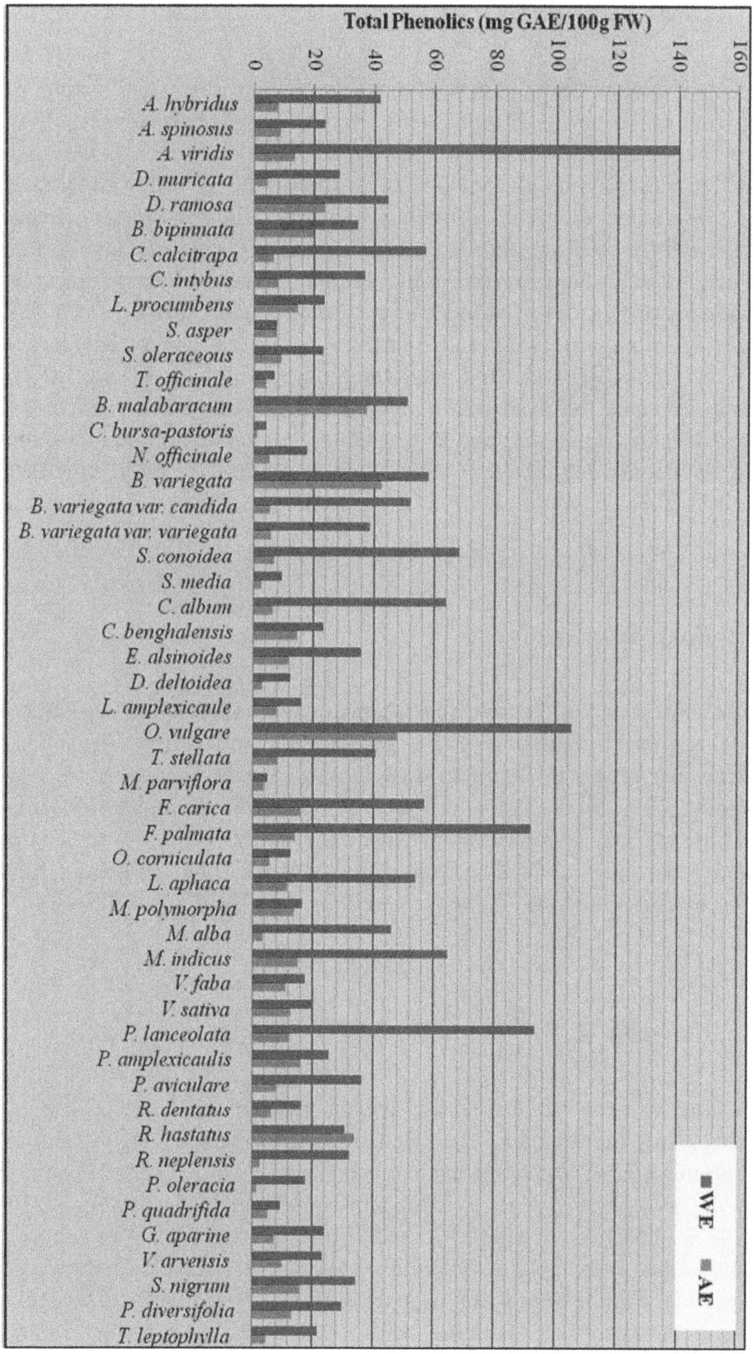

Fig. 7.51 Phenolic contents

7.4.2 Total Flavonoids

Comparative evaluation of flavonoid contents of both water and acetone extracts expressed as milligram Rutin equivalents on fresh weight basis (mg Rt/100 g, FW) is mentioned in Fig. 7.52, which shows that water extract of *Rumex neplensis* exhibited highest flavonoids contents (267.4 mg Rt/100 g), followed by *Origanum vulgare*, *Portulaca oleracea*, *Bauhinia variegata*, *Ficus palmata*, *Amaranthus viridis*, *Melilotus indicus*, and *Polyganum aviculare*, whereas lowest values of flavonoid contents were observed in the water extracts of *Capsella bursa-pastoris* (1.420 mg Rt/100 g), followed by *Discoria deltoidea*, *Stellaria media*, *Taraxacum officinale*, *Oxalis corniculata*, and *Ficus carica*. In the case of acetone extracts *Origanum vulgare* again showed maximum concentration (143.0 mg Rt/100 g) of flavonoids contents followed by *Rumex neplensis*, *Amaranthus viridis*, *Melilotus indicus*, *Polyganum avicular*, and *Bidens bipinnata*, while minimum concentration of total flavonoids was noted in *Capsella bursa-pastoris* (2.430 mg Rt/100 g), followed by *Taraxacum officinale*, *Malva parviflora*, and *Veronica arvensis*.

7.4.3 Total Flavonols

Flavonol contents in wild edible vegetables are also expressed as mg Rutin equivalent/100 g fresh weight (mg Rt/100 g, FW) as mentioned in Fig. 7.53, which indicates water extract of *Bauhinia variegata var. variegata* showed highest value of flavonols (199.1 mg Rt/100 g) followed by *Ficus palmata* and *Melilotus indicus*, whereas lowest level was recorded in *Capsella bursa-pastoris*. Acetone extracts of *Origanum vulgare* contained maximum flavonol contents (168.8 mg Rt/100 g) followed by *Ficus palmata*, *Amaranthus viridis*, and *Melilotus indicus*.

7.4.4 Ascorbic Acid

Comparative assessment of ascorbic acid contents of wild edible vegetables is expressed as mg ascorbic acid equivalent/100 g fresh weight (mg AA/100 g, FW) as given in Fig. 7.54, which reveals that water extracts of *Melilotus indicus*, *Silene conoidea*, and *Ficus palmata* exhibited significant concentration 2.570 mg AAE/100 g, 2.510 mg AAE/100 g, and 2.160 mg AAE/100 g of ascorbic acid contents, respectively, while *Bauhinia variegata var. variegata* showed lowest concentration. Similarly maximum concentration of ascorbic acid was observed in the acetone extract of *Melilotus indicus* (2.106 mg AAE/100 g), followed by *Silene conoidea* and *Ficus palmata*. However, *Capsella bursa-pastoris* showed lowest concentration of ascorbic acid in acetone extracts (0.142 mg AAE/100 g).

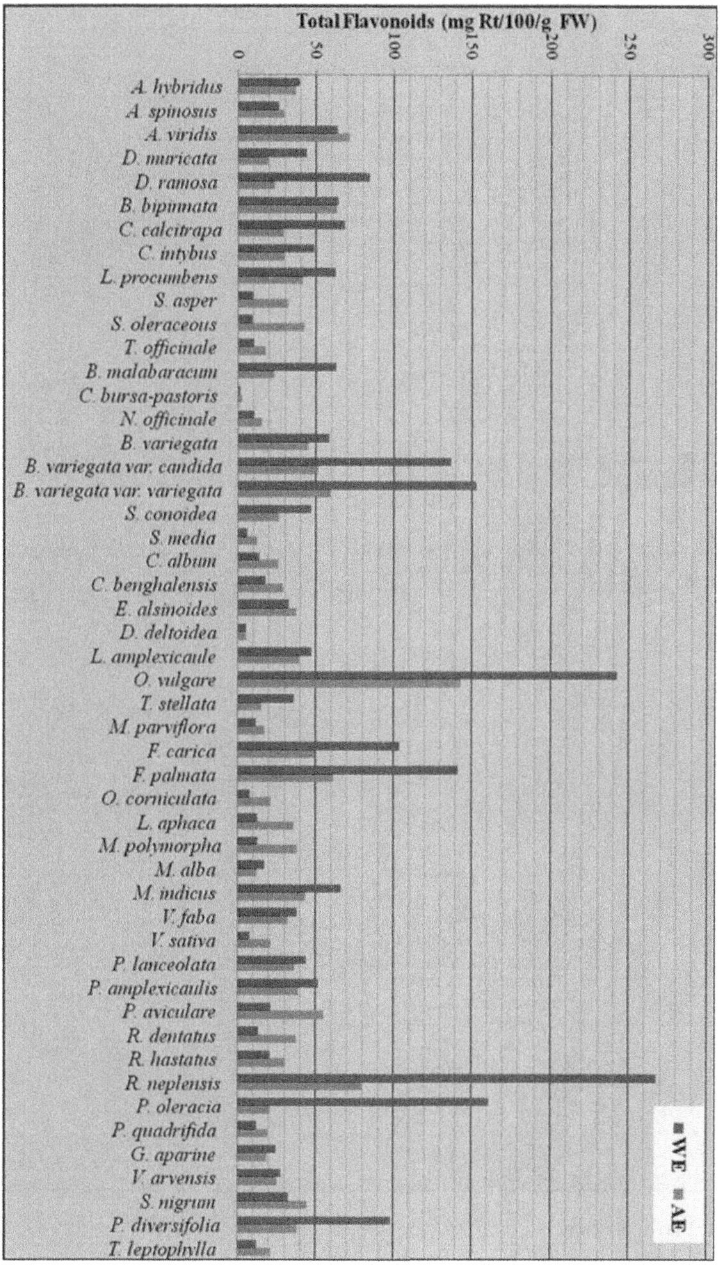

Fig. 7.52 Flavonoid contents

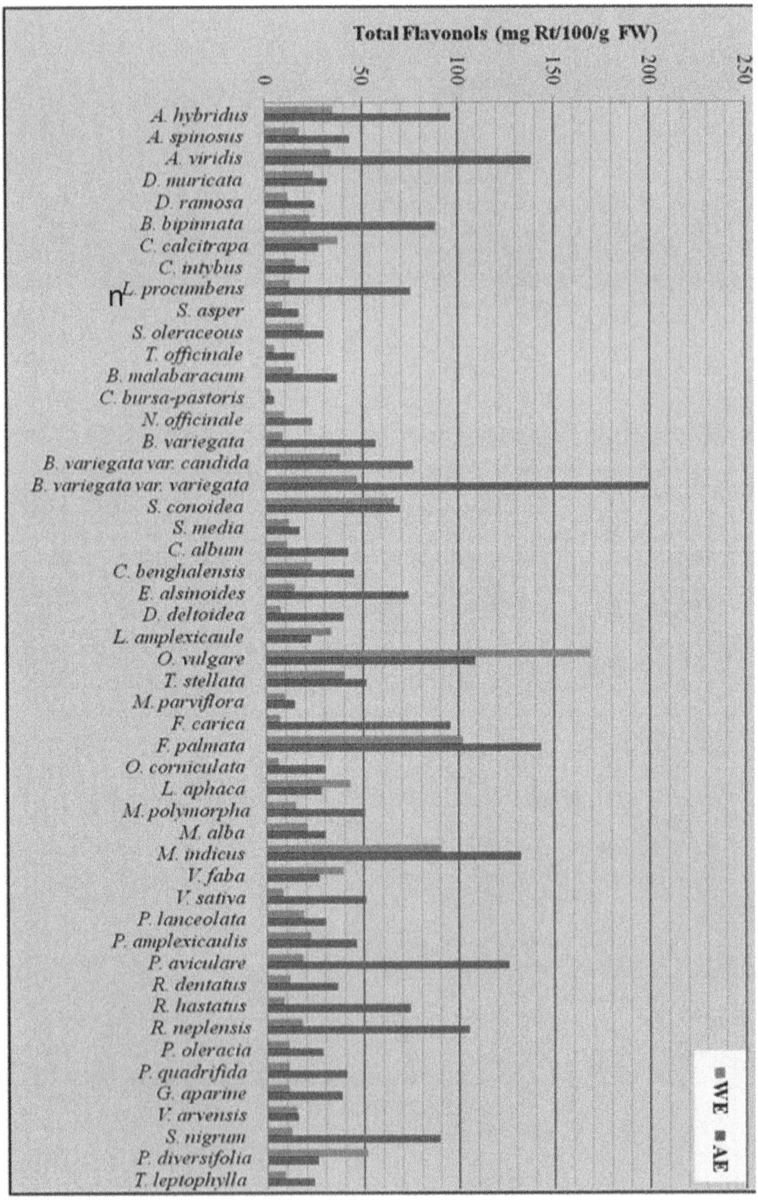

Fig. 7.53 Flavonols contents

Phytochemical constituents investigated in wild edible vegetables were mostly found in the range of reported literature; however, variations in the chemical composition may be due to prevailing environmental conditions, age of the plants at collection time, and uptake of nutrients from the soil.

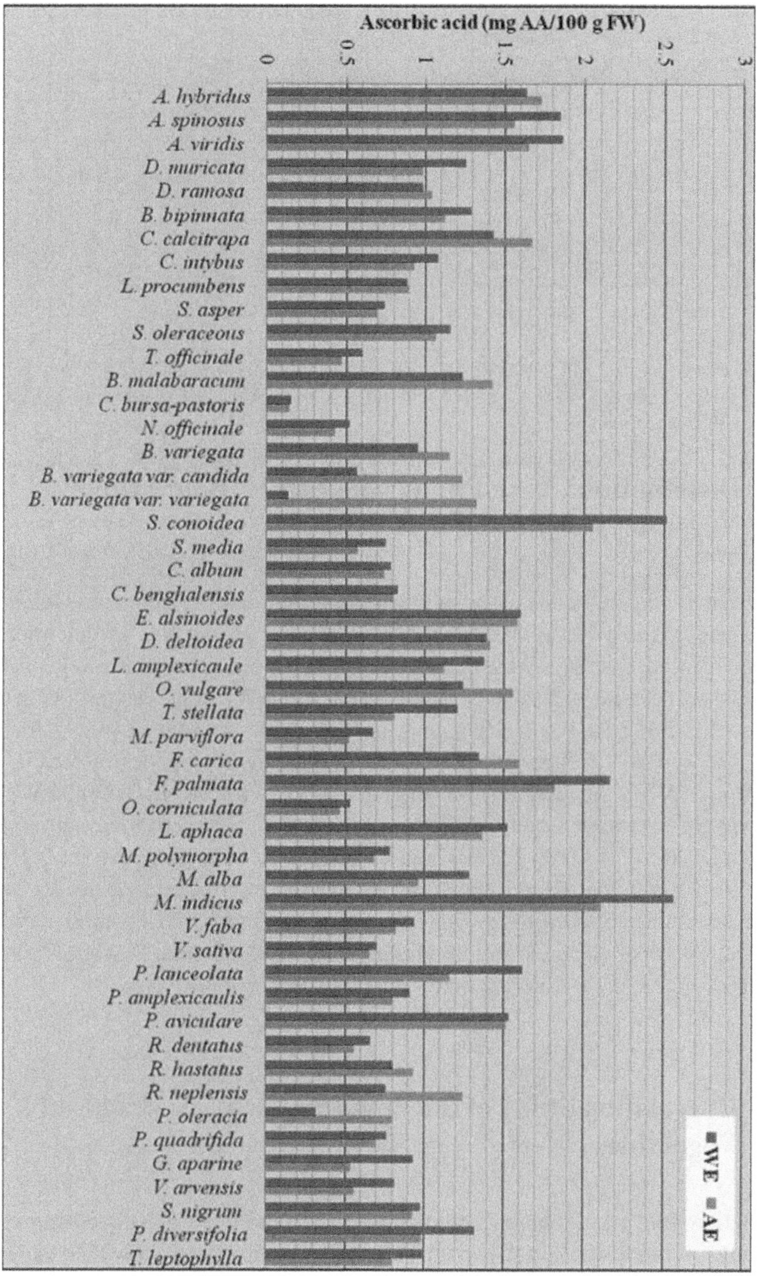

Fig. 7.54 Ascorbic acid contents

Table 7.1 Statistical distribution parameters for selected phytochemicals (fresh weight) in wild edible vegetables of lesser Himalayas

	Min	Max	Mean	Median	SD	SE	Kurtosis	Skewness
TP (WE)	3.872	139.9	36.74	30.14	27.52	3.892	3.41	1.643
TP (AE)	0.717	47.86	11.4	8.136	9.991	1.413	4.808	2.133
TF (WE)	1.416	267.5	52.12	34.11	57.74	8.166	4.891	2.137
TF (AE)	2.345	143	35.31	31.34	22.56	3.19	9.752	2.428
TFL (WE)	2.158	168.9	25.87	15.12	28.99	4.1	12.28	3.154
TFL (AE)	4.182	199.2	55.11	40.27	41.15	5.819	2.031	1.453
AA (WE)	0.14	2.567	1.1	0.98	0.523	0.074	0.958	0.799
AA (AE)	0.144	2.107	1.054	0.971	0.452	0.064	−0.457	0.422

TP total phenolics (mg GAE/100 g, FW), *TF* total flavonoids (mg Rt/100 g, FW), *TFL* total flavonols (mg Rt/100 g, FW), *AA* ascorbic acid (mg AA/100 g, FW), *WE* water extract, *AE* acetone extract, *FW* fresh weight

7.5 Distribution of the Phytochemicals in Wild Edible Vegetables

Basic statistical parameters associated with distribution of selected phytochemicals (Fresh weight) in wild edible vegetables of Lesser Himalayas are mentioned in Table 7.1. In the case of wild vegetables dominant mean levels were shown by total flavonols in acetone extract (55.11 mg Rt/100 g, FW), total flavonoids in water extract (52.12 mg Rt/100 g, FW), and total phenolics in water extract (36.74 mg GAE/100 g, FW), followed by total flavonoids in acetone extract (35.31 mg Rt/100 g, FW), total flavonols in water extract (25.87 mg Rt/100 g, FW), total phenolics in acetone extract (11.40 mg GAE/100 g, FW), ascorbic acid in water extract (1.100 mg AA/100 g, FW), and ascorbic acid in acetone extract (1.054 mg AA/100 g, FW). The decreasing trend of average phytochemical levels in wild edible vegetables shows following order total flavonols (AE) > total flavonoids (WE) > total phenolics (WE) > total flavonoids (AE) > total flavonols (WE) > total phenolics (AE) > ascorbic acid (WE) > ascorbic acid (AE).

7.6 Correlation Study of the Phytochemicals in Wild Edible Vegetables

Correlation coefficients among phytochemicals pairs were calculated to examine their inter-relationships in the wild vegetables samples as shown in the matrix (Table 7.2). Among wild edible vegetables very strong positive correlations were observed for following pairs; ascorbic acid in water extract-ascorbic acid in acetone extract ($r = 0.820$), total flavonoids in water extract-total flavonoids in acetone extract ($r = 0.727$), total flavonoids in acetone extract-total flavonols in water extract ($r = 0.685$), total phenolics in water extract-Ascorbic acid in acetone extract

Table 7.2 Correlation coefficient[a] matrix for phytochemicals in wild edible vegetables

	TP (WE)	TP (AE)	TF (WE)	TF (AE)	TFL (WE)	TFL (AE)	AA (WE)	AA (AE)
TP (WE)	1.000							
TP (AE)	0.428	1.000						
TF (WE)	0.425	0.274	1.000					
TF (AE)	0.588	0.513	**0.727**	1.000				
TFL (WE)	0.588	0.380	0.559	**0.685**	1.000			
TFL (AE)	0.538	0.231	0.527	**0.659**	0.480	1.000		
AA (WE)	0.584	0.160	0.033	0.220	0.481	0.318	1.000	
AA (AE)	**0.665**	0.271	0.406	0.478	0.558	**0.627**	**0.820**	1.000

TP total phenolics (mg GAE/100 g, FW), *TF* total flavonoids (mg Rt/100 g, FW), *TFL* total flavonols (mg Rt/100 g, FW), *AA* ascorbic acid (mg AA/100 g, FW), *WE* water extract, *AE* acetone extract, *FW* fresh weight

[a]Bold r-values are significant at $p < 0.01$

($r = 0.665$), total flavonoids in acetone extract-total flavonols in acetone extract ($r = 0.659$), total flavonols in acetone extract-ascorbic acid in acetone extract ($r = 0.627$). However, no negative correlation was observed among any pair.

References

Andarwulan, N., Batari, R., Sandrasari, D. A., Bolling, B., & Wijaya, H. (2010). Flavonoids content and antioxidant activity of vegetables from Indonesia. *Food Chemistry, 121*, 1231–1235.

Barros, L., Carvalho, A. M., Morais, J. S., & Ferreira, I. C. F. R. (2010). Strawberry-tree, blackthorn and rose fruits: Detailed characterization in nutrients and phytochemicals with antioxidant properties. *Food Chemistry, 120*, 247–254.

Beecher, G. R. (2003). Overview of dietary flavonoids: Nomenclature, occurrence and intake. *Journal of Nutrition, 133*, 3248–3254.

Bors, W., Heller, W., Michel, C., & Saran, M. (1990). Flavonoids as antioxidant: Determination of radical scavenging efficiencies. *Methods in Ezymology, 186*, 343–355.

Einhellig, F. A., Putnam, A. R., & Tang, C. S. (1986). *The science of allelopathy* (Vol. 163, pp. 171–189). New York: Wiley.

Formica, J. V., & Regelson, W. (1995). Review of the biology of quercetin and related bioflavonoids. *Food and Chemical Toxicology, 33*, 1061–1080.

Fotherby, M. D., Williams, J. C., Forster, L. A., Craner, P., & Ferns, G. A. (2000). Effect of vitamin C on ambulatory blood pressure and plasma lipids in older persons. *Journal of Hypertension, 18*, 411–415.

Gryglewski, R. J., Korbut, R., & Robak, J. (1987). On the mechanism of antithrombotic action of flavonoids. *Biochemical Pharmacology, 36*, 317–321.

Halliwell, B., & Gutteridge, J. M. C. (1990). Role of free radicals and catalytic metal ions in human disease: An overview. *Methods in Enzymology, 186*, 80–85.

Harborne, J. B. (1976). Functions of flavonoids in plants. In T. W. Goodwin (Ed.), *Chemistry and biochemistry of plant pigments* (Vol. 67, pp. 736–778). New York: Academic.
Harborne, J. B. (1993). *The flavonoids: Advances in research since 1986* (Vol. 134, pp. 543–553). London: Chapman & Hall.
Harborne, J. B., & Baxter, H. (1999). *The handbook of natural flavonoids, Vols. 1 & 2* (Vol. 96, pp. 612–632). Chichester: Wiley.
Herrmann, K. (1976). Flavonoids and flavones in food plants: A review. *Journal of Food Technology, 11*, 433–448.
Ji, L., Wu, L., Gao, W., Yang, J., & Guo, C. (2011). Antioxidant capacity of different fraction of vegetables and correlation with the contents of ascorbic acid, phenolics, and flavonoids. *Journal of Food Science, 76*, 1248–1257.
Kim, D., Jeond, S., & Lee, C. (2003). Antioxidant capacity of phenolic phytochemicals from various cultivars of plums. *Food Chemistry, 81*, 321–326.
Klein, B. P., & Perry, A. K. (1982). Ascorbic acid and vitamin A activity in selected vegetables from different geographical areas of the United States. *Journal of Food Science, 47*, 941–948.
Kukic, J., Petrovic, S., & Niketic, M. (2006). Antioxidant activity of four endemic Stachys taxa. *Biological and Pharmaceutical Bulletin, 29*, 725–729.
Kumaran, A., & Karunakaran, R. J. (2006). Antioxidant and free radical scavenging activity of an aqueous extract of *Coleus aromaticus. Food Chemistry, 97*, 109–114.
Lako, J., Trenerry, V. C., Wahlqvist, M., Wattanapenpaiboon, N., Sotheeswaran, S., & Robert, P. (2007). Phytochemical flavonols, carotenoids and the antioxidant properties of a wide selection of Fijian fruit, vegetables and other readily available foods. *Food Chemistry, 101*, 1727–1741.
Lin, L., Cui, C., Wen, L., Yang, B., Luo, W., & Zhao, M. (2011). Assessment of in vitro antioxidant capacity of stem and leaf extracts of *Rabdosia serra* (MAXIM.) HARA and identification of the major compound. *Food Chemistry, 126*, 54–59.
Manach, C., Scalbert, A., Morand, C., Remesy, C., & Jimenez, L. (2004). Polyphenols: Food sources and bioavailability. *American Journal of Clinical Nutrition, 79*, 727–747.
Mandal, S. M., Chakraborty, D., & Dey, S. (2010). Phenolic acids act as signaling molecules in plant microbe symbioses. *Plant Signaling and Behavior, 5*, 359–368.
Motamed, S. M., & Naghibi, F. (2010). Antioxidant activity of some edible plants of the Turkmen Sahra region in northern Iran. *Food Chemistry, 119*, 1637–1642.
Ramchoun, M., Harnafi, H., Alem, C., Benlys, M., Elrhaffari, L., & Amrani, S. (2009). Study on antioxidant and hypolipidemic effects of polyphenol rich extract from Thymus vulgaris and Lavendula multifida. *Pharmacognosy Research, 1*, 106–112.
Schorah, C. J., Downing, C., Piripitsi, A., Gallivan, L., Al-Hazaa, A. H., Sanderson, M. J., & Bodenham, A. (1996). Total vitamin C, ascorbic acid, and dehydroascorbic acid concentrations in plasma of critically ill patients. *American Journal of Clinical Nutrition, 63*(5), 760–765.
Sies, H., Stahl, W., & Sundquist, A. R. (1992). Antioxidant function of vitamins, vitamins E and C, beta-carotene, and other carotenoids. *Annals of the New York Academy of Science, 669*, 7–20.
Tomas-Barberan, F., & Espin, J. C. (2001). Phenolic compounds and related enzymes as determinants of quality of fruits and vegetables. *Journal of the Science of Food and Agriculture, 81*, 853–876.

Chapter 8
Antioxidant Properties of Wild Edible Vegetables

8.1 Methodology

8.1.1 Sampling

Approximately, 5 kg fresh edible parts of wild vegetables were collected from several plants of each species. Collected samples were washed with tap water followed by distilled water and dried at room temperature to remove residual moisture, then placed in paper envelope and dried at 70 °C for 48 h in electric oven, following the method as described by Abuye et al. (2003) and Wahab et al. (2008). The dried samples were ground into a fine powder using a pestle and mortar and sieved through 20-mesh sieve. The powder samples were placed in cleaned, labelled plastic bottles and stored at room temperature in desiccators for antioxidant analysis.

8.1.2 Extraction

The freshly collected vegetable samples were washed with tap water, followed by deionized water, then cut into small pieces and dried at 70 °C for 48 h in an electric oven. The dried samples were ground into fine powder. A two-step extraction process as reported by Ji et al. (2011) was adopted to assess the hydrophilic (water) and hydrophobic (acetone) contributions. First the samples were extracted with water (hydrophilic part) and then with acetone (hydrophobic part). About 1 g of dried powered sample was mixed with 10 mL of distilled water and this homogenate was centrifuged at 6,000 rpm for 15 min. This extraction procedure is repeated three times; the solid residue is further extracted thrice in acetone (1:10 w/v).

A.M. Abbasi et al., *Wild Edible Vegetables of Lesser Himalayas*,
DOI 10.1007/978-3-319-09543-1_8

8.2 Antioxidant Assays

Antioxidant properties of water and acetone extracts of wild edible vegetables were measured by using protocols as mentioned below.

8.2.1 DPPH Scavenging Activity

DPPH scavenging activity of the wild edible vegetables was determined according to the method as described by Yu et al. (2002) with some modification. This method is based on the ability of an antioxidant to scavenge the DPPH cation radical. Briefly 2.0 mL of the sample extract or standards was added to the 5 mL of DPPH solution (0.1 mM in methanol) and vortexes vigorously; then incubated in dark for 30 min at room temperature and the decolorization of DPPH was measured against blank at 517 nm. Results were expressed as Gallic acid equivalents and % inhibition was calculated by following relationship:

$$\text{Inhibition } (\%) = \frac{\left(A_{\text{Blank}} - A_{\text{Sample}}\right)}{\left(A_{\text{Blank}}\right)} \times 100$$

8.2.2 Hydroxyl Radical Scavenging Activity

The hydroxyl radical scavenging activity of water and acetone extracts was calculated as described by Yu et al. (2004) with some modifications. This assay is based on Fenton reaction. Briefly 2.0 mL of 0.2 M phosphate buffer (pH 7.2), 0.04 mL ferrous sulphate (0.02 M), 2 mL of extract, and 1 mL of 1, 10-phenanthroline (0.04 M) were delivered into the test tube. The Fenton reaction was initiated by addition of 0.1 mL of 7 mM H_2O_2. Absorbance was measured at 560 nm after 5 min incubation at room temperature. The relative hydroxyl radical scavenging activity (%) was calculated as:

$$\text{Scavenging activity } (\%) = \frac{\left(A_{\text{Blank}} - A_{\text{Sample}}\right)}{\left(A_{\text{Blank}}\right)} \times 100$$

8.2.3 Hydrogen Peroxide Scavenging Activity

Hydrogen peroxide scavenging activity of the hydrophilic and hydrophobic extracts was determined by Aeyigoro and Okoh (2010) method. The extract (4 mL) was mixed with 2.4 mL of 4 mM H_2O_2 solution prepared in phosphate buffer (0.1 M, pH 7.4) and incubated for 10 min at room temperature. The absorbance was measured at 230 nm against blank, containing the extract without H_2O_2. Scavenging activity (%) was calculated by:

$$\text{Scavenging activity }(\%) = \frac{\left(A_{\text{Blank}} - A_{\text{Sample}}\right)}{\left(A_{\text{Blank}}\right)} \times 100$$

8.2.4 Fe^{2+} Chelating Activity

The extracts were assessed for their ability to compete with ferrozine for iron (II) in solution. The chelating ability of ferrous ion of various fractions was estimated by using the method of Dinis et al. (1994). About 2.0 mL of water or acetone extract was added to the 2.0 mL of ferrous sulphate (0.125 mM). The reaction was initiated by addition of 2 mL of 0.3125 mM ferrozine. The mixture was shaken vigorously and left standing at room temperature for 10 min. Absorbance of the solution was measured at 562 nm against blank prepared in same manner using ferrous chloride and water. EDTA (0.625–5.0 mg) served as positive control and sample without extract or EDTA served as negative control. The percentage inhibition of ferrozine-Fe (II) complex was calculated using formula:

$$\text{Chelating activity }(\%) = \frac{\left(A_{\text{Control}} - A_{\text{Sample}}\right)}{\left(A_{\text{Control}}\right)} \times 100$$

8.2.5 Ferric Ion Reducing Antioxidant Power (FRAP Assay)

The ferric ion reducing power of the extracts was calculated by the method of Hazra et al. (2008) with slight modification. Briefly 2.0 mL of sample was mixed with the 2.0 mL of phosphate buffer (0.2 M, pH 6.6) and 2 mL of 0.1 % potassium ferricyanide, followed by incubation at 50 °C in water bath for 20 min. Afterward the reaction was stopped with addition of 2 mL of trichloroacetic acid (10 %). The upper portion of solution (2 mL) was mixed with 2 mL of distilled water and 2 mL of 0.01 % ferric chloride and left for 20 min at room temperature and absorbance was measured at 700 nm against blank. A higher absorbance of reaction mixture indicated greater reducing power. Rutin, Gallic acid, or ascorbic acid can be used as positive control.

8.2.6 Phosphomolybdenum Complex Assay

The total antioxidant capacity (TAC) of the vegetable extracts was measured by Phosomolybdenum complex assay (PMA) as described by Prieto et al. (2006). Briefly 2.0 mL of sample solution in water or acetone extract was added to the 6.6 mL of reagent solution (0.6 mol/L sulphuric acid, 28 mol/L sodium phosphate, and 4 mol/L ammonium molybdate), capped and incubated at 95 °C for 90 min. After cooling to room temperature, the absorbance was measured at 695 nm against blank containing 1 mL of reagent solution and same volume of water instead of extract solution and then subjected to the same experimental condition. The results from three independent experiments, each run in triplicate, are expressed as the mean of relative antioxidant activity (RAA) compared to ascorbic acid (Vitamin C).

8.3 Antioxidant Properties of Wild Vegetables

8.3.1 *Amaranthus hybridus* L.

Measured levels of antioxidant properties in the leaves of *A. hybridus* are given in Fig. 8.1, which indicates that comparatively water extracts exhibited higher antioxidant activity than acetone extract. Percentage DPPH scavenging activity was found higher in acetone extract than the corresponding water extract. However, comparatively high levels were measured for OH^- radical scavenging, hydrogen peroxide (H_2O_2) radical scavenging, and ferrous ion (Fe^{2+}) chelating activities in water extracts than corresponding acetone extracts. Similarly water extract showed high value in Ferric ion reducing antioxidant power (FRAP), expressed as μM GAE/100 g FW, compared to acetone extract on fresh weight basis. However, in the case of Phosphomolybdenum complex assay (PMA) acetone extract showed higher concentration, articulated as μM AAE/100 g, than the corresponding water extract.

8.3.2 *Amaranthus spinosus* L.

Results of antioxidant properties of *A. spinosus* leaves indicate that in general, water extracts show better potential as compared to acetone extracts in DPPH scavenging, OH^- radical scavenging, H_2O_2 radical scavenging, and Fe^{2+} chelating activities as mentioned in Fig. 8.2. Similarly high level was noted for FRAP assay, expressed as μM GAE/100 g, FW, in water extract than corresponding acetone extract. However, in the case of total antioxidant activity (PM assay) acetone extract showed significant level, articulated as μM AAE/100 g, FW, than water extract on fresh weight basis.

8.3.3 *Amaranthus viridis* L.

Results of antioxidant activity in the leaves of *A. viridis* are mentioned in Fig. 8.3, which indicates that in majority of the cases water extracts show greater values than acetone extracts. Percentage levels of DPPH scavenging activity, H_2O_2 radical scavenging activity, and Fe^{2+} chelating activity were found more in water extracts than acetone extracts. Similarly OH^- radical scavenging activity was measured only in water extract, whereas in acetone extract it was below detection limit.

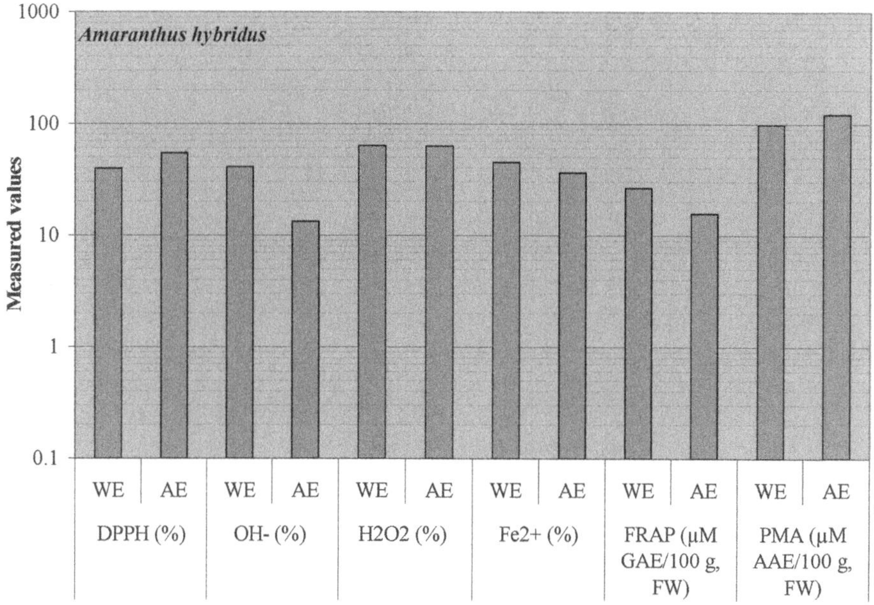

Fig. 8.1 Antioxidant activities of *Amaranthus hybridus*

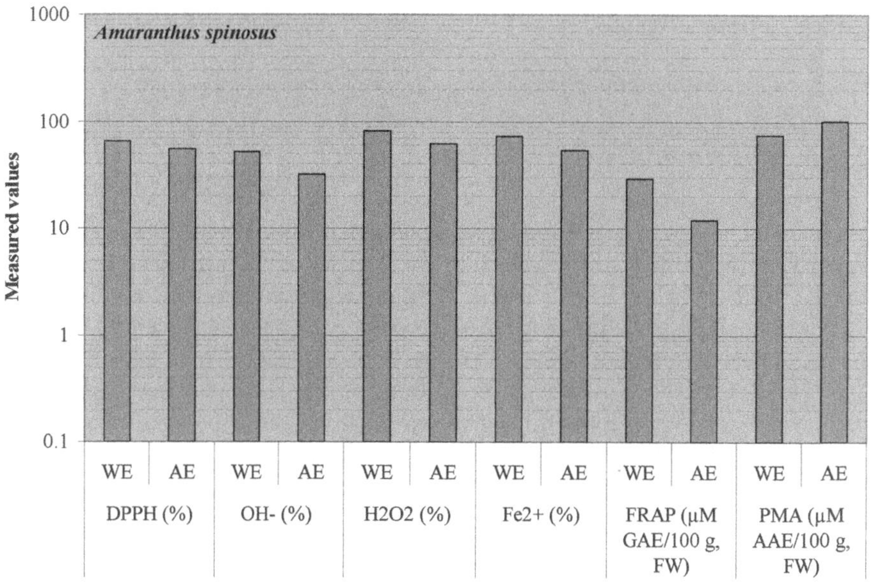

Fig. 8.2 Antioxidant activities of *Amaranthus spinosus*

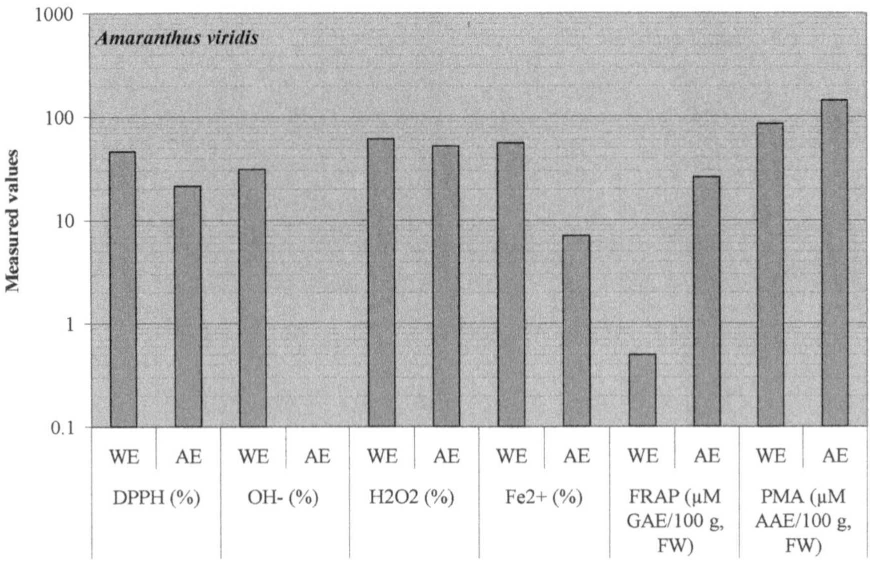

Fig. 8.3 Antioxidant activities of *Amaranthus viridis*

However, acetone extracts show significantly higher values during FRAP assay (µM GAE/100 g, FW) and PMA (µM AAE/100 g, FW) as compared to water extract on fresh weight basis.

8.3.4 *Digera muricata (L.) Mart.*

Antioxidant potential of *D. muricata* is shown in Fig. 8.4. DPPH scavenging activity was found higher in water extract as compared to the corresponding acetone extract. In the case of OH^- radical scavenging activity, H_2O_2 radical scavenging activity and Fe^{2+} chelating activity acetone extracts show comparatively more potential than corresponding water extracts. Similarly, FRAP values, expressed as µM GAE/100 g, FW, were found to be higher in the acetone extract than water extract. Nevertheless, elevated level of total antioxidant activity in term of PM assay (µM AAE/100 g, FW) was found in acetone extract compared to water extract on fresh weight basis.

8.3.5 *Dryopteris ramosa (Hope) C. Chr.*

Measured levels of antioxidant activities in the leaves of *D. ramosa* are mentioned in Fig. 8.5, which indicates that in general, water extracts show better response

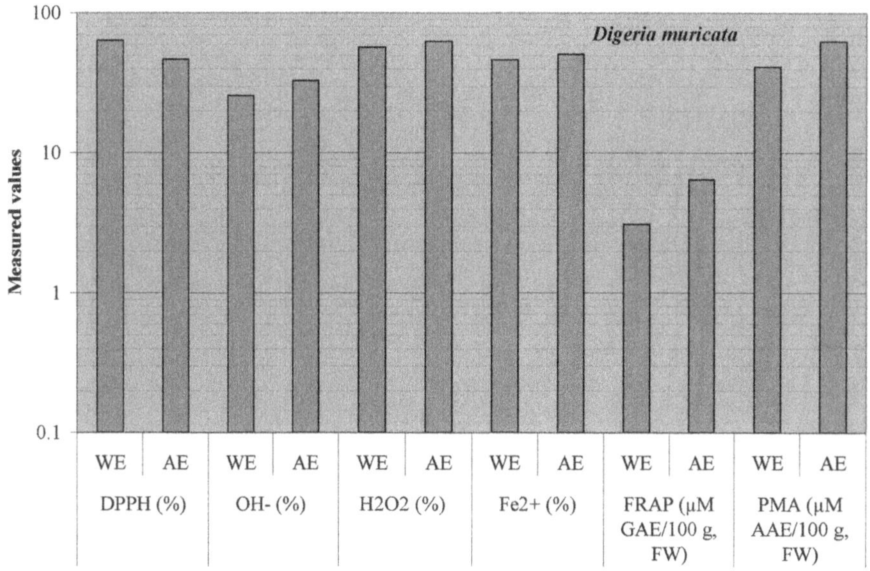

Fig. 8.4 Antioxidant activities of *Digera muricata*

compared to acetone extracts. Percentage levels of DPPH scavenging, OH^- radical scavenging, H_2O_2 radical scavenging, and Fe^{2+} chelating activities were found higher in water extracts as compared to the corresponding levels in acetone extracts. During FRAP and PM assays acetone extracts show better response, articulated as μM GAE/100 g, FW and μM AAE/100 g, FW, respectively, than water extracts on fresh weight basis.

8.3.6 *Bidens bipinnata* L.

Antioxidant activities measured in the leaves of *B. bipinnata* are illustrated in Fig. 8.6, which demonstrates that water extract exhibited high DPPH scavenging activity than acetone extract. OH^- radical scavenging activity was measured only in acetone extract, whereas it was below detection limit in water extract. Similarly acetone extract showed higher level during H_2O_2 radical scavenging activity compared to water extract. In the case of ferrous ion (Fe^{2+}) chelating activity elevated level was observed in water extract as compared to acetone extract. However, acetone extract exhibited higher value in the case of FRAP assay, expressed as μM GAE/100 g, FW, than water extract. In the case of total antioxidant activity (PM assay) significantly higher concentration was recorded for both acetone and water extracts, expressed as μM AAE/100 g, FW, on fresh weight basis.

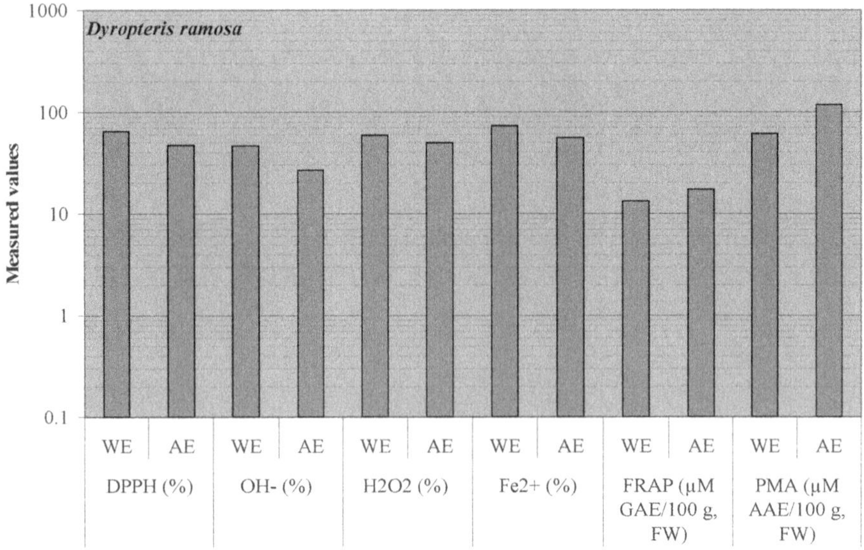

Fig. 8.5 Antioxidant activities of *Dryopteris ramosa*

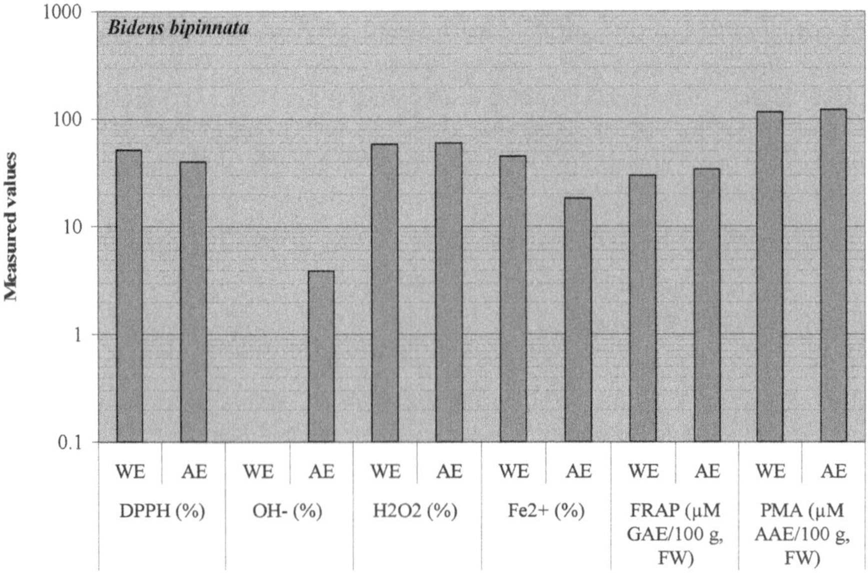

Fig. 8.6 Antioxidant activities of *Bidens bipinnata*

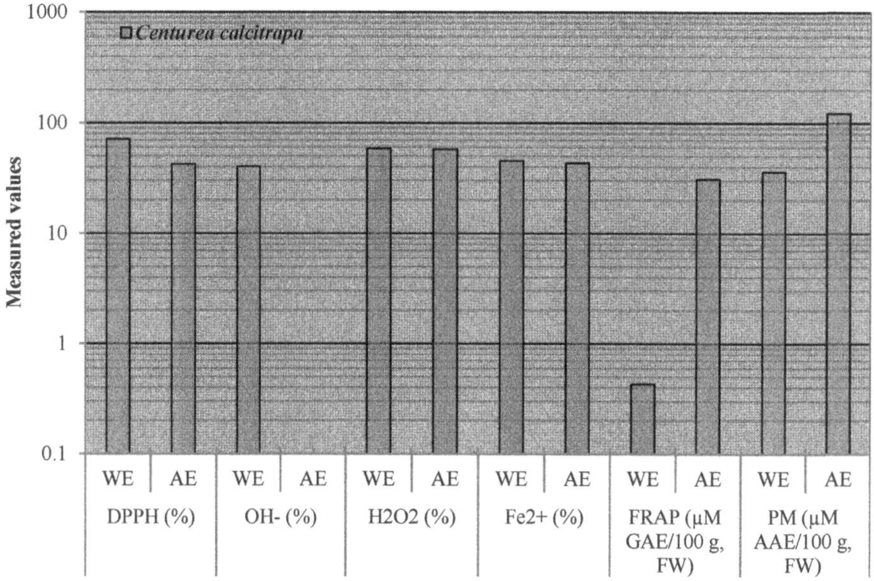

Fig. 8.7 Antioxidant activities of *Centaurea calcitrapa*

8.3.7 Centaurea calcitrapa L.

Measured levels of antioxidant properties in *C. calcitrapa* are given in Fig. 8.7, which revealed that percentage levels of DPPH scavenging activity, OH^- radical scavenging activity, H_2O_2 radical scavenging activity, and Fe^{2+} chelating activity were comparatively higher in the water extracts than corresponding acetone extracts. However, total antioxidant activity in terms of FRAP and PM complex assays, articulated as μM GAE/100 g and μM AA/100 g, on fresh weight basis were found significantly high in acetone extracts than water extracts.

8.3.8 Cichorium intybus L.

Acetone extract of the leaves of *C. intybus* shows more DPPH scavenging activity, while in water extract it was found below detection limit. Similarly in the case of OH^- radical scavenging activity, H_2O_2 radical scavenging activity and Fe^{2+} chelating activity acetone extracts were found more active than corresponding water extracts. However, during FRAP assay it was noted that water extract showed high value, expressed as μM GAE/100 g, FW, than acetone extract. In term of total antioxidant activity (PM assay) almost same behavior was observed both in acetone and water extracts and expressed as μM AAE/100 g, FW on fresh weight basis (Fig. 8.8).

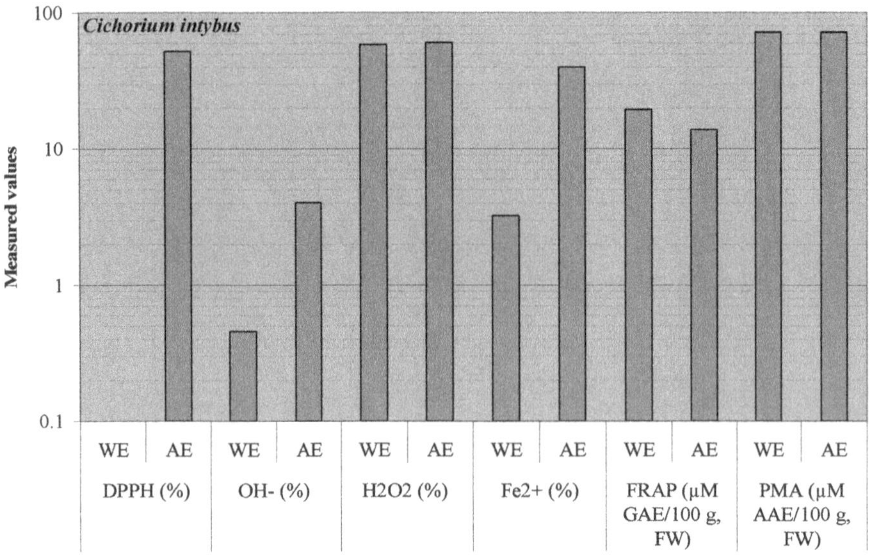

Fig. 8.8 Antioxidant activities of *Cichorium intybus*

8.3.9 *Launaea procumbens (Roxb).*

In the case of *L. procumbens* measured levels for DPPH scavenging activity, OH^{-} radical scavenging activity, H_2O_2 radical scavenging activity, and Fe^{+2} chelating activity were high in water extracts compared to the acetone extracts (Fig. 8.9). Similarly, water extract exhibited higher value in FRAP and PM assays, expressed as μM GAE/100 g, FW and μM AAE/100 g, FW, than acetone extracts on fresh weight basis.

8.3.10 *Sonchus asper L.*

The antioxidant potential in the leaves of *S. asper* as given in the Fig. 8.10 reveals that in majority of the cases measured levels were high in water extracts than acetone extracts. DPPH, OH^{-}, and H_2O_2 radical scavenging activities were found higher in water extracts. Similarly higher scavenging was recorded in water extract during Fe^{2+} chelating activity while ferric ion reducing power was higher in acetone extracts compared to water extracts. Nevertheless, in the case of total antioxidant activity acetone extract showed elevated level as expressed in (μM AAE/100 g, FW) than water extract on fresh weight basis.

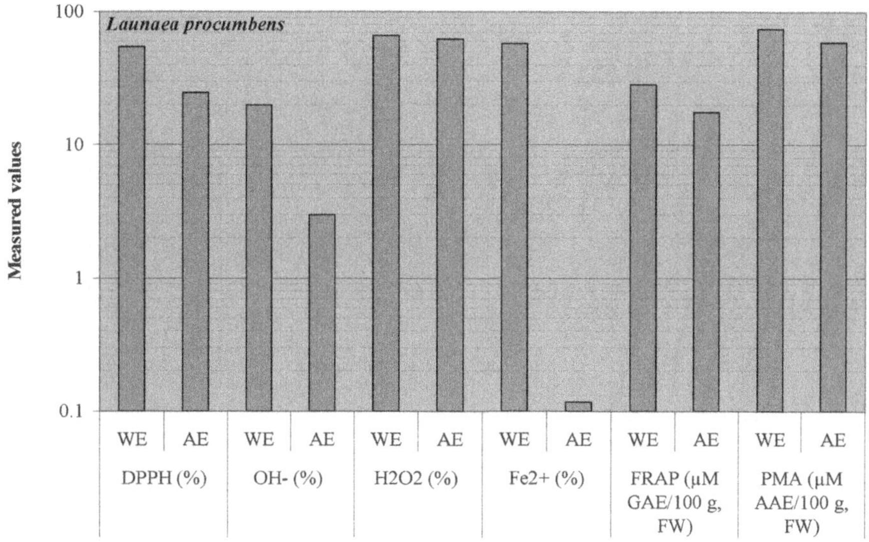

Fig. 8.9 Antioxidant activities of *Launaea procumbens*

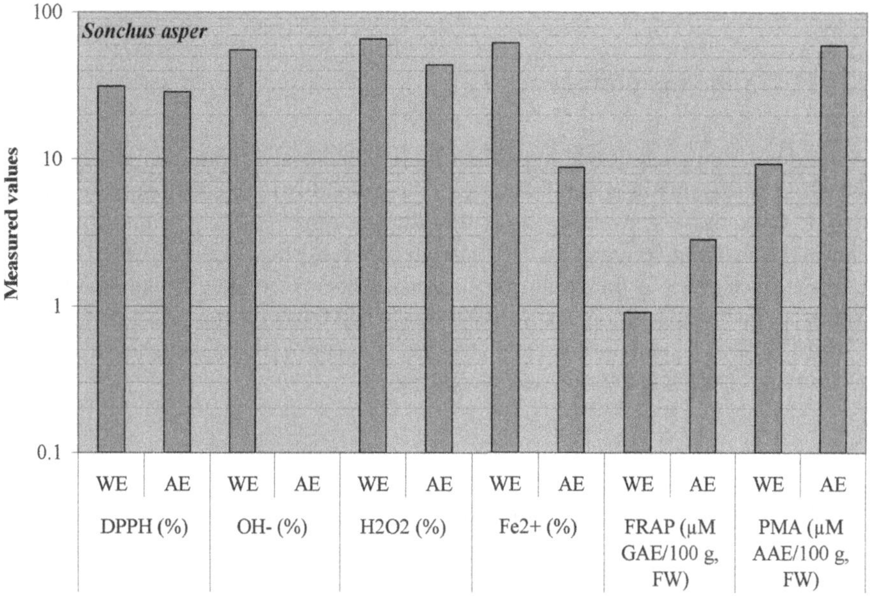

Fig. 8.10 Antioxidant activities of *Sonchus asper*

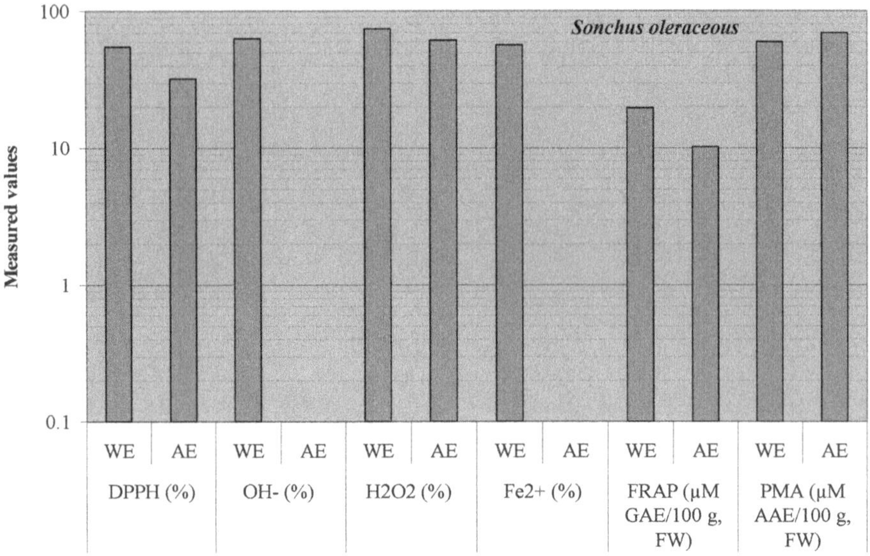

Fig. 8.11 Antioxidant activities of *Sonchus oleraceous*

8.3.11 *Sonchus oleraceous* L.

Antioxidant activities of edible parts of *S. oleraceous* are mentioned in Fig. 8.11, which indicates that in general water extracts exhibit better performance in DPPH and H_2O_2 radical scavenging activities than acetone extracts. Similarly, OH^- radical scavenging activity and Fe^{+2} chelating activity were also observed only in water extract, whereas their values in acetone extracts were below detection limits. Water extract showed higher value during FRAP assay compared to acetone extract. However, total antioxidant activity (PM assay) was found higher in acetone extract than water extract on fresh weight basis.

8.3.12 *Taraxacum officinale* L.

Antioxidant activity assessment of *T. officinale* shown in Fig. 8.12 reveals that water extracts show better results in OH^-, H_2O_2 radicals scavenging and Fe^{2+} chelating activity than the corresponding acetone extracts. Similarly, water extract showed more value illustrated as μM GAE/100 g, FW in FRAP assay than acetone extract. However, in PM complex assay total antioxidant activity was measured high in acetone extract than water extract on fresh weight basis and expressed as μM AAE/100 g, FW.

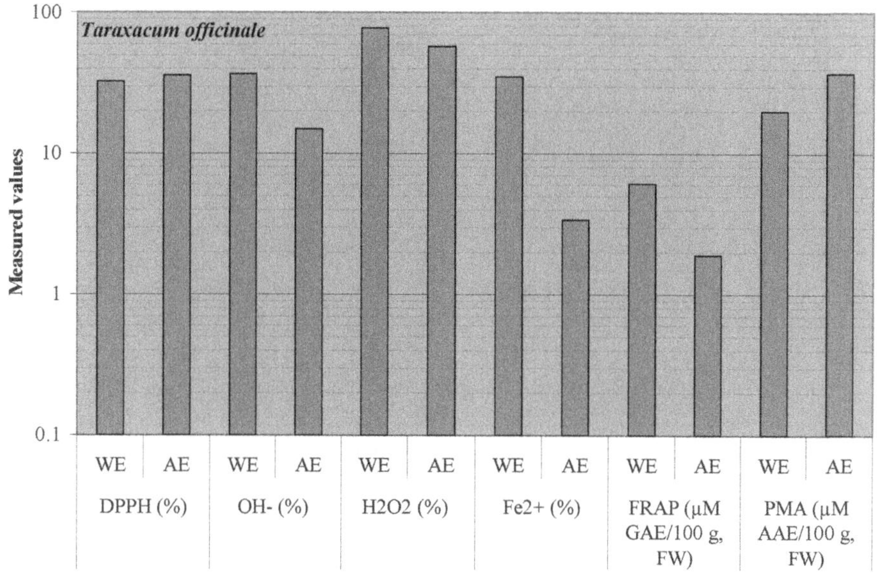

Fig. 8.12 Antioxidant activities of *Taraxacum officinale*

8.3.13 *Bombax malabaracum* DC.

Antioxidant activity profile of the flowering buds of *B. malabaracum* is given in Fig. 8.13, which shows that acetone extract exhibited more DPPH scavenging activity compared to water extract. OH^- radical scavenging activity was measured only in acetone extract, while in water extract it was below detection limit. Nevertheless, water extract exhibited significantly high potential during H_2O_2 radical scavenging activity as compared to acetone extract. Ferric ion reducing antioxidant power was found high in water extract than acetone extract. However, in total antioxidant activity (PM assay) acetone extract showed higher concentration than water extract on fresh weight basis.

8.3.14 *Capsella bursa-pastoris* (L.) Medic.

Measured levels of antioxidant properties in the leaves of *C. bursa-pastoris* are mentioned in Fig. 8.14, which indicates that water extracts show elevated levels in DPPH and OH^- radical scavenging and Fe^{2+} chelating activities than corresponding acetone extracts. However, in the case of H_2O_2 radical scavenging activity, FRAP assay, and PM complex assay measured levels were high in acetone extracts than corresponding water extracts on fresh weight basis.

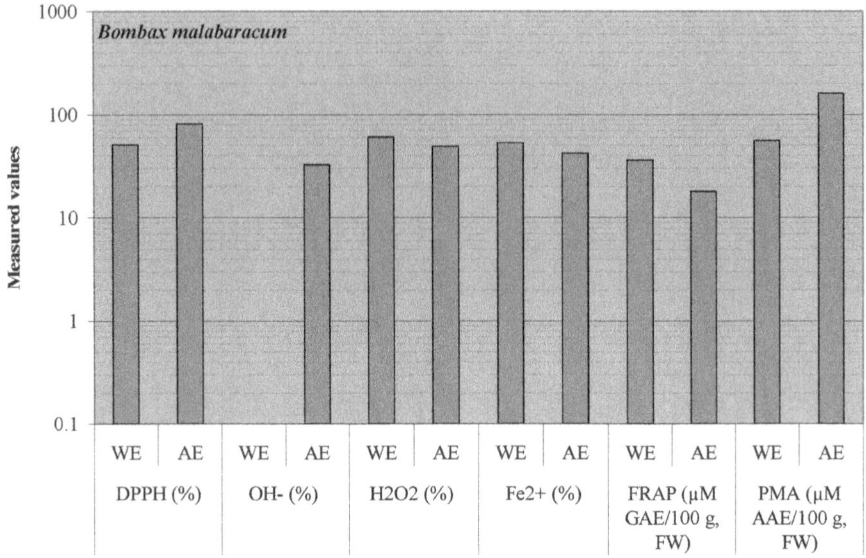

Fig. 8.13 Antioxidant activities of *Bombax malabaracum*

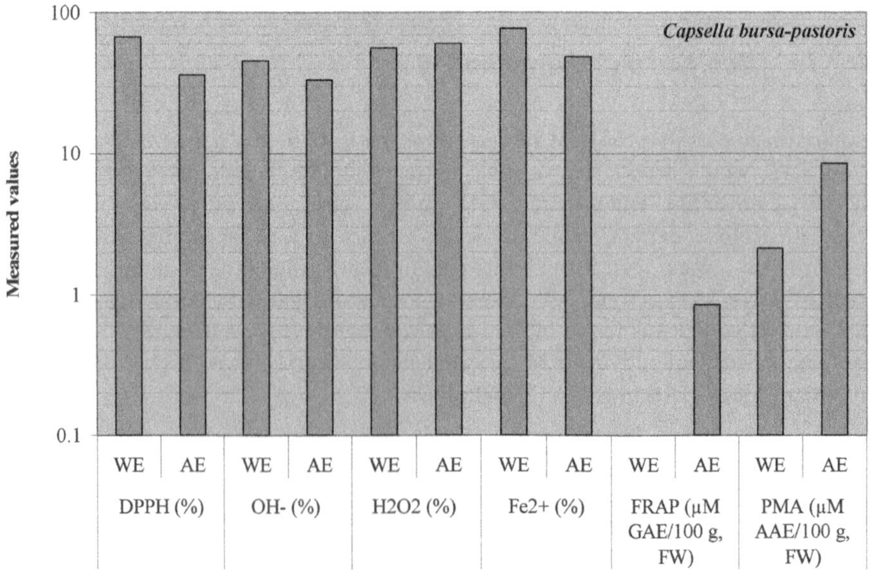

Fig. 8.14 Antioxidant activities of *Capsella bursa-pastoris*

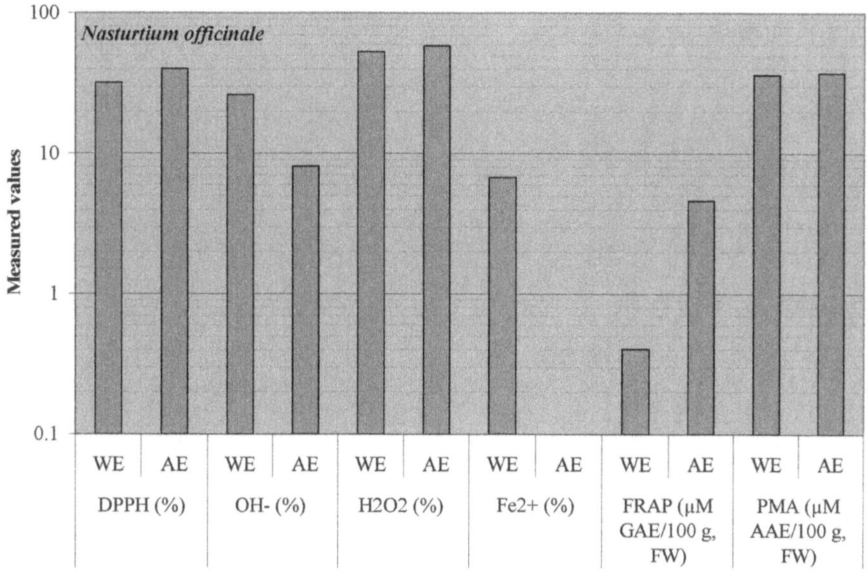

Fig. 8.15 Antioxidant activities of *Nasturtium officinale*

8.3.15 *Nasturtium officinale R. Br.*

Antioxidant activities in the leaves of *N. officinale* are shown in Fig. 8.15, which demonstrates that measured levels in DPPH and H_2O_2 radicals scavenging activities were higher in acetone extracts compared to water extracts. However, in the case of OH^- radical scavenging activity and Fe^{2+} chelating activity water extracts were found more active than corresponding acetone extracts. Nevertheless, during FRAP and PM assays acetone extracts were found more active than corresponding water extracts as expressed in μM GAE/100 g, FW and μM AAE/100 g, FW, respectively.

8.3.16 *Bauhinia variegata L.*

Flowering buds of *B. variegata* show high antioxidant potential as given in Fig. 8.16. DPPH scavenging activity was found higher in acetone extract than water extract. However, water extracts show more OH^- and H_2O_2 radicals scavenging activity and Fe^{2+} chelating activity as compared to acetone extract. On the other hand, acetone extracts show high antioxidant potential in terms of ferric ion reducing antioxidant power (FRAP) and phosphomolybdenum assay compared to water extracts.

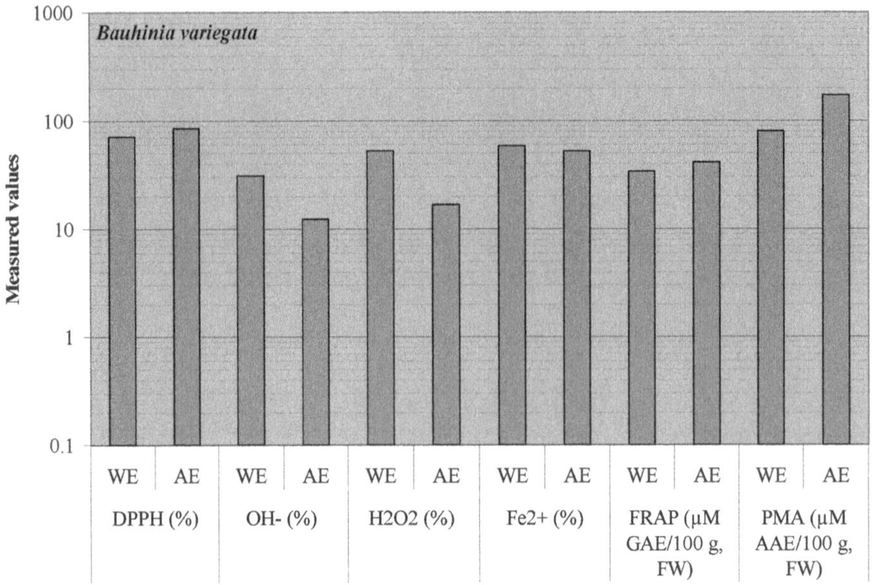

Fig. 8.16 Antioxidant activities of *Bauhinia variegata*

8.3.17 *Bauhinia variegata var. candida Voigt*

Antioxidant activity levels measured in the flowering buds of *B. variegata* var. *candida* as shown in the Fig. 8.17 indicated that acetone extracts show more concentrations in the case of DPPH scavenging and Fe^{2+} chelating activities compared to water extracts. Similarly, during FRAP and PM assays acetone extracts were found more active than corresponding water extracts as expressed in μM GAE/100 g, FW and μM AAE/100 g, FW, respectively. However, measured levels for OH^- and H_2O_2 radicals scavenging activities were found more in water extracts than acetone extracts.

8.3.18 *Bauhinia variegata L. var. variegata*

Flowering buds of *B. variegata* var. *variegata* showed high antioxidant potential as given in Fig. 8.18. DPPH scavenging activity was found higher in acetone extract than water extract. However, water extracts showed high levels in H_2O_2 scavenging and Fe^{2+} chelating activity, but low level in OH^- radical scavenging activity than their corresponding extracts. Similarly water extract showed high antioxidant potential in terms of ferric ion reducing antioxidant power (FRAP). However, acetone extract showed high value in PMA compared to water extract.

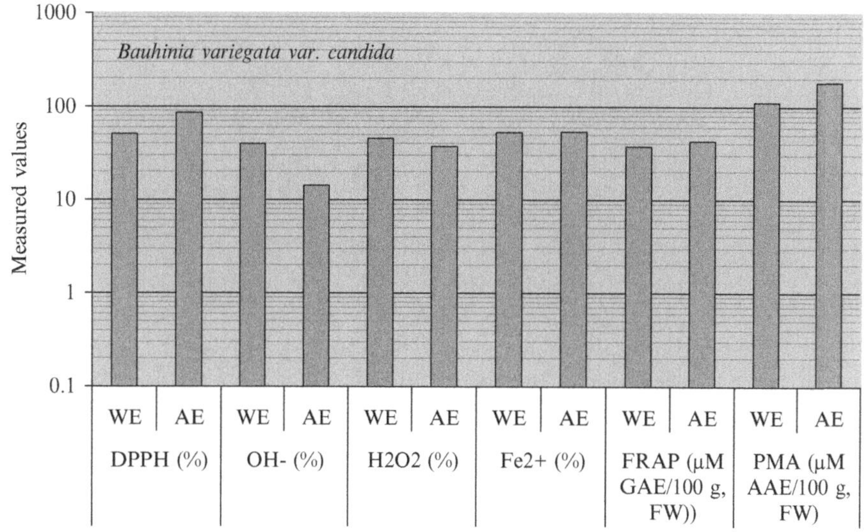

Fig. 8.17 Antioxidant activities of *Bauhinia variegata var. candida*

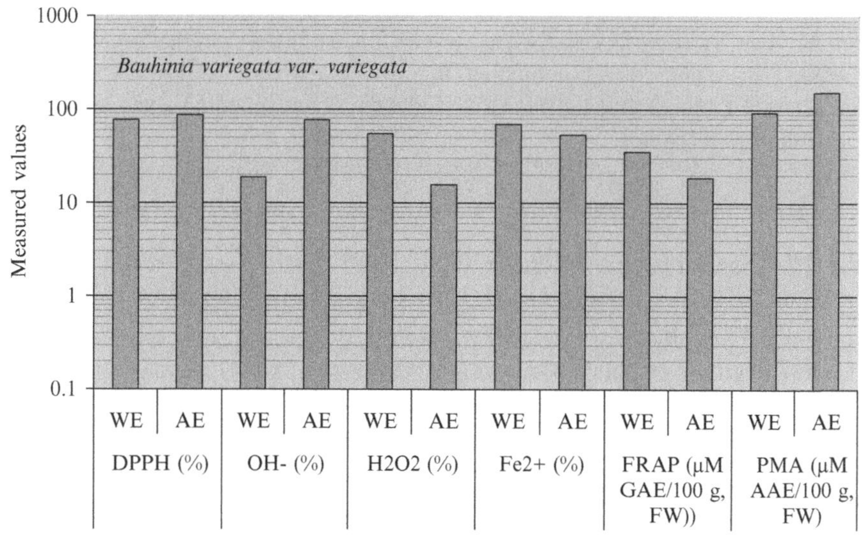

Fig. 8.18 Antioxidant activities of *Bauhinia variegata var. variegata*

8.3.19 Silene conoidea L.

Antioxidant properties of *S. conoidea* are given in Fig. 8.19, which indicates that DPPH scavenging activity was found comparatively higher in water extract than acetone extract. Acetone extract exhibits better result during OH^- radical scavenging activity, while in the case of water extract it was below detection limit. Similarly acetone extract shows higher potential during H_2O_2 radical scavenging

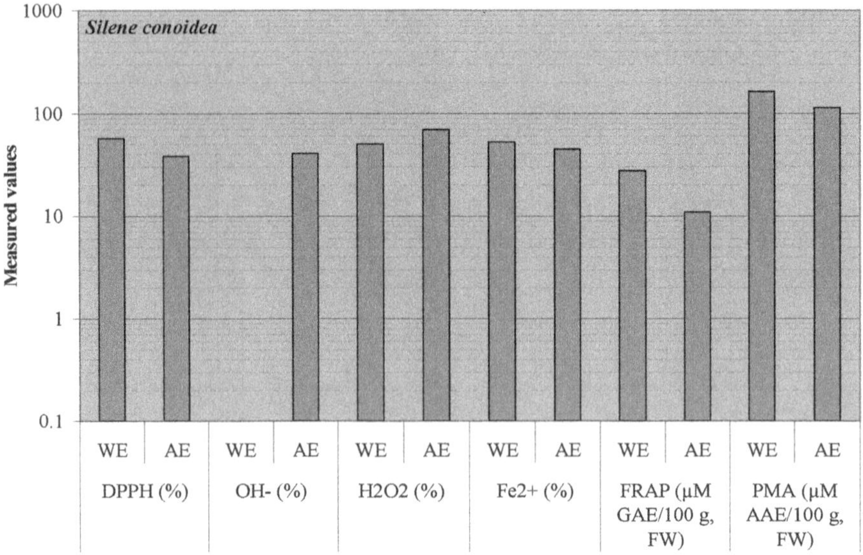

Fig. 8.19 Antioxidant activities of *Silene conoidea*

activity than water extract. In the case of Fe^{2+} chelating assay higher potential was recorded in water extract compared to acetone extract. Comparatively water extracts show more values in FRAP and PM than corresponding acetone extracts on fresh weight basis.

8.3.20 *Stellaria media (L.) Cyr.*

Antioxidant potential in the leaves of *S. media* is given in Fig. 8.20. It was observed that water extracts exhibited better results in the case of DPPH scavenging activity, OH^- radical scavenging activity, and ferrous ion chelating activity as compared to acetone extracts. Significantly, higher H_2O_2 radical scavenging activity was noted in acetone extract as compared to water extract. Measured levels of ferric ion antioxidant power (FRAP) and total antioxidant activity (PM assay) were found high in acetone extracts compared to water extracts on fresh weight basis.

8.3.21 *Chenopodium album L.*

The antioxidant properties in the leaves of *C. album* are given in Fig. 8.21, which indicates that water extract showed more DPPH scavenging activity than acetone extract. OH^- radical scavenging activity and Fe^{2+} chelating activity were measured

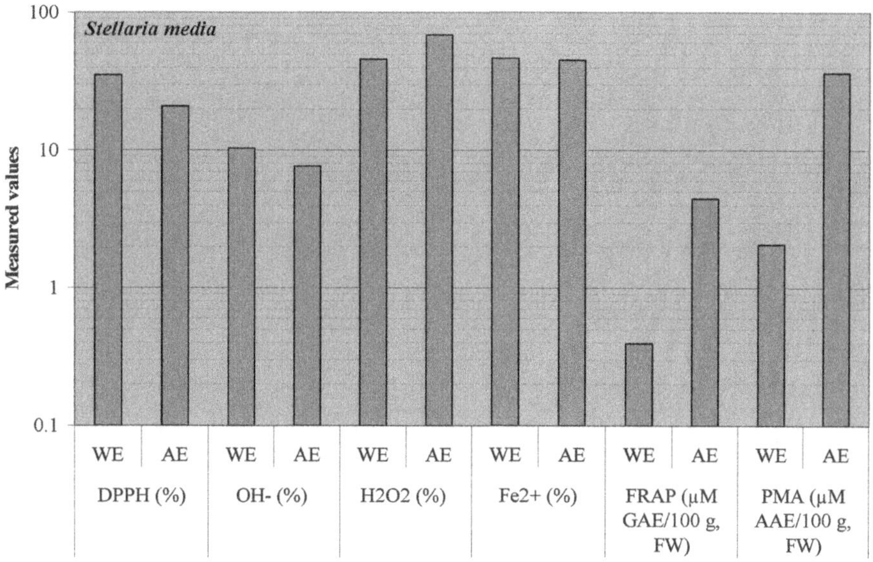

Fig. 8.20 Antioxidant activities of *Stellaria media*

only in water extracts, while in acetones extracts their levels were below detection limit. H_2O_2 radical scavenging activity was observed more in acetone extract than water extract. Nevertheless, acetone extracts show better values for FRAP and PMA assays than corresponding water extracts.

8.3.22 *Commelina benghalensis* L.

In the leaves of *C. benghalensis* acetone extract showed comparatively high DPPH scavenging activity than water extract. However, percentage values of OH^- radical scavenging activity, H_2O_2 radical scavenging activity, and Fe^{2+} chelating activity were high in water extracts than acetone extracts. Similarly FRAP value was also recorded high in water extract than acetone extract. However, in the case of PM assay acetone extract shows significantly higher value than water extract (Fig. 8.22).

8.3.23 *Evolvulus alsinoides* L.

Results of antioxidant activity in the leaves of *E. alsinoides* are presented in Fig. 8.23. It was observed that water extracts showed comparatively higher potential than acetone extracts. In the case of DPPH scavenging, OH^- radical scavenging

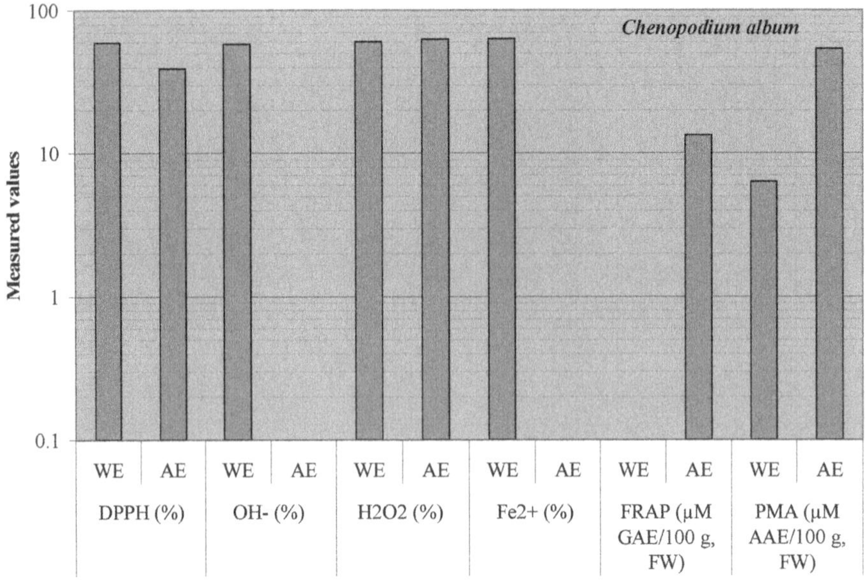

Fig. 8.21 Antioxidant activities of *Chenopodium album*

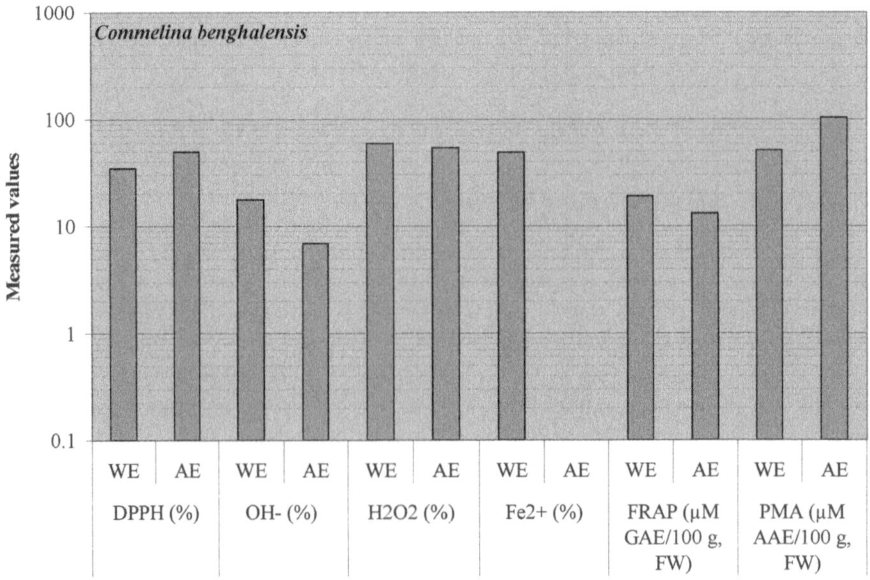

Fig. 8.22 Antioxidant activities of *Commelina benghalensis*

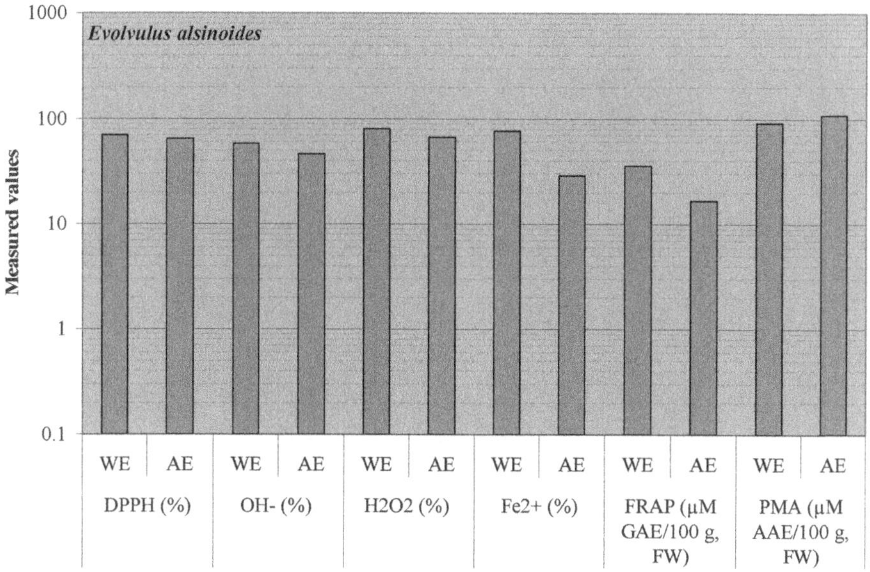

Fig. 8.23 Antioxidant activities of *Evolvulus alsinoides*

activity, H_2O_2 radical scavenging activity, and Fe^{2+} chelating activity water extracts exhibited higher value compared to acetone extract. Similarly FRAP value expressed (μM GAE/100 g, FW) was also found higher in water extract than acetone extract. However, in the case of PM assay acetone extract exhibited higher potential (μM AAE/100 g, FW) than water extract on fresh weight basis.

8.3.24 *Dioscorea deltoidea* Wall. ex, Griseb

Antioxidant property of the tubers of *D. deltoidea* is mentioned in Fig. 8.24, which indicates that percentage DPPH scavenging activity, H_2O_2 radical scavenging activity, and Fe^{2+} chelating activity were measured high in water extracts than their corresponding acetone extracts on fresh weight basis. OH^- radical scavenging activity was measured significantly higher in acetone extract compared to water extract. However, water extracts showed elevated levels in FRAP and PM complex assays as calculated in μM GAE/100 g, FW and μM AAE/100 g, FW than acetone extracts on fresh weight basis.

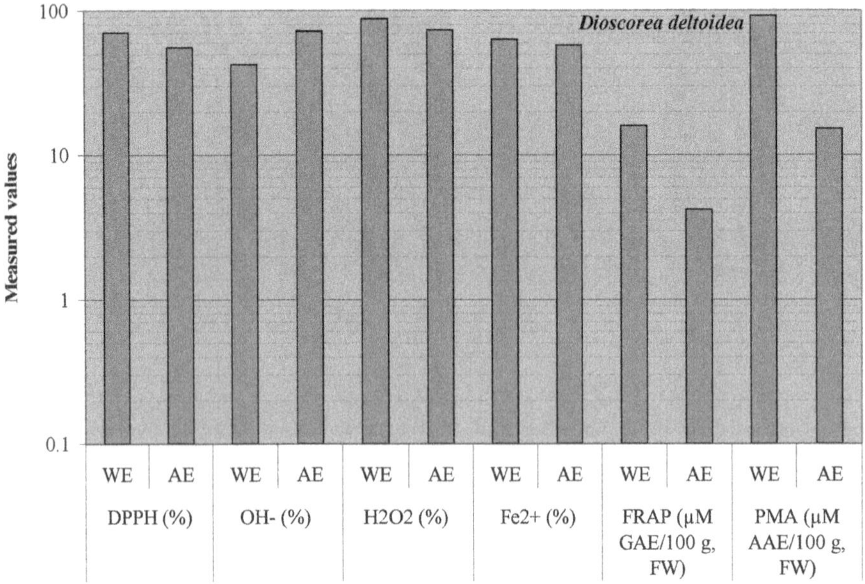

Fig. 8.24 Antioxidant activities of *Dioscorea deltoidea*

8.3.25 *Lamium amplexicaule* L.

Antioxidant activities measured in the leaves of *L. amplexicaule* are given in Fig. 8.25, which reveals that acetone extracts show comparatively better potential in DPPH scavenging activity and H_2O_2 radical scavenging activity than water extracts. In the case of OH^- radical scavenging activity water extract showed high value. Fe^{2+} chelating activity was observed only in water extract, as it was below detection limit in acetone extract. Water extract exhibited better result in FRAP assay (µM GAE/100 g, FW) than acetone extract. Similarly, water extract exhibited higher value (µM AAE/100 g, FW) in PM assay than acetone extract on fresh weight basis.

8.3.26 *Origanum vulgare* subsp. *Hirtum* (Link) Letsw

Measured levels for antioxidant activities in the leaves of *O. vulgare* are illustrated in Fig. 8.26. It was observed that water extract showed higher value during DPPH activity than acetone extract. During OH^- scavenging activity results were below detection limit both in water and acetone extracts. Acetone extract showed higher value during H_2O_2 radical scavenging activity, while below detection limit in water extract. Similarly Fe^{2+} chelating activity was found more in water extract, while noted below detection limit in acetone extract. Water extract showed more value in

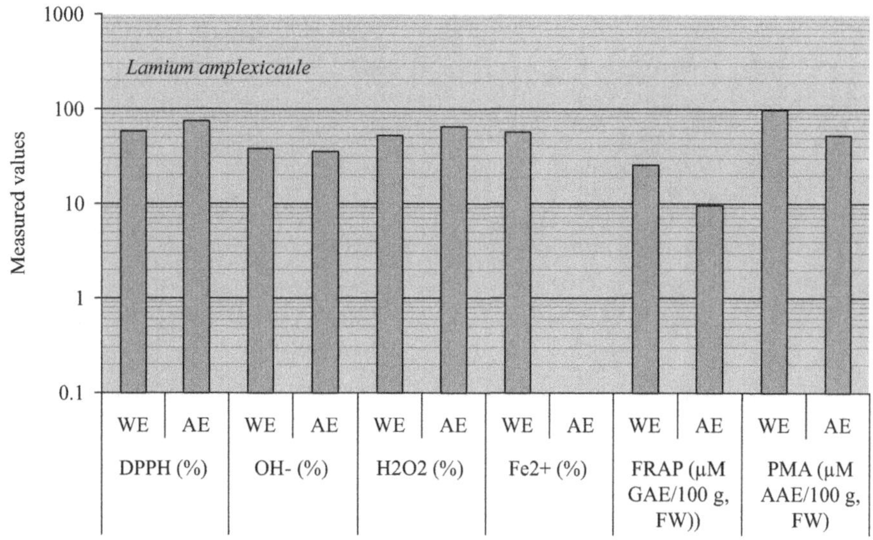

Fig. 8.25 Antioxidant activities of *Lamium amplexicaule*

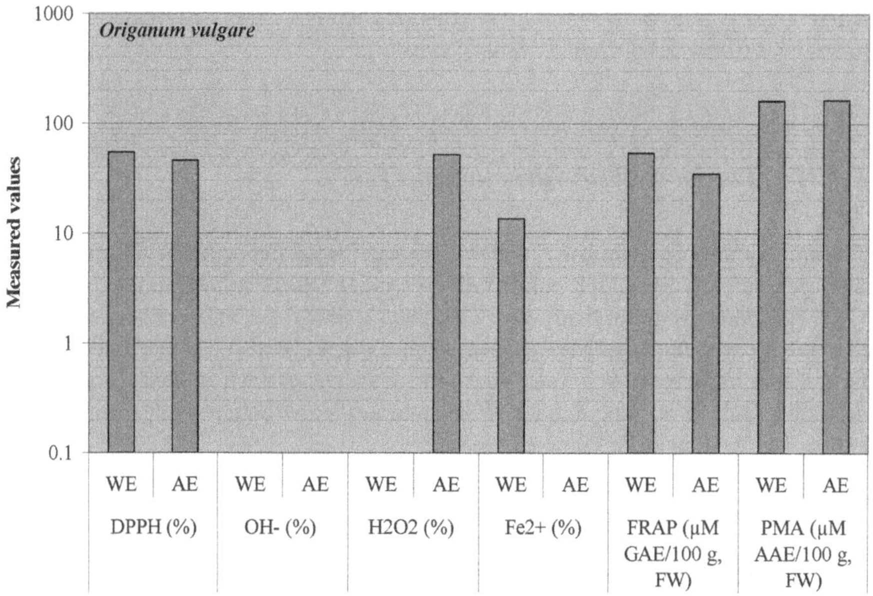

Fig. 8.26 Antioxidant activities of *Origanum vulgare* subsp. *Hirtum*

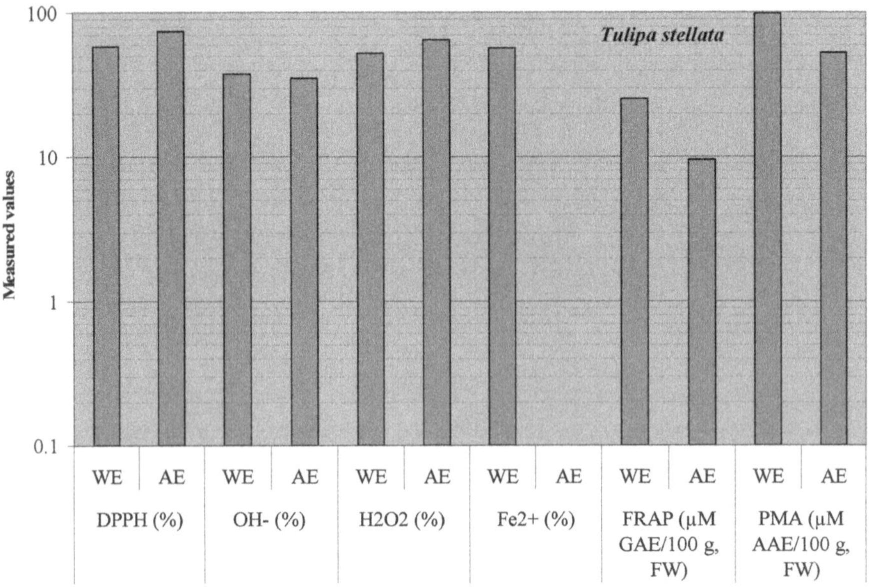

Fig. 8.27 Antioxidant activities of *Tulipa stellata var. clusiana*

FRAP assay than acetone extract. Nevertheless, acetone extract shows more potential value PM assay than water extract.

8.3.27 *Tulipa stellata var. clusiana Hk. f.*

The antioxidant properties of *T. stellata* are mentioned in Fig. 8.27, which reveals that acetone extracts exhibit comparatively higher DPPH scavenging activity and H_2O_2 radical scavenging than water extracts. Nevertheless, OH^- radical scavenging activity was found to be higher in water extract compared to acetone extract. Fe^{2+} chelating activity was measured only in water extract, while it was below detection limit in acetone extract. Water extracts show better result than acetone extract in FRAP and PM complex assays.

8.3.28 *Malva parviflora L.*

Antioxidant properties measured in the leaves of *M. parviflora* are given in Fig. 8.28, which shows that acetone extract showed significantly higher value for DPPH scavenging activity than water extract. However, measured levels of OH^- radical scavenging activity and H_2O_2 radical scavenging were found high in water

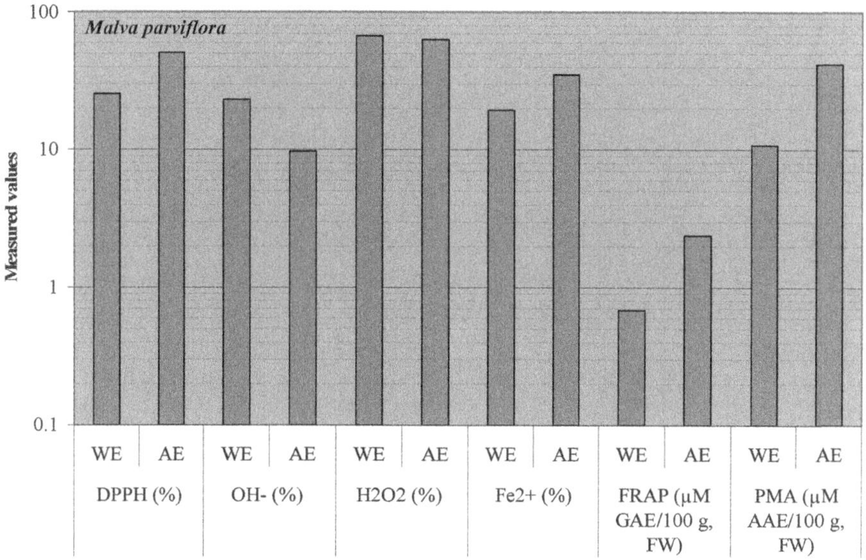

Fig. 8.28 Antioxidant activities of *Malva parviflora*

extracts as compared to acetone extract. Fe^{2+} chelating activity was estimated more in acetone extract than water extract. In FRAP and PM assays acetone extracts again show high levels than water extracts.

8.3.29 Ficus carica L.

Antioxidant activities of *F. carica* are given in Fig. 8.29, which reveals that measured levels of DPPH and OH^- radicals scavenging activities were high in water extracts compared to acetone extracts. However, acetone extract exhibited more value in H_2O_2 radical activity than water extract. Fe^{2+} chelating activity was found only in the water extract, while it was below detection limit in acetone extract. Nevertheless, water extract exhibited more value in FRAP assay than acetone extract, whereas in PM complex assay acetone extract showed high value than water extract.

8.3.30 Ficus palmata Forssak.

Antioxidant activity measured in the leaves of *F. palmata* demonstrates that water extract showed higher DPPH scavenging activity than acetone extract. However, in the case of OH^- and H_2O_2 radical scavenging assays water extracts showed no

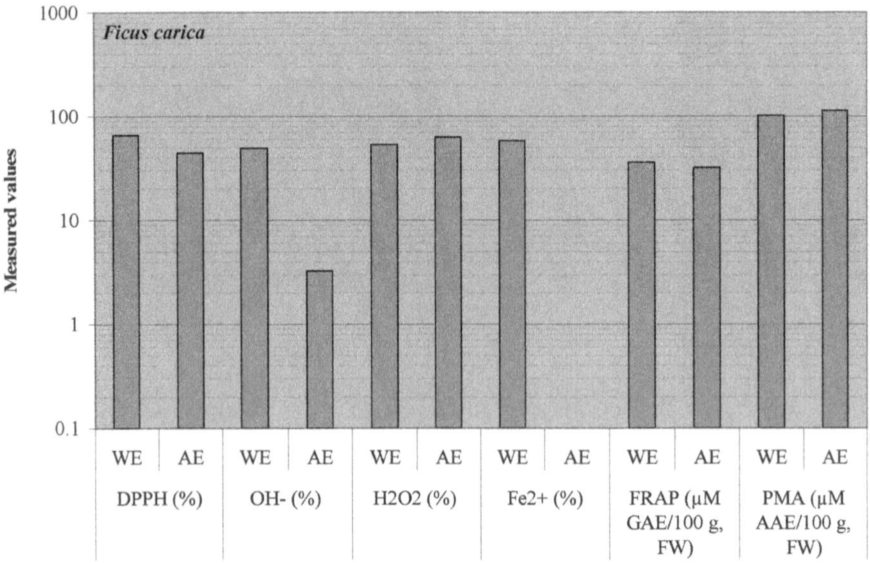

Fig. 8.29 Antioxidant activities of *Ficus carica*

response but acetone extract exhibited significantly higher value. Fe^{2+} chelating activity was also measured more in acetone extract than water extract. Nevertheless, water extract shows significant value during FRAP assay than acetone extract, whereas in PM assay acetone extract shows higher value (than water extract on fresh weight basis) (Fig. 8.30).

8.3.31 Oxalis corniculata L.

Antioxidant potential measured in the leaves of *O. corniculata* is mentioned in Fig. 8.31, which indicates that water extracts exhibited significantly higher DPPH scavenging activity, H_2O_2 radical scavenging activity, and Fe^{2+} chelating activity than acetone extracts. Similarly OH^- radical scavenging activity was also measured only in the water extract, whereas it was below detection limit in acetone extract. Measured levels for FRAP assay were also found high in water extract than acetone extract. However, PM complex activity was found more in acetone extract, whereas in water extract it was below detection limit.

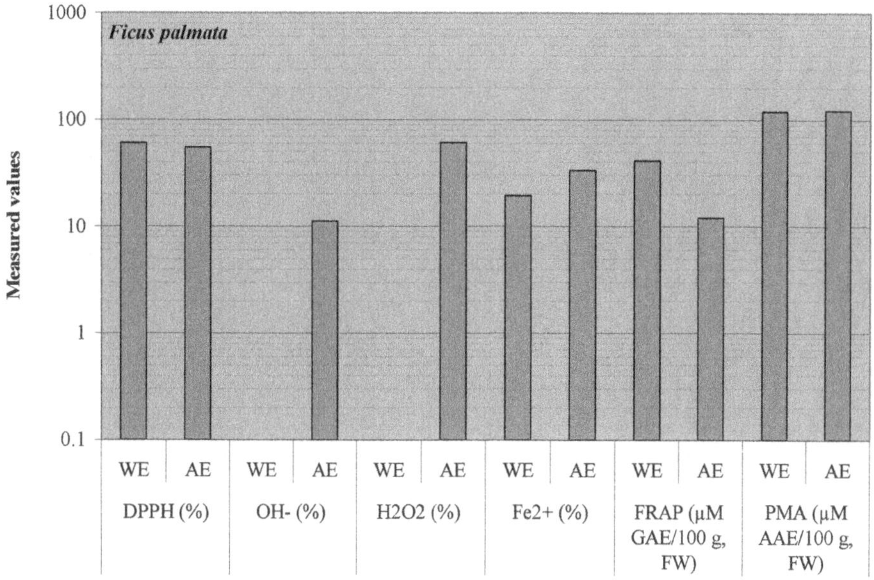

Fig. 8.30 Antioxidant activities of *Ficus palmata*

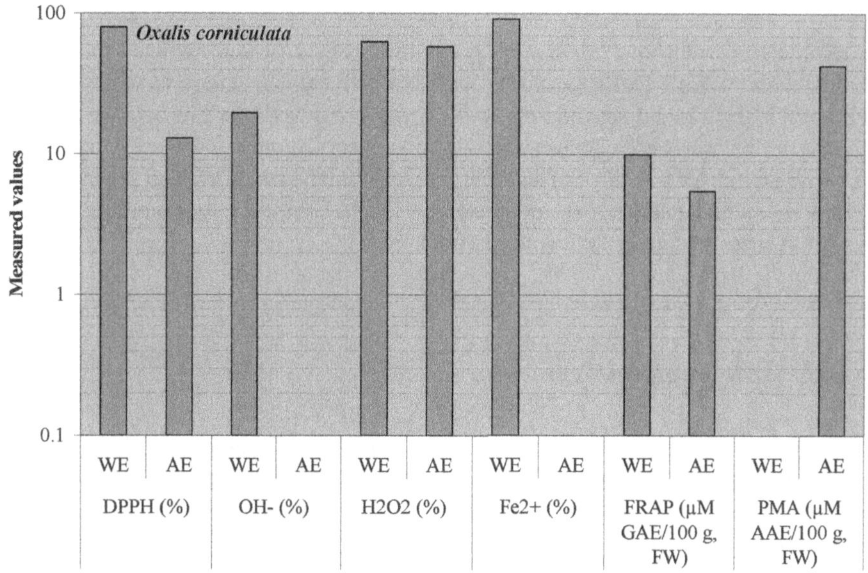

Fig. 8.31 Antioxidant activities of *Oxalis corniculata*

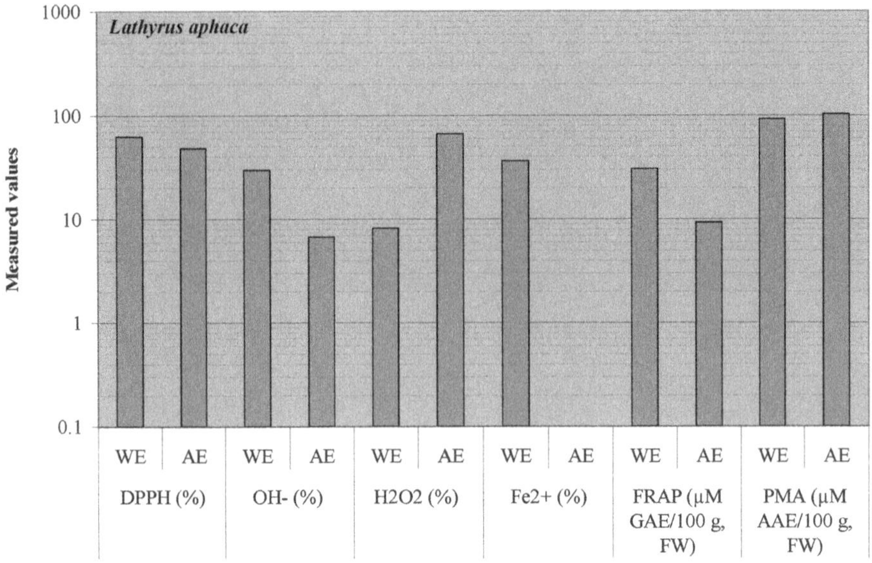

Fig. 8.32 Antioxidant activities of *Lathyrus aphaca*

8.3.32 *Lathyrus aphaca* L.

Antioxidant activity of *L. aphaca* is given in Fig. 8.32, which revealed that water extracts show high DPPH and OH^- radical scavenging activities than acetone extracts. H_2O_2 radical scavenging activity was measured more in acetone extract than water extract. Fe^{2+} chelating activity was observed only in water extract while it was below detection limit in acetone extract. Water extract showed high level in FRAP assay than acetone extract. However, in PM complex assay more value was found in acetone extract than water extract.

8.3.33 *Medicago polymorpha* L

Results of antioxidant activity in the leaves of *M. polymorpha* are mentioned in Fig. 8.33, which indicates that percentage values for DPPH scavenging activity, OH^- radical scavenging activity, H_2O_2 radical scavenging activity, and Fe^{2+} chelating activity were higher in water extracts than corresponding acetone extracts. Similarly water extract exhibited better result in FRAP assay than acetone extract. However, measured value for PMA was noted comparatively higher in acetone extract than water extract.

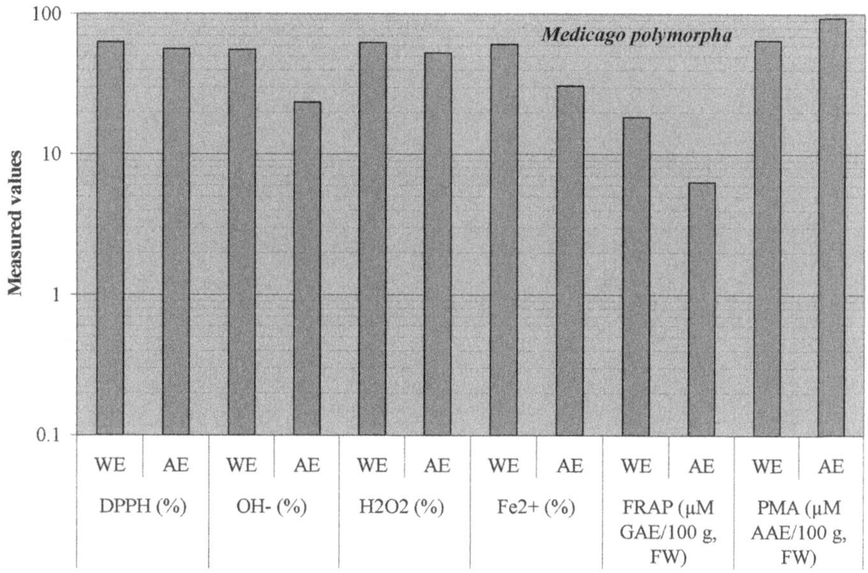

Fig. 8.33 Antioxidant activities of *Medicago polymorpha*

8.3.34 *Melilotus alba* Ledeb.

Antioxidant properties in the leaves of *M. alba* are mentioned in Fig. 8.34, which revealed that DPPH activity was found high in water extract than corresponding acetone extract. OH^- radical scavenging activity was measured only in acetone extract, while in water extract it was below detection limit. Acetone extract exhibited higher contribution in H_2O_2 radical scavenging activity compared to water extract. Fe^{2+} chelating activity was recorded higher in water extract compared to acetone extract. Water extracts showed significantly higher concentrations in FRAP and PM complex assays compared to acetone extracts.

8.3.35 *Melilotus indicus* (L.) All.

Antioxidant activities in the leaves of *M. indicus* are presented in Fig. 8.35, which shows that DPPH scavenging activity was more in acetone extract than water extract. Comparatively higher level was noted for water extract in OH^- radical scavenging activity compared to acetone extract. Nevertheless, acetone extract was found more active in H_2O_2 radical scavenging activity than water extract. Fe^{2+} chelating activity was noted only in water extract while in acetone extract it was below detection limit. Water extracts showed better result in FRAP and Phosphomolybdenum assays than their corresponding acetone extracts.

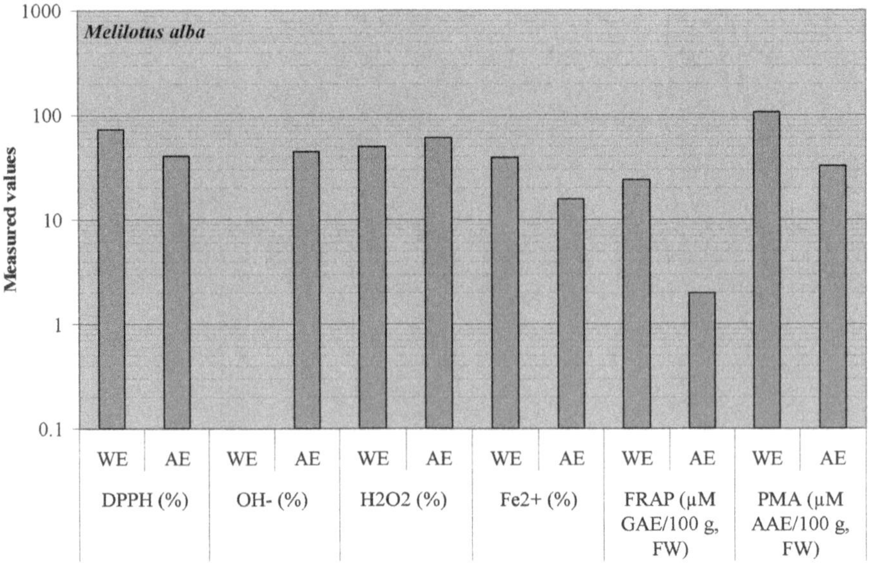

Fig. 8.34 Antioxidant activities of *Melilotus alba*

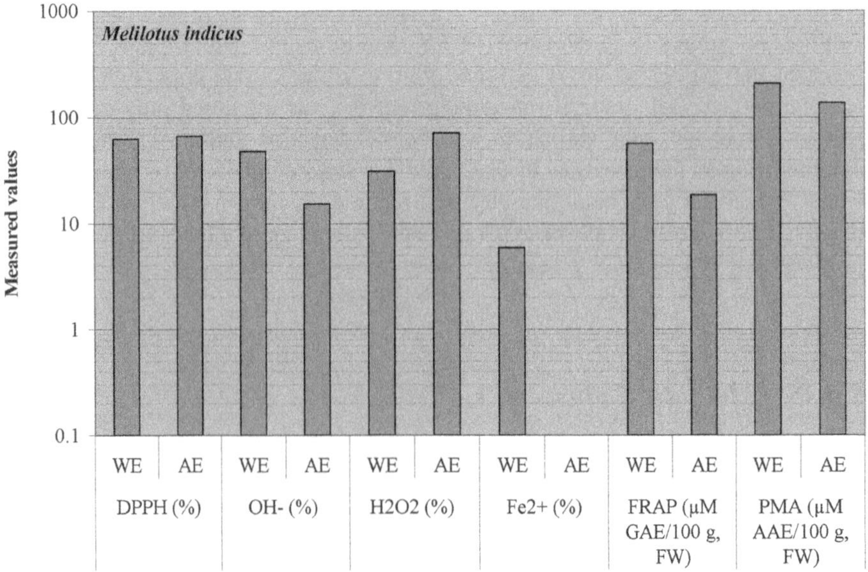

Fig. 8.35 Antioxidant activities of *Melilotus indicus*

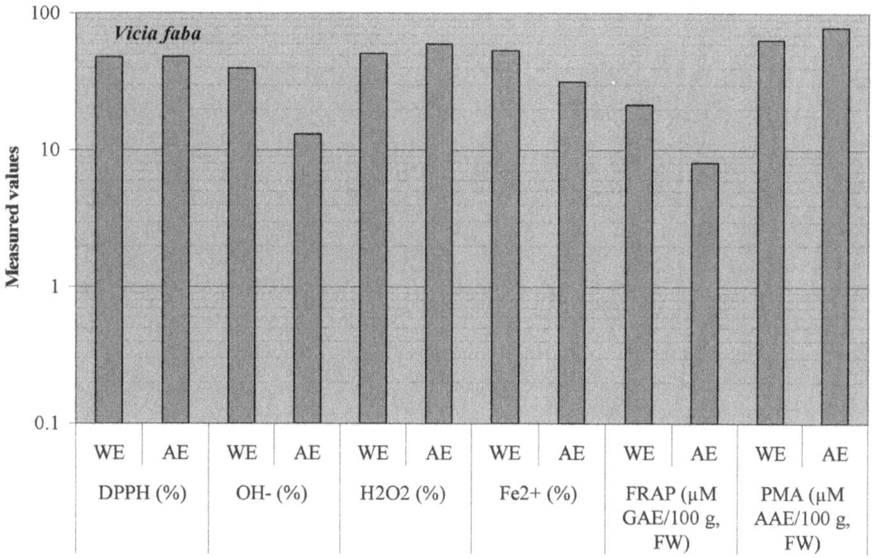

Fig. 8.36 Antioxidant activities of *Vicia faba*

8.3.36 *Vicia faba L*

Antioxidant properties measured in the edible parts of *V. faba* are given in Fig. 8.36, which revealed that both acetone and water extracts exhibit almost same DPPH scavenging activity. However, water extract shows more values during OH^- radical scavenging activity and Fe^{2+} chelating activity than acetone extracts. However, acetone extract shows higher potential in H_2O_2 radical scavenging activity compare to water extract. In FRAP assay water extract showed comparatively more value than acetone extract, whereas in PM assay measured value for acetone extract was found higher as compared to water extract.

8.3.37 *Vicia sativa L.*

The antioxidant properties of *V. sativa* are shown in Fig. 8.37, which indicates that acetone extract showed higher value during DPPH assay compared to water extract, whereas OH^- and H_2O_2 radicals scavenging activities were measured high in water extracts than acetone extracts. Fe^{2+} chelating activity was measured only in water extract, while it was found below detection limit in acetone extract. FRAP and PM assays indicated that acetone extracts exhibited higher values than water extracts.

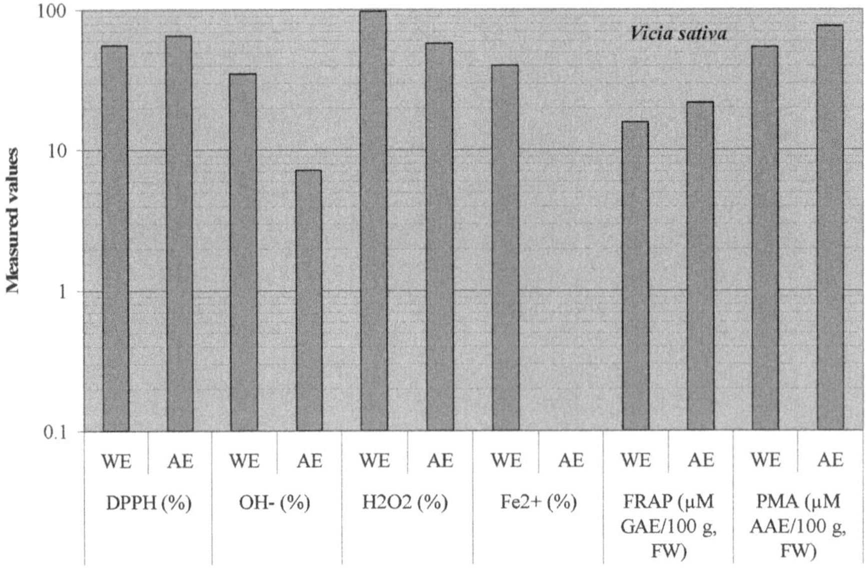

Fig. 8.37 Antioxidant activities of *Vicia sativa*

8.3.38 *Plantago lanceolata* L.

Water extract of *P. lanceoplata* showed comparatively higher DPPH scavenging activity than acetone extract. However, in the case of OH^- radical scavenging activity acetone extract showed better contribution than water extract. Approximately similar value was noted for both water extract and acetone extract during H_2O_2 radical scavenging activity. In the case of Fe^{2+} chelating activity acetone extract was found more active than water extract. Water extract showed better level in the case of ferric ion reducing antioxidant power (FRAP) than acetone extract. Nevertheless, in the case of PM assay acetone extract showed more contribution compared to water extract on fresh weight basis (Fig. 8.38).

8.3.39 *Persicaria amplexicaulis* (D. Don) Ronse Decr

Results of antioxidant properties measured in the leaves of *P. amplexicaulis* are given in Fig. 8.39, which indicates that water extracts show high percentage levels in DPPH scavenging activity, OH^- radical scavenging activity, H_2O_2 radical scavenging activity, and Fe^{2+} chelating activity than corresponding acetone extracts. Similarly significant level was noted for water extract during FRAP assay than acetone extract. However, in the case of PM assay measured value was high in acetone extract compared to corresponding water extract.

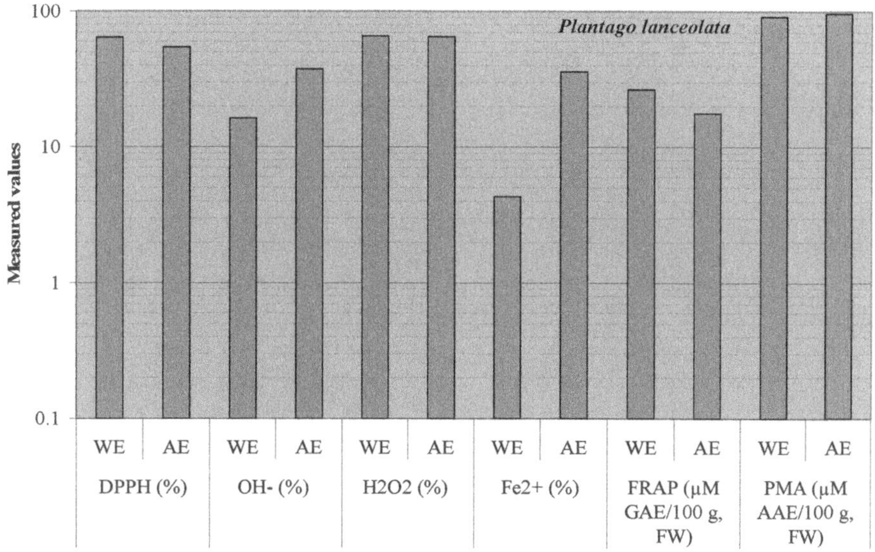

Fig. 8.38 Antioxidant activities of *Plantago lanceolata*

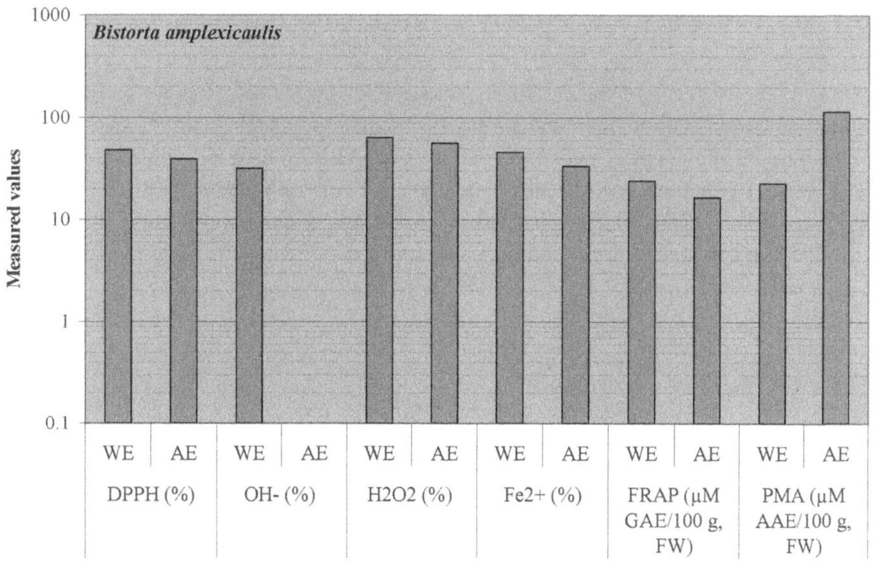

Fig. 8.39 Antioxidant activities of *Persicaria amplexicaulis*

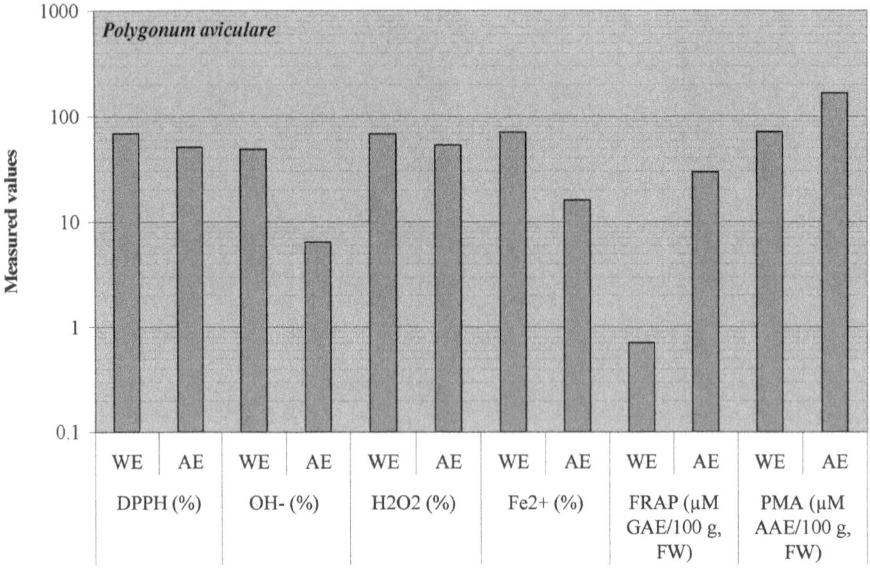

Fig. 8.40 Antioxidant activities of *Polygonum aviculare*

8.3.40 *Polygonum aviculare* L.

Antioxidant properties measured in the leaves of *P. aviculare* are given in Fig. 8.40, which revealed that in the case of DPPH scavenging, OH^- radical scavenging, H_2O_2 radical scavenging, and Fe^{2+} chelating activities measured values were high in water extracts than corresponding acetone extracts on fresh weight basis. However, in FRAP and PM complex assays measured values were found higher in acetone extracts than corresponding water extracts.

8.3.41 *Rumex dentatus* L.

Results of antioxidant properties in the leaves of *R. dentatus* are shown in Fig. 8.41, which indicates that water extract showed higher value in DPPH scavenging activity than corresponding acetone extract. OH^- radical scavenging activity was measured in water extract while it was below detection limit in acetone extract. H_2O_2 radical scavenging activity and Fe^{2+} chelating activity were found higher in acetone extracts than water extracts. Nevertheless, acetone extracts show more values in FRAP and PMAs compared to water extracts.

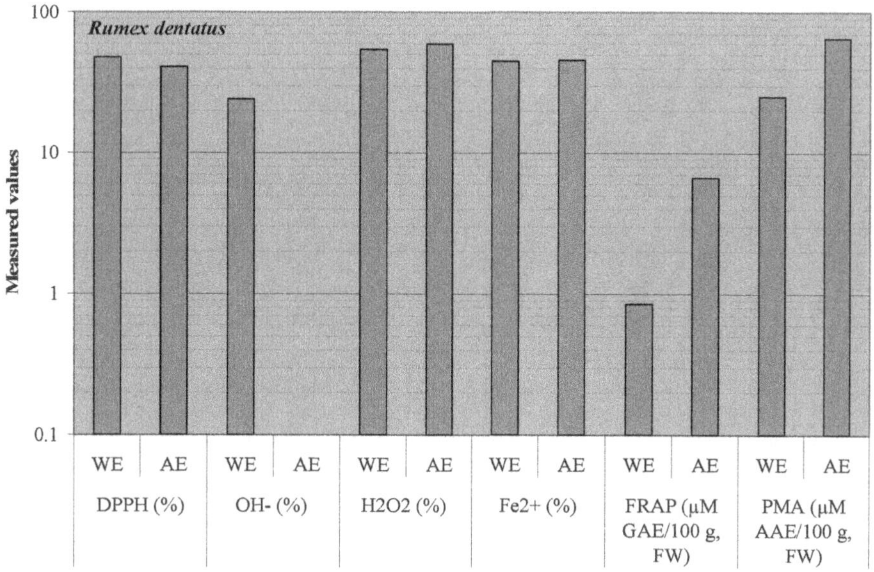

Fig. 8.41 Antioxidant activities of *Rumex dentatus*

8.3.42 *Rumex hastatus D. Don, Prodor.*

Antioxidant activities in the leaves of *R. hastatus* are given in Fig. 8.42, which shows that water extracts show significant levels in DPPH scavenging activity, H_2O_2 radical scavenging activity, and Fe^{2+} chelating activity than corresponding acetone extracts. OH^- radical scavenging activity was measured only in water extract, while it was below detection limit in acetone extract. However, acetone extracts show higher value in FRAP and PM complex assays compared to water extracts.

8.3.43 *Rumex nepalensis Spreng.*

Antioxidant properties shown by the leaves of *R. neplensis* are given in Fig. 8.43, which revealed that measured levels for DPPH and H_2O_2 radicals scavenging activities and Fe^{2+} chelating activity were higher in water extracts compared to acetone extracts. In the case of OH^- radical scavenging activity measured levels were below detection limit in both water and acetone extracts. However, in the case of FRAP and PM complex assays, values were high in acetone extracts than water extracts.

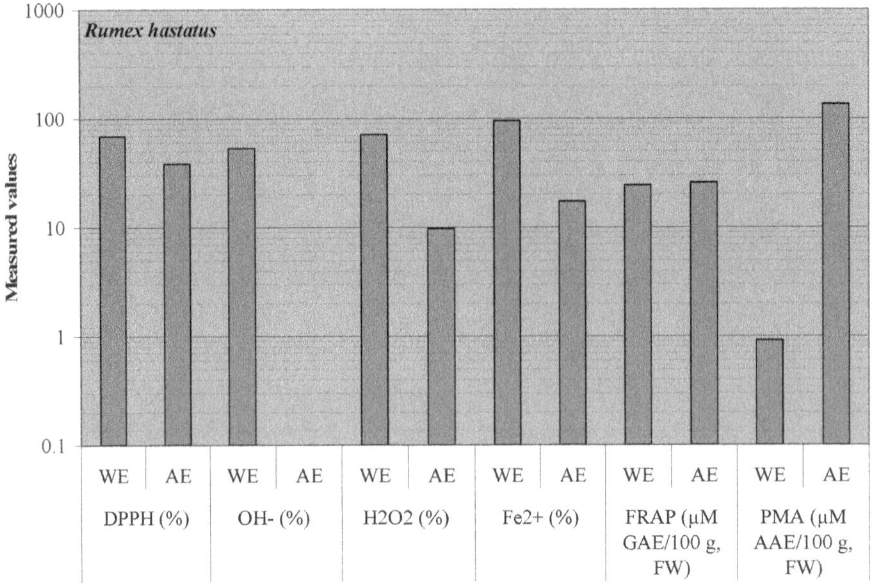

Fig. 8.42 Antioxidant activities of *Rumex hastatus*

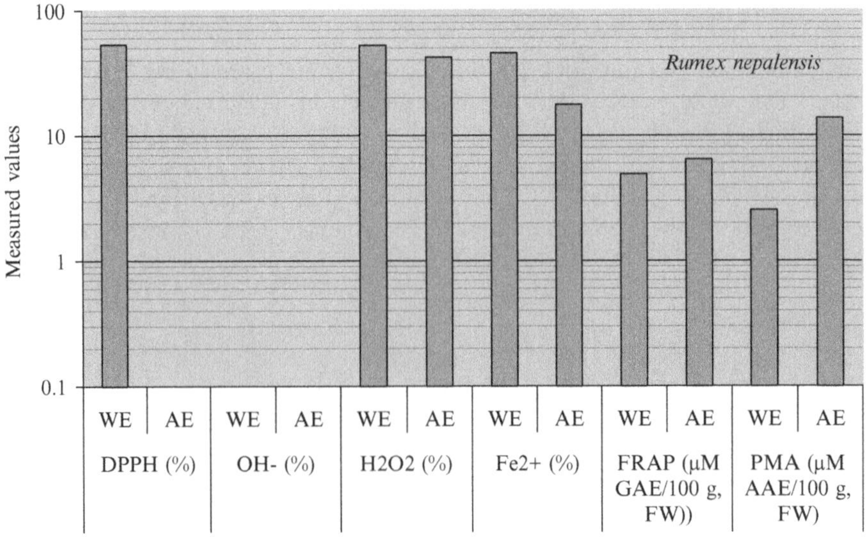

Fig. 8.43 Antioxidant activities of *Rumex nepalensis*

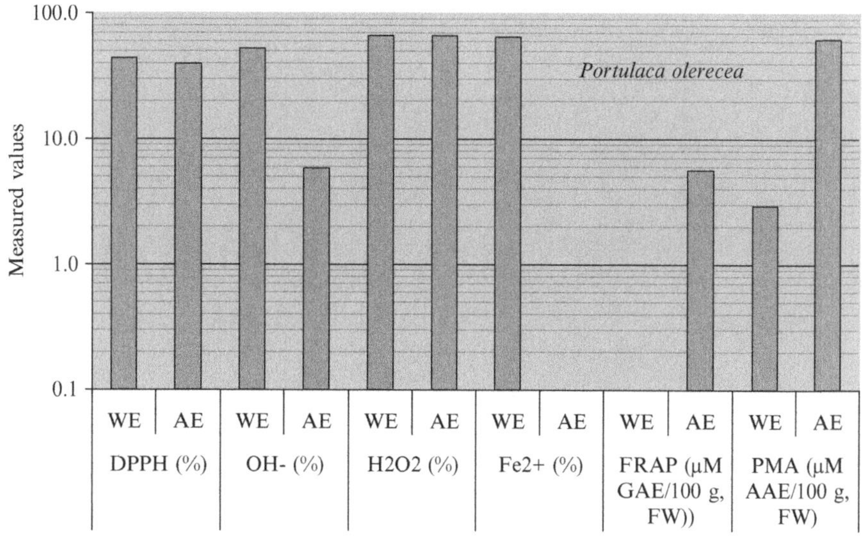

Fig. 8.44 Antioxidant activities of *Portulaca oleracea*

8.3.44 *Portulaca oleracea* L.

Antioxidant activities in the leaves of *P. oleracea* are given in Fig. 8.44, which shows that measured values for DPPH and OH^- radicals scavenging activities and Fe^{2+} chelating activity were higher in water extracts compared to acetone extracts. In the case of H_2O_2 radicals scavenging activity acetone extract showed more concentration than water extract. However, in the case of FRAP and PM complex assays, measured levels were high in acetone extracts than water extracts.

8.3.45 *Portulaca quadrifida* L.

Antioxidant activities determined in the edible parts of *P. quadrifida* are mentioned in Fig. 8.45. DPPH scavenging activity was found high in acetone extract compared to water extract. Measured levels of OH^- and H_2O_2 radicals scavenging activities and Fe^{2+} chelating activity were more in water extracts than corresponding acetone extracts. Nevertheless, acetone extract showed more contribution during FRAP assay than water extract. In the case of PM assay acetone extract showed significant value than water extract on fresh weight basis.

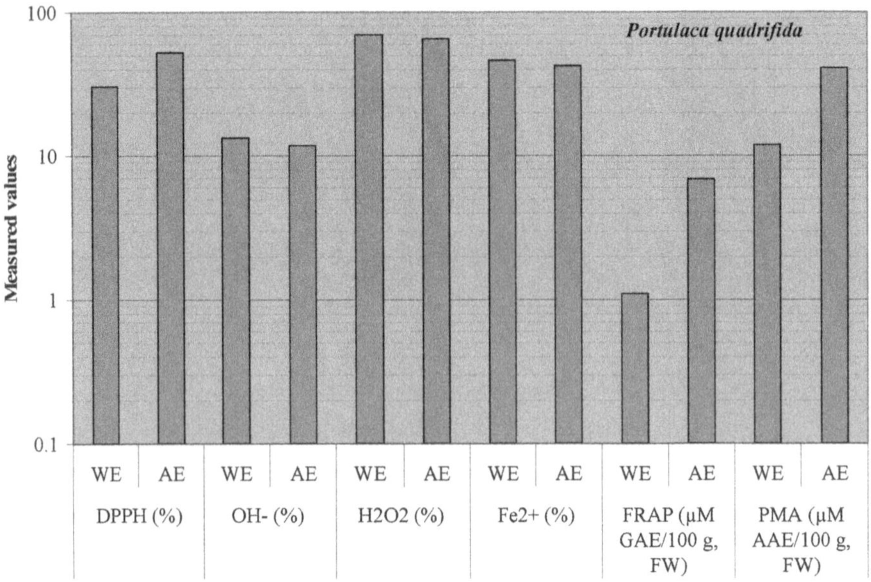

Fig. 8.45 Antioxidant activities of *Portulaca quadrifida*

8.3.46 *Galium aparine* L.

The antioxidant activities measured in the leaves of *G. aparine* are presented in Fig. 8.46, indicating that water extract showed higher potential in DPPH and OH^{-} radical scavenging activity compared to acetone extracts. However, acetone extract showed better results during H_2O_2 radical scavenging activity compared to water extract. Ferrous ion chelating activity was below detection limit both in the case of water as well as acetone extracts. Measured levels for ferric ion reducing antioxidant power and PMAs were high in water extracts than acetone extracts.

8.3.47 *Veronica arvensis* L.

Antioxidant properties determined in the leaves of *V. arvensis* are mentioned in Fig. 8.47, which indicates that values for DPPH, H_2O_2 radical scavenging activities, and Fe^{2+} chelating activity were high in acetone extracts compared to water extract. OH^{-} radical scavenging activity was measured only in water extract, whereas it was below detection limit in acetone extract. Water extract showed higher concentration in ferric ion reducing antioxidant power activity compared to acetone extract. However, in the case of PM assay acetone extract exhibited more value compared to corresponding water extract.

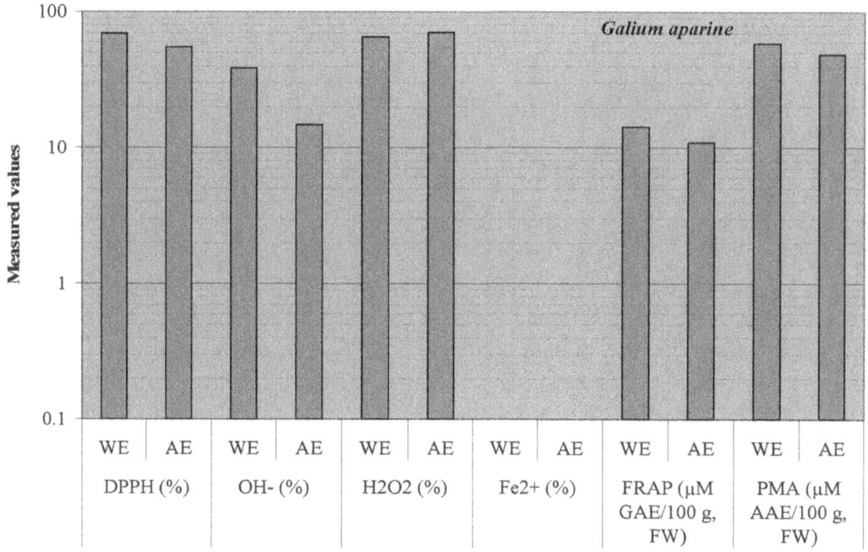

Fig. 8.46 Antioxidant activities of *Galium aparine*

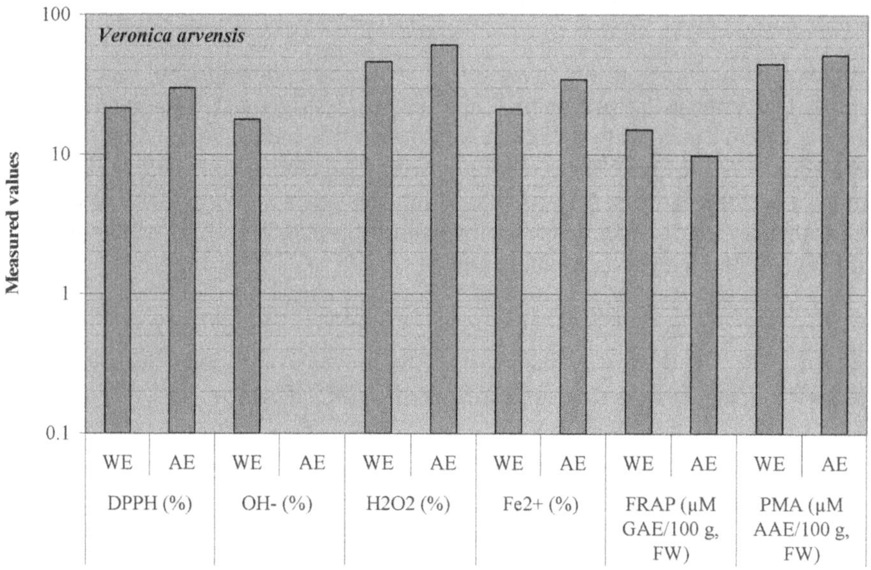

Fig. 8.47 Antioxidant activities of *Veronica arvensis*

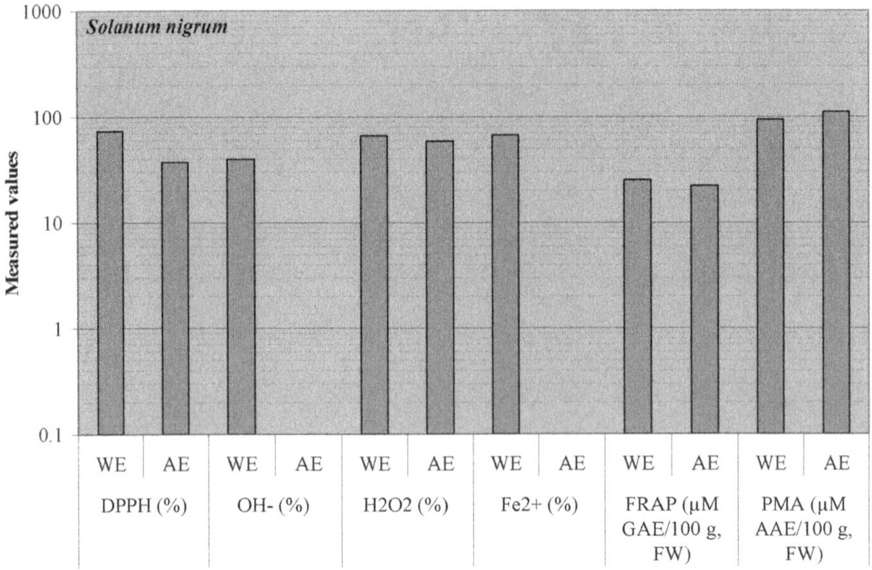

Fig. 8.48 Antioxidant activities of *Solanum nigrum*

8.3.48 Solanum nigrum auct.

Antioxidant activities in the leaves of *S. nigrum* are shown in Fig. 8.48, which shows that water extracts exhibited high levels in DPPH and H_2O_2 radical scavenging activity than acetone extracts. Similarly OH^- radical scavenging and Fe^{2+} chelating activities were measured only in water extracts, while below-detection limit in acetone extracts. Measured level in water extract during FRAP assay was more than acetone extract. However, in PM complex assay acetone extract showed significantly high level than water extract.

8.3.49 Pimpinella diversifolia (Wall.) DC. Prodor.

Antioxidant properties of *P. diversifolia* are given in Fig. 8.49, which shows that acetone extracts exhibit high value in DPPH and H_2O_2 radical scavenging activity compared to water extracts. OH^- radical scavenging activity was measured only in acetone extract, while it was below detection limit in water extract. In the case of Fe^{2+} chelating activity higher value was noted in water extract than acetone extract. Similarly measured level in water extract was high for FRAP assay compared to acetone extract. Nevertheless, acetone extract showed higher value during PM assay compared to water extract.

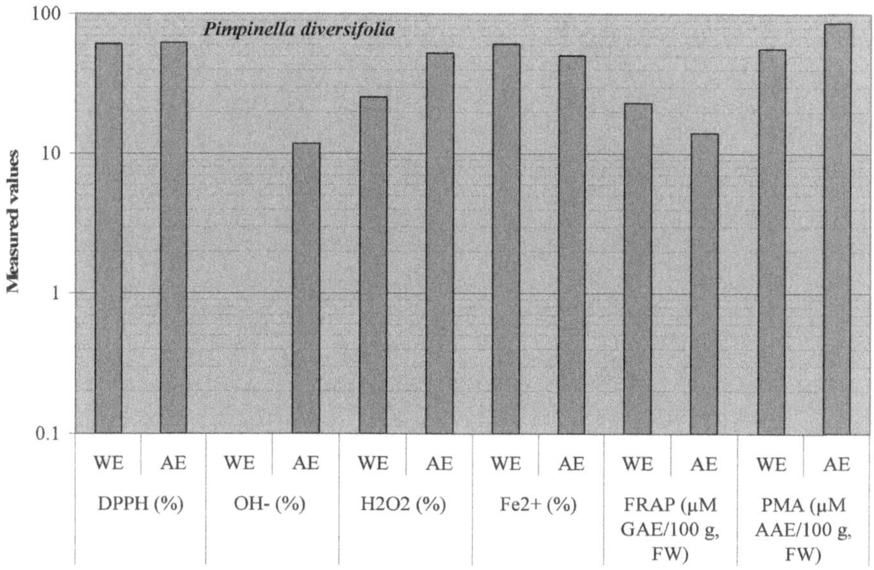

Fig. 8.49 Antioxidant activities of *Pimpinella diversifolia*

8.3.50 *Torilis leptophylla (L.) Reichb.f.*

Results of antioxidant activity studied in the leaves of *T. leptophylla* are mentioned in Fig. 8.50. It was observed that acetone extract exhibited higher concentration during DPPH scavenging activity compared to water extract. Measured levels in OH^-, H_2O_2 radicals scavenging activities, and Fe^{2+} chelating activity were found higher in water extracts than acetone extracts. Similarly water extract showed higher value in FRAP assay, expressed as μM GAE/100 g, FW, compared to acetone extract. In the case of PM assay acetone extract showed more value expressed as μM AAE/100 g, FW than water extract on fresh weight basis.

8.4 Comparative Evaluation of Antioxidant Properties

Antioxidant properties of wild edible vegetables used by the local inhabitants of Lesser Himalayas–Pakistan were investigated six diverse types of antioxidant assays including DPPH (2,2-diphenyl-1-picrylhydrazyl) scavenging activity, OH^- radical scavenging activity, H_2O_2 scavenging activity, Fe(II) ion chelating activity, FRAP (ferric ion reducing antioxidant power), and Phosphomolybdenum assay.

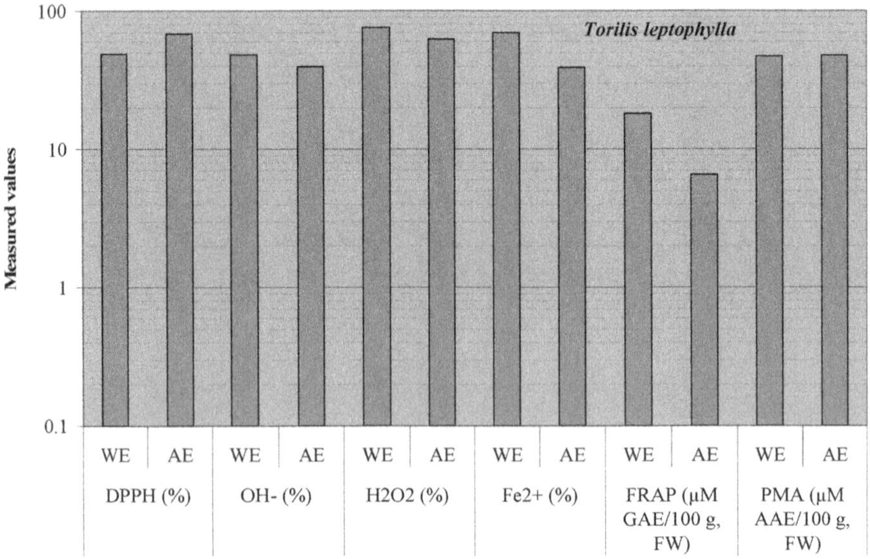

Fig. 8.50 Antioxidant activities of *Torilis leptophylla*

8.4.1 DPPH Scavenging Activity

The stable radical DPPH has been used widely for the determination of primary antioxidant activity of pure antioxidant compounds, fruits, and other foods (Wong et al. 2006). DPPH is a stable free radical at room temperature and accepts an electron or hydrogen radical to form a stable diamagnetic molecule. DPPH radical is scavenged by antioxidants through the donation of a proton forming the reduced DPPH. The color changes from purple to yellow after reduction, which can be quantified by its decrease in absorbance at wavelength 517 nm. Radical scavenging activity increased with increasing percentage of the free radical inhibition (Huang et al. 2005). DPPH scavenging activity of both water and acetone extracts of wild edible vegetables is given in Fig. 8.51, which indicates that among wild vegetables water extracts of *Oxalis corniculata* showed maximum DPPH scavenging activity (79.84 %), followed by *Bauhinia variegata var. variegata* (77.30 %), *Solanum nigrum* (73.16 %), and *Melilotus alba* (72.90 %), whereas *Veronica arvensis* showed lowest value (21.50 %). DPPH scavenging activity was ranged from 25.50 to 71.43 % in the water extracts of all other vegetables samples. The acetone extracts of *Bauhinia variegata var. variegata* exhibited highest DPPH activity (87.21 %), followed by *Bauhinina variegata var. candida* (85.50 %), *Bauhinia variegata* (85.30 %), *Bombax malabaracum* (81.59 %), and *Tulipa stellata* (74.41 %), whereas minimum activity was observed in the acetone extract of *Oxalis corniculata* (12.95 %). Acetone extracts of all other vegetables samples showed DPPH activity ranged from 20.90 to 68.33 %. Previously it was reported that in acetone, methanol, and water extracts of *Amaranthus spinosus, Amaranthus viridis,*

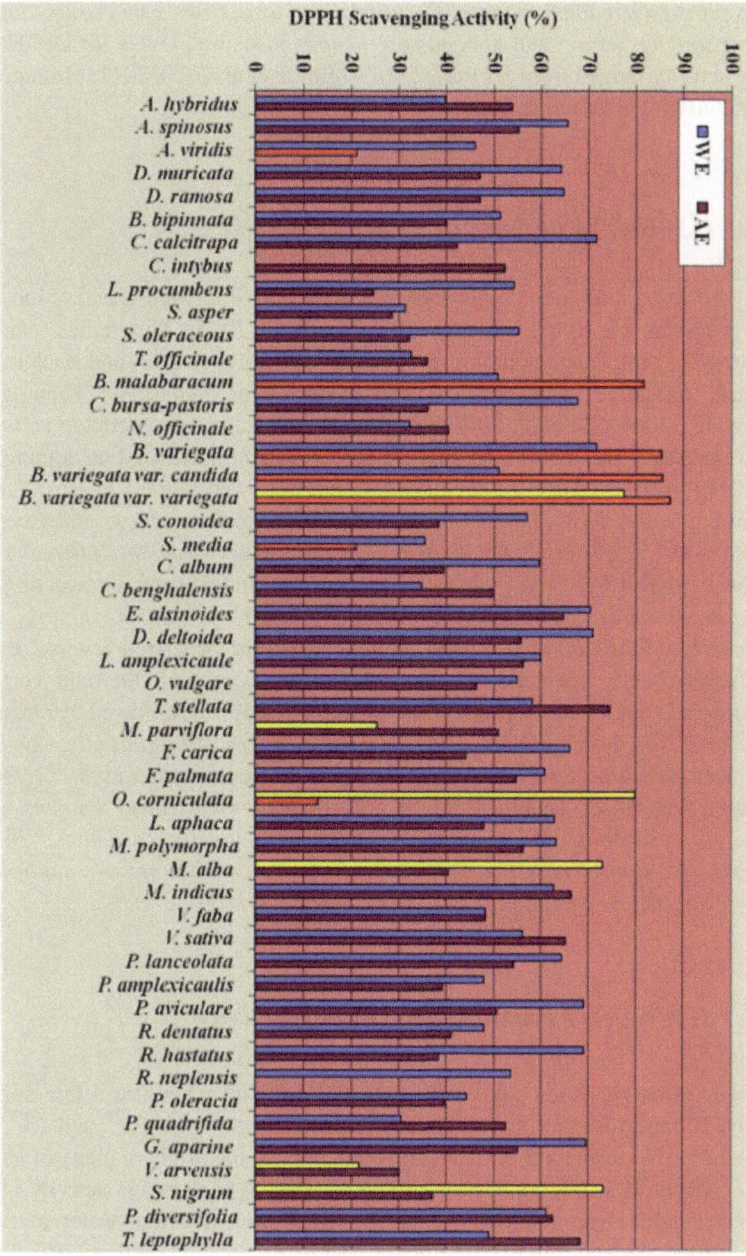

Fig. 8.51 DPPH scavenging activity in wild vegetables

Bidens pilosa, Origanum vulgare, Portulaca oleracea, Rumex tuberosus, Solanum nigrum, Sonchus asper, and *Sonchus oleraceous* measured levels for DPPH scavenging activity ranged from 56.1 to 92.90 % (Jimoh et al. 2010, 2011; Motamed and Naghibi 2010; Odhav et al. 2007).

8.4.2 Hydroxyl Radical Scavenging Activity

The hydroxyl radical is the most active in the reactive oxygen species, and it can cause most damage to the organism (Balaban et al. 2005). This radical can react with array of molecules including proteins, lipids, polypeptides, and nucleic acids to cause oxidative damages and cell necrosis or mutations (Klaunig and Kamendulis 2004; Shi et al. 2004). Results of OH^- radical scavenging activity of both vegetables are mentioned in Fig. 8.52, which indicated that among wild vegetables water extracts of *Sonchus oleraceous* showed highest hydroxyl radical scavenging activity (63.28 %) followed by *Chenopodium album* and *Evolvulus alsinoides* (57.97 %) each and *Medicago polymorpha* (55.69 %), whereas lowest value was observed in *Cichorium intybus* (0.466 %). In the water extracts of *Bidens bipinnata, Bombax malabaracum, Silene conoidea, Origanum vulgare, Ficus palmata, Melilotus alba,* and *Pimpinella diversifolia* measured levels for OH^- activity were below detection limit. The acetone extract of *Bauhinia variegata var. variegata* exhibited maximum OH^- activity (77.30 %), followed by *Dioscorea deltoidea* (72.24 %), *Evolvulus alsinoides* (45.77 %), and *Melilotus alba* (44.48 %), while acetone extract of *Launaea procumbens* showed lowest activity (2.990 %). Measured levels for acetone extracts of *Amaranthus viridis, Sonchus asper, Sonchus oleraceous, Chenopodium album, Origanum vulgare, Bistorta amplexicaulis Rumex hastatus, Rumex dentatus, Veronica arvensis,* and *Solanum nigrum* were below detection limit.

8.4.3 Hydrogen Peroxide Scavenging Activity

Hydrogen peroxide is a weak oxidizing agent and can inactivate a few enzymes directly. It can cross cell membranes quickly and react with Fe^{2+} and Cu^{2+} ions within the cell to form hydroxyl radical which may be the origin of many of its toxic effects (Miller et al. 1993). Measured levels of H_2O_2 scavenging activity of wild edible vegetables are given in Fig. 8.53, which revealed that water extract of *Vicia sativa* showed highly significant H_2O_2 scavenging activity (97.74 %), followed by *Dioscorea deltoidea* (87.55 %) and *Amaranthus viridis* (82.26 %), whereas *Lathyrus aphaca* exhibited lowest value (8.230 %). In the water extracts of other vegetables H_2O_2 activity was ranged from 25.36 to 80.68 %. However, in the case of *Origanum vulgare* and *Ficus palmata* results were below detection limit. H_2O_2 scavenging activity was found highest in the acetone extract of

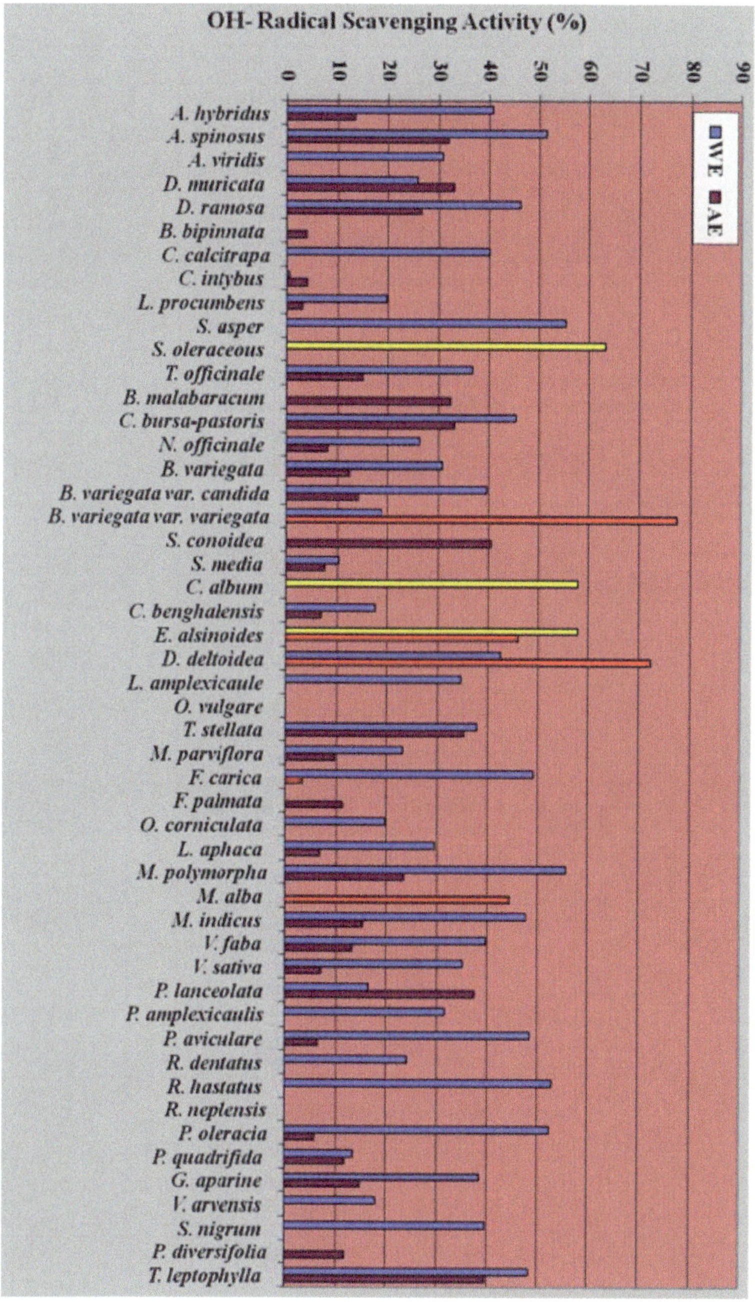

Fig. 8.52 Hydroxyl radical scavenging activity in wild vegetables

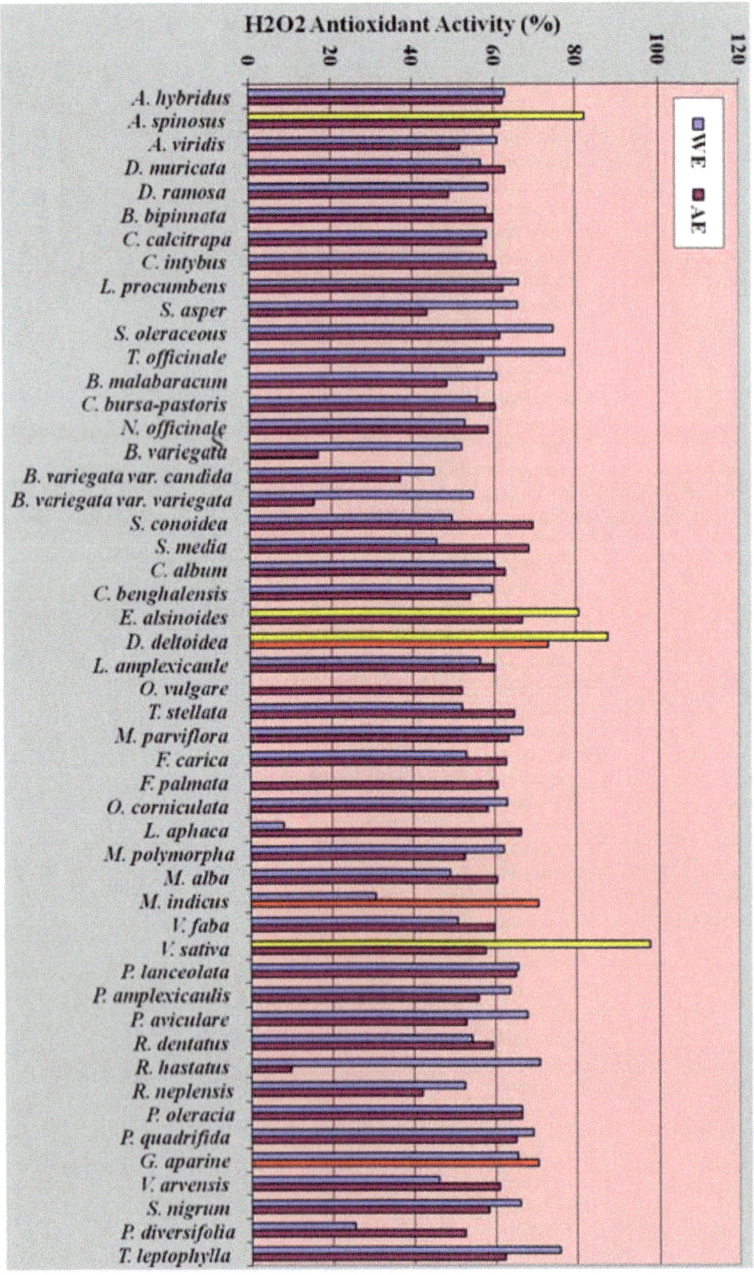

Fig. 8.53 H_2O_2 scavenging activity in wild vegetables

Dioscorea deltoidea (73.30 %) and lowest in *Rumex hastatus* (9.730 %), whereas in all other acetone extracts of analyzed samples measured values were ranged from 16.70 to 70.90 %.

8.4.4 Fe^{2+} Chelating Activity

The chelating capacity of the extract indicates how efficiently its compounds can compete with ferrozine for ferrous ion (Fe^{2+}). The iron-ferrozine complex has highest absorbance at 562 nm and a large decline in absorbance indicates strong chelating power. An extract with high chelating power reduces the free ferrous ion concentration by forming a stable iron (II) chelate and decreases the level of Fenton reaction which is implicated in many diseases (Halliwell and Gutteridge 1990). Fe (II) ions chelating activity in water and acetone extracts of wild edible vegetables is given in Fig. 8.54. Measured levels in the water extract of *Rumex hastatus* showed maximum (94.86 %) potential, followed by *Oxalis corniculata* (91.35 %) and *Capsella bursa-pastoris* (77.30 %). *Cichorium intybus* exhibited lowest value (0.324 %), while value for *Galium aparine* was below detection limit. In the case of acetone extract significant chelating power was observed in *Dioscorea deltoidea, Dyropteris ramose* and *Amaranthus spinosus* (57.68 %, 55.69 %, and 53.81 %), respectively, whereas Fe (II) ions chelating activity was found below detection limit in the acetone extracts of *Sonchus oleraceous, Nasturtium officinale, Commelina benghalensis, Lamium amplexicaule, Origanum vulgare, Tulipa stellata, Ficus carica, Lathyrus aphaca, Vicia sativa, Galium aparine,* and *Solanum nigrum.*

8.4.5 Ferric Ion Reducing Antioxidant Power (FRAP) Assay

FRAP is a simple test to measure antioxidant capability. This method was primarily developed to assay plasma antioxidant capacity, but can be used with plant extracts too (Gourine et al. 2010). FRAP assay measures the reducing capacity of antioxidants against oxidative effects of reactive oxygen species. The antioxidative activity is estimated by measuring the increase in absorbance caused by the formation of ferrous ions from FRAP reagent containing TPTZ (2,4,6-Tri (2-pyridyl)-s-triazine) and $FeCl_3{\cdot}6H_2O$. The absorbance is measured at 595 nm (Benzie and Strain 1996). Results of ferric ion reducing antioxidant power (FRAP) of the wild edible vegetables are presented in Fig. 8.55, which indicates that measured levels in water extracts of *Melilotus indicus*, *Origanum vulgare,* and *Ficus palmata* were highest in concentrations: 56.55 μM GAE/100 g, 54.04 μM GAE/100 g, and 41.08 μM GAE/100 g, respectively, whereas *Chenopodium album* exhibited lowest value 0.070 μM GAE/100 g. The acetone extract of *Bauhinia variegata var. candida* showed maximum value (42.03 μM GAE/100 g), followed by *Bauhinia variegata* (41.02 μM GAE/100 g), *Origanum vulgare* (34.80 μM

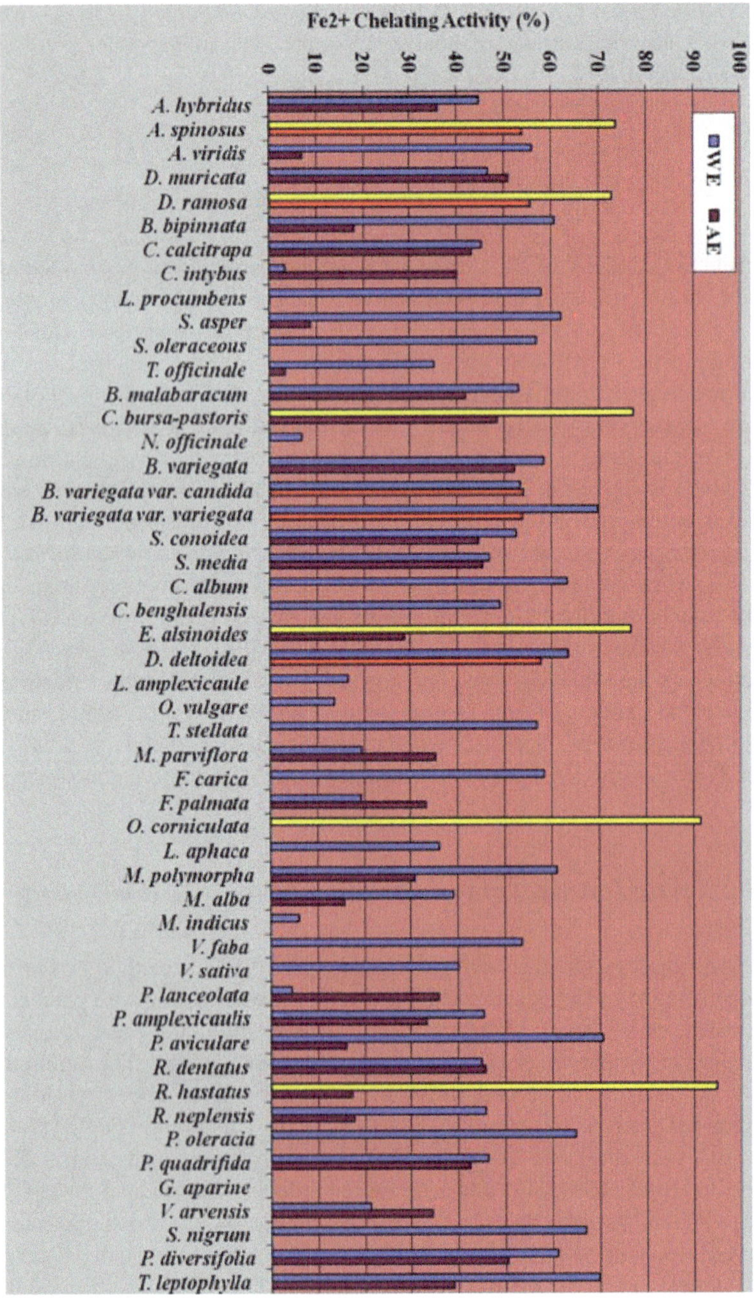

Fig. 8.54 Ferrious ion chelating activity in wild vegetables

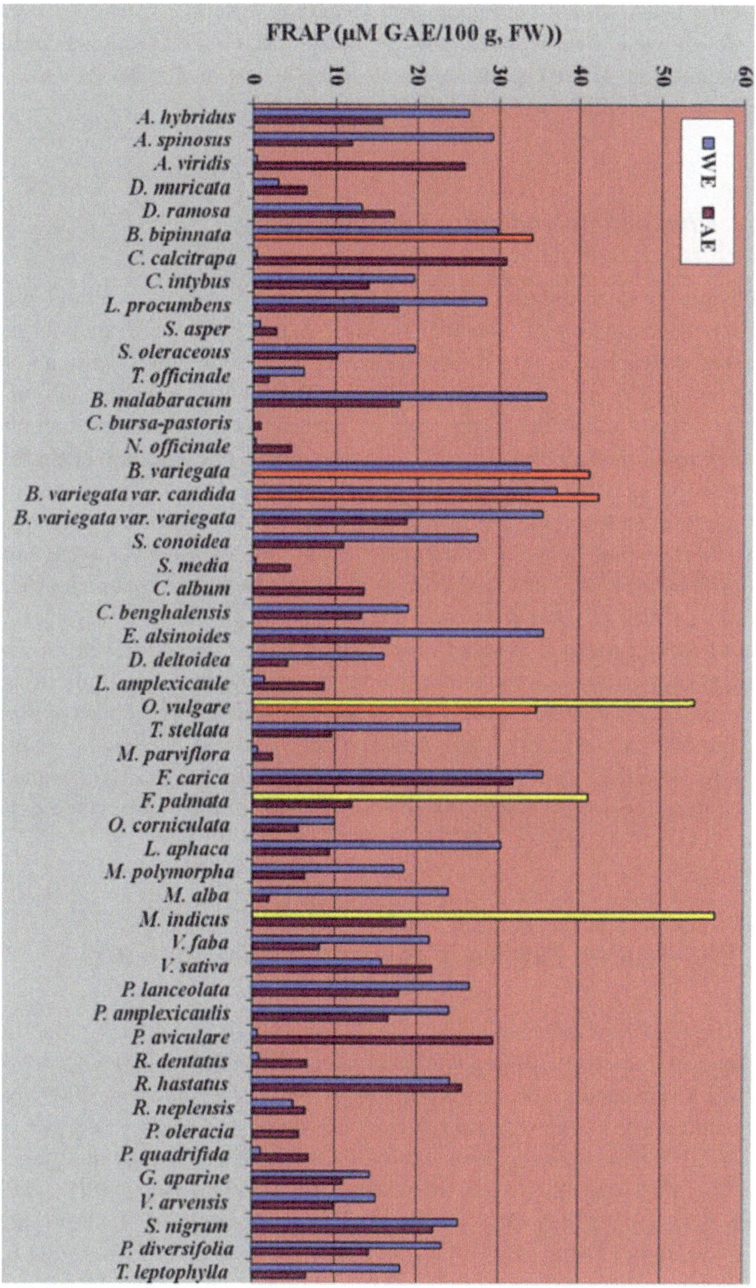

Fig. 8.55 FRAP values in wild vegetables

GAE/100 g), and *Bidens bipinnata* (34.19 μM GAE/100 g), while *Taraxacum officinale* showed lowest concentration (1.910 μM GAE/100 g). According to Andarwulan et al. (2010) measured level in the leaves of *Portulaca oleracea* for FRAP assay was 24.60 μM TE/g FRAP.

8.4.6 Phosphomolybdenum Complex Assay

The TAC of the samples was measured by Phosphomolybdenum method, which is based on the reduction of Mo (VI) to Mo (V) by the sample analyte and the successive formation of green phosphate/Mo (V) compounds with an utmost absorption at 695 nm (Prieto et al. 1999). The phosphomolybdenum method generally detects antioxidants such as ascorbic acid, some phenolics, tocopherols, and carotenoids (Miladi and Damak 2008). Results of Phosphomolybdenum assay was expressed as μM AAE/100 g, FW Tables 30. The comparative assessment of both water and acetone extracts of wild edible vegetables is mentioned in Fig. 8.56, which indicates that in water extracts significant values for PMA were found in *Melilotus indicus* (208.7 μM AAE/100 g), followed by *Silene conoidea* (163.0 μM AAE/100 g) and *Origanum vulgare* (160.8 μM AAE/100 g). However, water extract of *Rumex hastatus* showed lowest value (0.914 μM AAE/100 g), while in the case of *Oxalis corniculata* measured level was below detection limit. In the case of acetone extracts elevated levels were measured in the flowering buds of *Bauhinia variegata var. candida* (181.7 μM AAE/100 g), followed by *Bauhinia variegata* (173.5 μM AAE/100 g), *Polygonum aviculare* (166.2 μM AAE/100 g), and *Origanum vulgare* (164.0 μM AAE/100 g), whereas *Capsella bursa-pastoris* exhibited minimum value (8.500 μM AAE/100 g) on fresh weight basis.

8.5 Distribution Pattern of the Antioxidant Activities

Basic statistical parameters related to the distribution of antioxidant properties are mentioned in Table 8.1. Among wild vegetables dominant mean levels, 87.85 μM AAE/100 g, FW and 65.91 μM AAE/100 g, FW, were shown by PMA both in acetone and water extracts, respectively, followed by H_2O_2 assay in water extract (59.81 %), H_2O_2 assay in acetone extract (56.78 %), DPPH activity in water extract (55.58 %), Fe^{2+} chelating assay in water extract (49.60 %), DPPH activity in acetone extract (48.89 %), OH^- activity in water extract (36.03 %), Fe^{2+} chelating assay in acetone extract (35.03 %), OH^- activity in acetone extract (21.62 %), FRAP assay in water extract (19.00 μM GAE/100 g, FW), and FRAP assay in acetone extract (14.51 μM GAE/100 g, FW). The decreasing trend of average antioxidant activity in wild edible vegetables showed following order PMA (AE) > PMA (WE) > H_2O_2 (WE) > H_2O_2 (AE) > DPPH (WE) > Fe^{2+} (WE) > DPPH (AE) > OH^- (WE) > Fe^{2+} (AE) > FRAP (WE) > FRAP (AE).

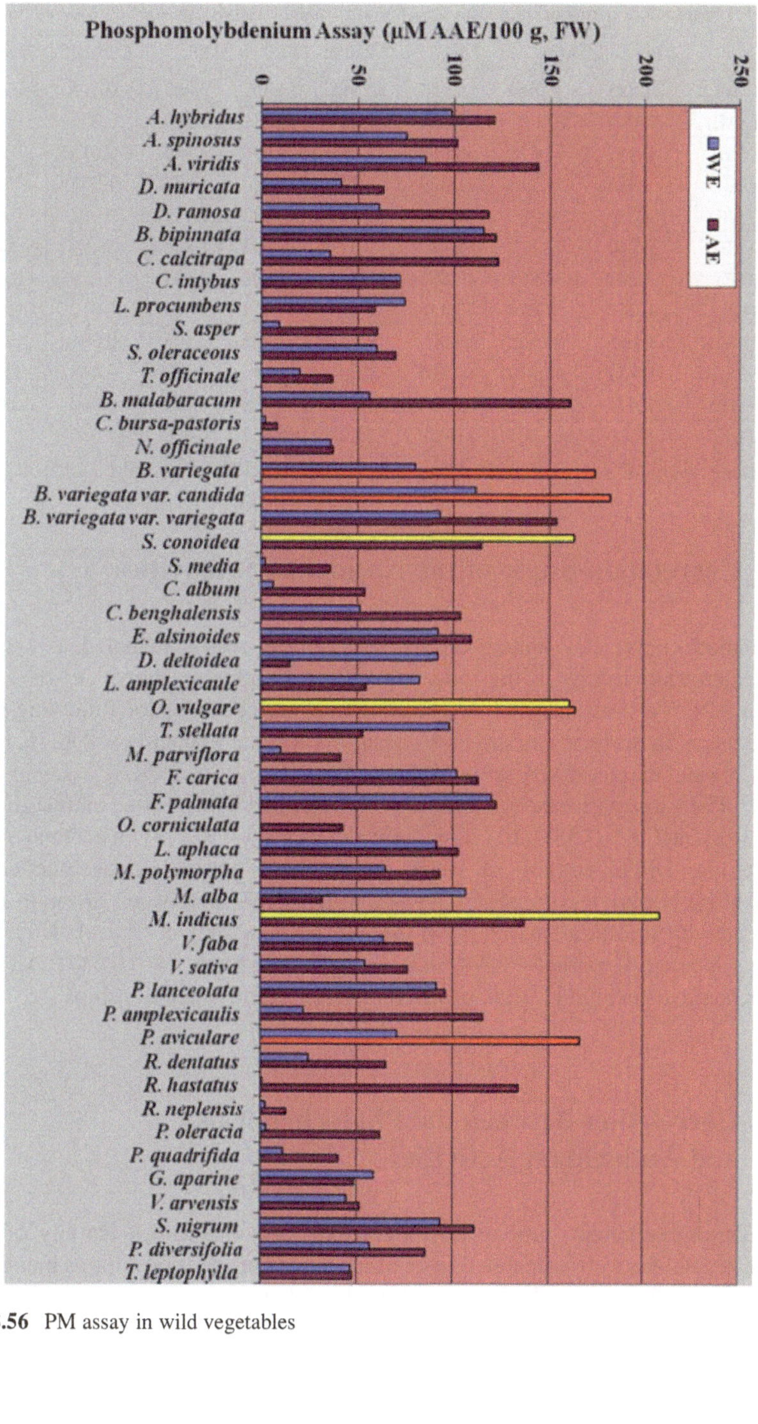

Fig. 8.56 PM assay in wild vegetables

Table 8.1 Statistical distribution parameters for antioxidant activity in wild edible vegetables

	Min	Max	Mean	Median	SD	SE	Kurtosis	Skewness
DPPH (WE)	21.495	79.840	55.587	58.211	14.452	2.065	−0.415	−0.605
DPPH (AE)	12.949	87.207	48.894	47.738	16.731	2.390	0.250	0.431
OH^- (WE)	0.455	63.278	36.039	38.164	14.930	2.304	−0.615	−0.266
OH^- (AE)	2.985	77.313	21.629	13.731	18.580	3.097	1.854	1.411
H_2O_2 (WE)	8.226	97.736	59.819	59.887	14.932	2.155	2.982	−0.676
H_2O_2 (AE)	9.730	73.307	56.786	60.322	13.092	1.852	5.299	−2.191
Fe^{2+} (WE)	3.243	94.865	49.603	53.243	22.006	3.144	−0.078	−0.495
Fe^{2+} (AE)	0.117	57.679	35.036	37.515	16.412	2.815	−0.672	−0.612
FRAP (WE)	0.065	56.552	19.003	19.533	14.735	2.105	−0.306	0.386
FRAP (AE)	0.852	42.349	14.516	12.003	10.453	1.478	0.418	1.016
PMA (WE)	0.914	208.683	65.915	64.177	45.436	6.491	0.960	0.698
PMA (AE)	8.500	181.701	87.850	82.721	45.452	6.428	−0.856	0.270

WE water extract, *AE* acetone extract, *FRAP* (μM GAE/100 g, FW), *PMA* (μM AAE/100 g, FW)

8.6 Correlation Study of the Antioxidant Activities

Correlation coefficients among antioxidant activity were calculated to examine their inter-relationships in the wild vegetables samples as shown in the matrix (Table 8.2). Very strong positive correlations were observed for following pairs; FRAP assay in acetone extract–PMA assay in acetone extract ($r = 0.827$), FRAP assay in water extract–PMA assay in water extract ($r = 0.768$), FRAP assay in water extract–PMA assay in acetone extract ($r = 0.596$), DPPH in acetone extract–Fe^{2+} in acetone extract ($r = 0.538$), PM assay in water extract–PM assay in acetone extract ($r = 0.517$), DPPH activity in water extract–OH^- activity in acetone extract ($r = 0.501$). Negative correlations were also observed in between different pairs; H_2O_2 activity in acetone extract–PM assay in acetone extract ($r = -0.477$), OH^- activity in acetone extract–FRAP assay in acetone extract ($r = -0.447$). All other phytochemicals exhibited weak positive or weak negative relationships.

8.7 Correlation Between the Phytochemicals and Antioxidant Activities

Correlation coefficients among phytochemicals and antioxidant activity of wild edible vegetables were calculated to examine their inter-relationships as mentioned in the matrix (Table 8.3). In the case of wild edible vegetables very strong positive correlations were observed for following pairs: ascorbic acid contents in acetone

Table 8.2 Correlation coefficient[a] matrix for antioxidant activities in wild edible vegetables

	DPPH (WE)	DPPH (AE)	OH^- (WE)	OH^- (AE)	H_2O_2 (WE)	H_2O_2 (AE)	Fe^{2+} (WE)	Fe^{2+} (AE)	FRAP (WE)	FRAP (AE)	PMA (WE)	PMA (AE)
DPPH (WE)	1.000											
DPPH (AE)	0.230	1.000										
OH^- (WE)	0.307	0.081	1.000									
OH^- (AE)	**0.501**	0.346	0.171	1.000								
H_2O_2 (WE)	−0.001	−0.040	0.255	0.223	1.000							
H_2O_2 (AE)	−0.177	−0.309	−0.037	−0.142	0.026	1.000						
Fe^{2+} (WE)	0.409	−0.072	0.492	0.318	0.276	−0.320	1.000					
Fe^{2+} (AE)	0.263	**0.538**	−0.133	0.374	−0.209	0.016	0.039	1.000				
FRAP (WE)	0.308	0.466	0.153	0.123	−0.213	−0.129	−0.100	0.156	1.000			
FRAP (AE)	0.292	0.317	0.102	−0.231	−0.048	**−0.447**	0.080	0.039	0.476	1.000		
PMA (WE)	0.376	0.328	0.066	0.195	−0.263	0.145	−0.268	0.110	**0.768**	0.420	1.000	
PMA (AE)	0.258	0.419	0.159	−0.069	−0.155	**−0.477**	0.069	0.109	**0.596**	**0.827**	**0.517**	1.000

[a]Bold r-values are significant at $p < 0.01$

Table 8.3 Correlation coefficient[a] matrix for phytochemicals and antioxidant activity in wild edible vegetables

	TP (WE)	TP (AE)	TF (WE)	TF (AE)	TFL (WE)	TFL (AE)	AA (WE)	AA (AE)
DPPH (WE)	0.234	0.166	0.105	0.098	0.090	0.292	0.227	0.333
DPPH (AE)	0.102	0.227	0.275	0.075	0.167	0.257	0.086	0.289
OH^- (WE)	0.012	0.112	0.016	0.089	−0.001	0.081	0.197	0.264
OH^- (AE)	0.069	−0.166	−0.005	−0.198	0.030	0.163	0.061	0.203
H_2O_2 (WE)	−0.264	−0.045	−0.248	−0.186	**−0.578**	−0.043	−0.168	−0.150
H_2O_2 (AE)	−0.093	**−0.466**	−0.300	−0.303	0.018	−0.338	0.346	−0.004
Fe^{2+} (WE)	−0.248	0.010	−0.136	−0.167	−0.393	0.058	−0.251	−0.098
Fe^{2+} (AE)	−0.091	−0.014	0.040	−0.316	0.139	−0.180	−0.030	0.085
FRAP (WE)	0.431	**0.556**	0.424	0.447	**0.582**	0.479	0.426	**0.570**
FRAP (AE)	**0.543**	**0.614**	0.369	**0.575**	0.287	**0.505**	0.205	0.479
PMA (WE)	**0.571**	0.296	0.243	0.433	**0.657**	**0.532**	**0.704**	**0.723**
PMA (AE)	**0.599**	**0.642**	0.332	**0.556**	0.440	**0.596**	0.368	**0.639**

TP total phenolics (mg GAE/100 g, FW), *TF* total flavonoids (mg Rt/100 g, FW), *TFL* total flavonols (mg Rt/100 g, FW), *AA* ascorbic acid (mg AA/100 g, FW), *FRAP* (μM GAE/100 g, FW), *PMA* phosphomolybdenum assay (μM AAE/100 g, FW), *WE* water extract, *AE* acetone extract, *FW* fresh weight

[a]Bold *r*-values are significant at $p < 0.01$

extract–PMA assay in water extract ($r = 0.723$), Ascorbic acid in water extract–PMA assay in water extract ($r = 0.704$), PMA assay in water extract–total flavonols in water extract ($r = 0.657$), total phenolics in acetone extract–PM assay in acetone extract ($r = 0.642$), PMA assay in acetone extract–Ascorbic acid in acetone extract ($r = 0.639$), FRAP assay in acetone extract–total phenolics in acetone extract ($r = 0.614$), PM assay in acetone extract–total phenolics in acetone extract ($r = 0.599$), PM assay in acetone extract–total flavonols in acetone extract ($r = 0.596$), FRAP water extract–total flavonols in water extract ($r = 0.582$), total phenolics in extract–FRAP assay in acetone extract ($r = 0.571$), FRAP water extract–Ascorbic acid acetone extract ($r = 0.570$), FRAP water extract–total phenolics in acetone extract ($r = 0.556$), PM assay in acetone extract–total flavonoids in acetone extract ($r = 0.556$), FRAP assay in acetone extract–total phenolics in water extract (0.543), PMA water extract–total flavonols in acetone extract ($r = 0.532$). Negative correlation, representing the contrasting distribution, was also observed in between different pairs; total flavonols in water extract–H_2O_2 activity in water extract ($r = -0.578$) and total phenolics in acetone extract–H_2O_2 activity in acetone extract ($r = -0.466$).

Table 8.4 Correlation coefficient[a] matrix for nutrients, phytochemicals, and antioxidant activity of wild edible vegetables

	MC	DM	CP	CF	Carb	Fib	Ash	EV
TP (WE)	**−0.511**	**0.511**	−0.070	−0.017	0.196	−0.154	−0.211	0.246
TP (AE)	−0.293	0.293	−0.350	0.034	0.385	0.076	−0.344	0.255
TF (WE)	**−0.616**	**0.616**	−0.235	−0.012	0.319	−0.004	−0.318	0.261
TF (AE)	**−0.523**	**0.523**	−0.230	−0.006	0.254	−0.127	−0.114	0.156
TFL (WE)	**−0.541**	**0.541**	−0.194	0.014	0.270	−0.110	−0.218	0.239
TFL (AE)	**−0.599**	**0.599**	−0.205	−0.082	0.314	−0.197	−0.215	0.253
AA (WE)	**−0.705**	**0.705**	−0.224	−0.115	0.283	−0.118	−0.161	0.160
AA (AE)	**−0.790**	**0.790**	−0.262	−0.136	0.343	−0.153	−0.204	0.207
DPPH (WE)	−0.292	0.292	−0.398	0.101	**0.407**	−0.009	−0.287	0.269
DPPH (AE)	−0.240	0.240	−0.090	−0.146	0.277	−0.071	−0.369	0.300
OH^- (WE)	−0.252	0.252	−0.153	−0.083	0.107	0.031	0.010	−0.047
OH^- (AE)	−0.227	0.227	−0.225	−0.172	0.387	−0.372	−0.224	0.321
H_2O_2 (WE)	−0.017	0.017	−0.089	−0.314	0.042	−0.140	0.215	−0.196
H_2O_2 (AE)	−0.108	0.108	0.226	−0.124	−0.205	−0.086	0.180	−0.140
Fe^{2+} (WE)	0.114	−0.114	−0.237	0.050	0.147	0.037	0.011	−0.014
Fe^{2+} (AE)	−0.047	0.047	0.142	−0.371	−0.023	−0.107	−0.017	−0.019
FRAP (WE)	**−0.529**	**0.529**	−0.335	−0.032	**0.449**	−0.067	−0.398	0.353
FRAP (AE)	−0.441	**0.441**	−0.271	−0.063	0.322	−0.026	−0.242	0.195
PMA (WE)	**−0.668**	**0.668**	−0.209	0.028	0.363	−0.148	−0.373	0.390
PMA (AE)	−0.494	**0.494**	−0.315	−0.055	0.327	0.099	−0.264	0.154

TP total phenolics (mg GAE/100 g, FW), *TF* total flavonoids (mg Rt/100 g, FW), *TFL* total flavonols (mg Rt/100 g, FW), *AA* ascorbic acid (mg AA/100 g, FW), *FRAP* (μM GAE/100 g, FW), *PMA* phosphomolybdenum assay (μM AAE/100 g, FW), *FW* fresh weight, *WE* water extract, *AE* acetones extract, *MC* moisture content, *DM* dry matter, *CP* crude proteins, *CF* crude fats, *Carb* carbohydrates, *Fib* fibers, *EV* energy value

[a]Bold r-values are significant at $p < 0.01$

8.8 Correlation Study of the Nutrients, Phytochemicals, and Antioxidant Activities

Correlation coefficients among nutrients, phytochemicals, and antioxidant activity of wild edible vegetables were carried out to study their inter-relationships as shown in the matrix (Table 8.4). In the majority of cases dry matter contents of wild vegetables showed very strong positive correlations with phytochemicals and antioxidant activity; Dry matter–AA in acetone extract ($r = 0.790$), Dry matter–AA in water extract ($r = 0.705$), Dry matter–PMA in water extract ($r = 0.668$),

Dry matter–total flavonoids in water extract ($r = 0.616$), Dry matter–total flavonols in acetone extract ($r = 0.599$), Dry matter–total flavonols in water extract ($r = 0.541$), Dry matter–FRAP in water extract ($r = 0.529$), Dry matter–total flavonoids in acetone extract ($r = 0.523$), Dry matter–total phenolics in water extract ($r = 0.511$), and Dry matter–PMA in acetone extract ($r = 0.494$), whereas, significantly positive correlations were also noted among carbohydrates–FRAP in water extract ($r = 0.449$) and carbohydrates–DPPH in water extract ($r = 0.407$). Negative correlations, representing the contrasting distribution, were mostly showed by moisture contents with AA in water extract, AA in acetone extract, PMA in water extract, total flavonols in acetone extract, total flavonols in water extract and FRAP in water extract ($r = -0.790$), ($r = -0.705$), ($r = -0.668$), ($r = -0.599$), ($r = -0.541$), and ($r = -0.529$), respectively. All other relationships between nutrient, phytochemicals, and antioxidant activity were found either weak positive or weak negative.

8.9 Correlation Study of the Metals, Phytochemicals, and Antioxidant Activities

Correlation coefficients among metals, phytochemicals, and antioxidant activity of wild edible vegetables were calculated to examine their inter-relationships as shown in Table 8.5. Significant positive correlations were observed for following pairs; Cu–Fe^{2+} activity in water extract ($r = 0.497$), Sr–OH^- activity in water extract ($r = 0.420$), Na–OH^- activity in water extract ($r = 0.361$), Ca–OH^- activity in water extract ($r = 0.329$), Sr–H_2O_2 activity in water extract ($r = 0.308$), and Cu–H_2O_2 activity in water extract ($r = 0.302$). Negative correlations were also observed in between different pairs; Fe–PMA assay in water extract ($r = -0.387$), Cu–H_2O_2 activity in acetone extract ($r = -0.378$), and Fe–total phenolics in acetone extracts ($r = -0.345$). All other relationships were either weak positive or weak negative.

Table 8.5 Correlation coefficient[a] matrix for metals, phytochemicals, and antioxidant activity in wild edible vegetables

	Ca	Mg	Na	K	Sr	Fe	Zn	Cu	Mn	Li	Co	Cr	Cd	Pb
TP (WE)	−0.286	0.225	−0.151	0.118	−0.285	−0.275	0.012	−0.285	−0.155	−0.049	**0.205**	−0.127	−0.011	−0.125
TP (AE)	−0.193	0.069	−0.032	0.012	−0.166	−0.345	−0.033	0.061	−0.042	−0.049	0.050	−0.131	−0.060	−0.116
TF (WE)	−0.246	0.011	−0.164	−0.010	−0.288	−0.159	−0.064	−0.317	0.009	−0.058	0.063	−0.145	−0.018	−0.154
TF (AE)	−0.092	**0.250**	−0.014	0.145	−0.152	−0.233	0.124	−0.165	0.108	−0.203	0.063	−0.080	−0.096	−0.086
TFL (WE)	−0.203	−0.011	−0.087	−0.021	−0.190	−0.128	−0.120	−0.449	−0.025	−0.128	−0.024	−0.097	0.032	−0.105
TFL (AE)	0.012	**0.257**	0.102	−0.110	0.014	−0.285	**0.266**	−0.177	−0.019	−0.046	−0.037	0.092	0.098	−0.007
AA (WE)	−0.054	−0.042	−0.024	−0.124	0.072	−0.148	−0.088	−0.286	−0.151	0.061	0.164	**0.255**	**0.276**	0.121
AA (AE)	−0.015	−0.024	−0.035	−0.143	0.110	−0.273	−0.064	−0.243	−0.148	0.031	0.100	0.236	**0.236**	0.097
DPPH (WE)	0.018	0.135	0.140	−0.188	0.199	−0.193	**0.218**	0.173	−0.048	−0.003	−0.204	0.140	0.036	0.111
DPPH (AE)	0.047	−0.263	−0.217	−0.312	0.138	−0.148	−0.261	−0.080	−0.230	**0.288**	0.046	0.151	−0.071	0.075
OH^- (WE)	**0.329**	−0.114	**0.361**	−0.342	**0.420**	−0.090	0.140	0.215	−0.065	0.110	−0.329	0.151	−0.048	0.088
OH^- (AE)	−0.036	−0.208	−0.026	−0.132	0.196	−0.102	−0.304	0.027	−0.315	0.073	0.001	0.215	0.094	−0.012
H_2O_2 (WE)	**0.270**	0.082	0.236	−0.139	**0.308**	−0.121	0.121	**0.302**	0.167	0.163	−0.124	0.125	−0.206	−0.044
H_2O_2 (AE)	0.091	−0.112	−0.143	−0.156	0.104	0.078	−0.134	−0.378	0.009	−0.017	−0.105	0.192	0.141	0.015
Fe^{2+} (WE)	0.224	0.077	0.206	−0.186	0.329	0.023	**0.220**	**0.497**	0.042	0.037	−0.192	0.098	−0.068	0.132
Fe^{2+} (AE)	−0.219	−0.268	−0.266	−0.020	0.057	0.052	−0.311	−0.184	−0.051	0.197	**0.299**	**0.278**	**0.296**	**0.230**
FRAP (WE)	−0.004	−0.157	0.069	−0.263	0.052	−0.324	−0.184	−0.185	−0.073	0.043	−0.049	0.121	0.199	−0.029
FRAP (AE)	−0.048	0.165	−0.039	−0.008	−0.089	−0.322	0.172	0.028	−0.020	0.003	0.176	0.009	0.020	−0.129
PMA (WE)	−0.115	0.056	−0.034	−0.081	−0.058	−0.387	−0.061	−0.308	−0.212	0.013	0.122	0.190	0.286	−0.096
PMA (AE)	−0.106	0.124	−0.003	−0.019	−0.017	−0.247	0.152	0.023	0.012	0.012	0.134	0.039	−0.045	0.038

[a]Bold r-values are significant at $p < 0.01$

References

Abuye, C., Urga, K., Knapp, H., Selmar, D., Omwega, A. M., & Imungi, J. K. (2003). A compositional study of *Moringa stenopetala* leaves. *East African Medical Journal, 80*, 247–252.

Aeyigoro, A. O., & Okoh, I. A. (2010). Preliminary phytochemical screening and in vitro antioxidant activities of the aqueous extract of *Helichrysum longifolium* DC. *BMC Complementary and Alternative Medicine, 10*, 1–8.

Andarwulan, N., Batari, R., Sandrasari, D. A., Bolling, B., & Wijaya, H. (2010). Flavonoids content and antioxidant activity of vegetables from Indonesia. *Food Chemistry, 121*, 1231–1235.

Balaban, R. S., Nemoto, S., & Finkel, T. (2005). Mitochondria, oxidants and aging. *Cell, 120*(4), 483–495.

Benzie, I. E. F., & Strain, J. J. (1996). The ferric reducing ability of plasma (FRAP) as a measure of antioxidant power: The FRAP assay. *Analytical Biochemistry, 239*, 70–76.

Dinis, T. C. P., Madeira, V. M. C., & Almeida, M. L. M. (1994). Action of phenolic derivates (acetoaminophen, salycilate and 5-aminosalycilate) as inhibitors of membrane lipid peroxidation and as peroxyl radical scavengers. *Archives of Biochemistry and Biophysics, 315*, 161–169.

Gourine, N., Yousfi, M., Bombarda, I., Nadjem, B., Stocker, P., & Gaydou, E. M. (2010). Antioxidant activities and chemical composition of essential oil of *Pistacia atlantica* from Algeria. *Industrial Crops and Products, 31*, 203–208.

Halliwell, B., & Gutteridge, J. M. C. (1990). Role of free radicals and catalytic metal ions in human disease. *Methods in Enzymology, 186*, 1–85.

Hazra, B., Biswas, S., & Mandal, N. (2008). Antioxidant and free radical scavenging activity of *Spondias pinnata. BMC Complementary and Alternative Medicine, 8*, 63–75.

Huang, D. J., Chen, H. J., Lin, C. D., & Lin, Y. H. (2005). Antioxidant and antiproliferative activities of water spinach (*Ipomea aquatica* Forsk) constituents. *Botanical Bulletin of Academia Sinica, 46*, 99–106.

Ji, L., Wu, L., Gao, W., Yang, J., & Guo, C. (2011). Antioxidant capacity of different fraction of vegetables and correlation with the contents of ascorbic acid, phenolics, and flavonoids. *Journal of Food Science, 76*, 1248–1257.

Jimoh, F. O., Adedapo, A. A., & Afolayan, A. J. (2010). Comparison of nutritional value and biological activity of the acetone, methanol and water extracts of the leaves of *Solanum nigrum* and *Leonotis leonorous. Food and Chemical Toxicology, 48*, 964–971.

Jimoh, F. O., Adedapo, A. A., & Afolayan, A. J. (2011). Comparison of the nutritive value, antioxidant and antibacterial activities of *Sonchus asper* and *Sonchus oleraceous. Records of Natural Products, 5*(1), 29–42.

Klaunig, J. E., & Kamendulis, L. M. (2004). The role of oxidative stress in carcinogenesis. *Annual Review of Pharmacology, 44*, 239–267.

Miladi, S., & Damak, M. (2008). *In vitro* antioxidant activities of *Aloe vera* leaf skin extracts. *Journal de la Societe Chimique de Tunisie, 10*, 101–109.

Miller, M. J., Sadowska-krowicka, H., Kakkis, J. L., & Clark, D. A. (1993). Amelioration of chronic ileitis by nitric oxidesynthase inhibition. *The Journal of Pharmacology and Experimental Therapeutics, 264*, 11–16.

Motamed, S. M., & Naghibi, F. (2010). Antioxidant activity of some edible plants of the Turkmen Sahra region in northern Iran. *Food Chemistry, 119*, 1637–1642.

Odhav, B., Beekrum, S., Akula, U., & Baijnath, H. (2007). Preliminary assessment of nutritional value of traditionally leafy vegetables in KwaZulu-Natal, South Africa. *Journal of Food Composition and Analysis, 20*, 430–435.

Prieto, P., Pineda, M., & Aguilar, M. (1999). Spectrophotometric quantitation of antioxidant capacity through the formation of a phosphomolybdenum complex: Specific application to the determination of vitamin E1. *Analytical Biochemistry, 269*, 337–341.

Prieto, P., Pineda, M., & Aguilar, M. (2006). Spectrophotometric quantitation of antioxidant capacity through the formation of a phosphomolybdenum complex: Specific application to the determination of vitamin E. *Analytical Biochemistry, 269*, 337–341.

Shi, H., Hudson, L. G., Ding, W., Wang, S. W., Cooper, K. L., Liu, S. M., et al. (2004). Arsenite causes DNA damage in keratinocytes via generation of hydroxyl radicals. *Chemical Research and Toxicology, 17*(7), 871–878.

Wahab, O. A., Rjoob, E., Massadeh, A. M., & Omari, M. N. (2008). Evaluation of Pb, Cu, Zn, Cd, Ni and Fe levels in *Rosmarinus officinalis* labaiatae (Rosemary) medicinal plant and soils in selected zones in Jordan. *Environmental Monitoring and Assessment, 140*, 61–68.

Wong, S. P., Lai, P. L., & Jen, H. W. K. (2006). Antioxidant activities of aqueous extracts of selected plants. *Food Chemistry, 99*, 775–783.

Yu, L., Haley, S., Perret, J., Harris, J. W., & Qian, M. (2002). Free radical scavenging properties of wheat extracts. *Journal of Agricultural and Food Chemistry, 50*, 1619–1624.

Yu, W., Zhao, Y., & Shu, B. (2004). The radical scavenging activities of radix puerariae isoflavanoids: A chemiluminescence study. *Food Chemistry, 86*, 525–529.

Annexure

Project Title ____

1. Date: ____________________ 2. Sheet No.: ____________________
3. Participants: ____
4. Name of the person contacted for the data: ____________________
5. Gender: Male/Female: ______________________________________
6. Habit: ______________________ 7. Habitat: __________________
8. Ethnic group/cast: _____________ 9. Language: ________________
10. Locality: _________________ 11. Altitude: __________________
12. Latitude: ________________ 13. Longitude: _________________
14. District: ____
15. Botanical name of the plant: ________________________________
16. Local name of the plant: ___________________________________
17. Is it familiar by this name: Yes/No ___________________________
18. In other places where it is found: ____________________________
19. Flowering and Fruiting period: _______________________________
20. Part used: Leaves, Branches, Flowers, Seeds, Roots ______________
21. Used as fruit/vegetable: ____________________________________
22. If fruit, then used as ripe, unripe, fresh, dried: __________________
23. Who collect it men, women, children: __________________________
24. Season of collection: _______________________________________
25. Taste: ____
26. Is it stored Yes/No: ____
27. If plant is used as vegetable then, which part is used? ______________
28. Eaten fresh or cooked as vegetable: ___________________________
29. If cooked, then how? _______________________________________
30. Is it favorite food? ____
31. Is it sold in the market Yes/No: ______________________________
32. If sold then price/kg: ______________________________________

A.M. Abbasi et al., *Wild Edible Vegetables of Lesser Himalayas*,
DOI 10.1007/978-3-319-09543-1

33. Traditional uses of plant:

34. Method of drug preparation:

35. Mode of application:
36. Diseases cured:
37. Other Ethnobotanical uses; fodder, fuel, furniture, fencing, tools handles, construction, ornamental purpose, decorative articles, brooms, baskets.

38. Status of the plant: Common, rare, endangered/ threatened.
39. Soil type:
40. Plant identified by:

41. Remarks if any:

____

____

____

____

Name of the Scientist

Index

A.M. Abbasi et al., *Wild Edible Vegetables of Lesser Himalayas*,
DOI 10.1007/978-3-319-09543-1

R

CPSIA information can be obtained at www.ICGtesting.com
Printed in the USA
LVOW02*1718191014

409493LV00001B/61/P

9 783319 095424